MATHEMATICAL THEORY OF HEMIVARIATIONAL INEQUALITIES AND APPLICATIONS

PURE AND APPLIED MATHEMATICS

A Program of Monographs, Textbooks, and Lecture Notes

MONOGRAPHS AND TEXTBOOKS IN PURE AND APPLIED MATHEMATICS

1. *K. Yano,* Integral Formulas in Riemannian Geometry (1970)
2. *S. Kobayashi,* Hyperbolic Manifolds and Holomorphic Mappings (1970)
3. *V. S. Vladimirov,* Equations of Mathematical Physics (A. Jeffrey, ed.; A. Littlewood, trans.) (1970)
4. *B. N. Pshenichnyi,* Necessary Conditions for an Extremum (L. Neustadt, translation ed.; K. Makowski, trans.) (1971)
5. *L. Narici et al.,* Functional Analysis and Valuation Theory (1971)
6. *S. S. Passman,* Infinite Group Rings (1971)
7. *L. Dornhoff,* Group Representation Theory. Part A: Ordinary Representation Theory. Part B: Modular Representation Theory (1971, 1972)
8. *W. Boothby and G. L. Weiss, eds.,* Symmetric Spaces (1972)
9. *Y. Matsushima,* Differentiable Manifolds (E. T. Kobayashi, trans.) (1972)
10. *L. E. Ward, Jr.,* Topology (1972)
11. *A. Babakhanian,* Cohomological Methods in Group Theory (1972)
12. *R. Gilmer,* Multiplicative Ideal Theory (1972)
13. *J. Yeh,* Stochastic Processes and the Wiener Integral (1973)
14. *J. Barros-Neto,* Introduction to the Theory of Distributions (1973)
15. *R. Larsen,* Functional Analysis (1973)
16. *K. Yano and S. Ishihara,* Tangent and Cotangent Bundles (1973)
17. *C. Procesi,* Rings with Polynomial Identities (1973)
18. *R. Hermann,* Geometry, Physics, and Systems (1973)
19. *N. R. Wallach,* Harmonic Analysis on Homogeneous Spaces (1973)
20. *J. Dieudonné,* Introduction to the Theory of Formal Groups (1973)
21. *I. Vaisman,* Cohomology and Differential Forms (1973)
22. *B.-Y. Chen,* Geometry of Submanifolds (1973)
23. *M. Marcus,* Finite Dimensional Multilinear Algebra (in two parts) (1973, 1975)
24. *R. Larsen,* Banach Algebras (1973)
25. *R. O. Kujala and A. L. Vitter, eds.,* Value Distribution Theory: Part A; Part B: Deficit and Bezout Estimates by Wilhelm Stoll (1973)
26. *K. B. Stolarsky,* Algebraic Numbers and Diophantine Approximation (1974)
27. *A. R. Magid,* The Separable Galois Theory of Commutative Rings (1974)
28. *B. R. McDonald,* Finite Rings with Identity (1974)
29. *J. Satake,* Linear Algebra (S. Koh et al., trans.) (1975)
30. *J. S. Golan,* Localization of Noncommutative Rings (1975)
31. *G. Klambauer,* Mathematical Analysis (1975)
32. *M. K. Agoston,* Algebraic Topology (1976)
33. *K. R. Goodearl,* Ring Theory (1976)
34. *L. E. Mansfield,* Linear Algebra with Geometric Applications (1976)
35. *N. J. Pullman,* Matrix Theory and Its Applications (1976)
36. *B. R. McDonald,* Geometric Algebra Over Local Rings (1976)
37. *C. W. Groetsch,* Generalized Inverses of Linear Operators (1977)
38. *J. E. Kuczkowski and J. L. Gersting,* Abstract Algebra (1977)
39. *C. O. Christenson and W. L. Voxman,* Aspects of Topology (1977)
40. *M. Nagata,* Field Theory (1977)
41. *R. L. Long,* Algebraic Number Theory (1977)
42. *W. F. Pfeffer,* Integrals and Measures (1977)
43. *R. L. Wheeden and A. Zygmund,* Measure and Integral (1977)
44. *J. H. Curtiss,* Introduction to Functions of a Complex Variable (1978)
45. *K. Hrbacek and T. Jech,* Introduction to Set Theory (1978)
46. *W. S. Massey,* Homology and Cohomology Theory (1978)
47. *M. Marcus,* Introduction to Modern Algebra (1978)
48. *E. C. Young,* Vector and Tensor Analysis (1978)
49. *S. B. Nadler, Jr.,* Hyperspaces of Sets (1978)
50. *S. K. Segal,* Topics in Group Kings (1978)
51. *A. C. M. van Rooij,* Non-Archimedean Functional Analysis (1978)
52. *L. Corwin and R. Szczarba,* Calculus in Vector Spaces (1979)

53. *C. Sadosky*, Interpolation of Operators and Singular Integrals (1979)
54. *J. Cronin*, Differential Equations (1980)
55. *C. W. Groetsch*, Elements of Applicable Functional Analysis (1980)
56. *I. Vaisman*, Foundations of Three-Dimensional Euclidean Geometry (1980)
57. *H. I. Freedan*, Deterministic Mathematical Models in Population Ecology (1980)
58. *S. B. Chae*, Lebesgue Integration (1980)
59. *C. S. Rees et al.*, Theory and Applications of Fourier Analysis (1981)
60. *L. Nachbin*, Introduction to Functional Analysis (R. M. Aron, trans.) (1981)
61. *G. Orzech and M. Orzech*, Plane Algebraic Curves (1981)
62. *R. Johnsonbaugh and W. E. Pfaffenberger*, Foundations of Mathematical Analysis (1981)
63. *W. L. Voxman and R. H. Goetschel*, Advanced Calculus (1981)
64. *L. J. Corwin and R. H. Szczarba*, Multivariable Calculus (1982)
65. *V. I. Istrățescu*, Introduction to Linear Operator Theory (1981)
66. *R. D. Järvinen*, Finite and Infinite Dimensional Linear Spaces (1981)
67. *J. K. Beem and P. E. Ehrlich*, Global Lorentzian Geometry (1981)
68. *D. L. Armacost*, The Structure of Locally Compact Abelian Groups (1981)
69. *J. W. Brewer and M. K. Smith, eds.*, Emily Noether: A Tribute (1981)
70. *K. H. Kim*, Boolean Matrix Theory and Applications (1982)
71. *T. W. Wieting*, The Mathematical Theory of Chromatic Plane Ornaments (1982)
72. *D. B. Gauld*, Differential Topology (1982)
73. *R. L. Faber*, Foundations of Euclidean and Non-Euclidean Geometry (1983)
74. *M. Carmeli*, Statistical Theory and Random Matrices (1983)
75. *J. H. Carruth et al.*, The Theory of Topological Semigroups (1983)
76. *R. L. Faber*, Differential Geometry and Relativity Theory (1983)
77. *S. Barnett*, Polynomials and Linear Control Systems (1983)
78. *G. Karpilovsky*, Commutative Group Algebras (1983)
79. *F. Van Oystaeyen and A. Verschoren*, Relative Invariants of Rings (1983)
80. *I. Vaisman*, A First Course in Differential Geometry (1984)
81. *G. W. Swan*, Applications of Optimal Control Theory in Biomedicine (1984)
82. *T. Petrie and J. D. Randall*, Transformation Groups on Manifolds (1984)
83. *K. Goebel and S. Reich*, Uniform Convexity, Hyperbolic Geometry, and Nonexpansive Mappings (1984)
84. *T. Albu and C. Năstăsescu*, Relative Finiteness in Module Theory (1984)
85. *K. Hrbacek and T. Jech*, Introduction to Set Theory: Second Edition (1984)
86. *F. Van Oystaeyen and A. Verschoren*, Relative Invariants of Rings (1984)
87. *B. R. McDonald*, Linear Algebra Over Commutative Rings (1984)
88. *M. Namba*, Geometry of Projective Algebraic Curves (1984)
89. *G. F. Webb*, Theory of Nonlinear Age-Dependent Population Dynamics (1985)
90. *M. R. Bremner et al.*, Tables of Dominant Weight Multiplicities for Representations of Simple Lie Algebras (1985)
91. *A. E. Fekete*, Real Linear Algebra (1985)
92. *S. B. Chae*, Holomorphy and Calculus in Normed Spaces (1985)
93. *A. J. Jerri*, Introduction to Integral Equations with Applications (1985)
94. *G. Karpilovsky*, Projective Representations of Finite Groups (1985)
95. *L. Narici and E. Beckenstein*, Topological Vector Spaces (1985)
96. *J. Weeks*, The Shape of Space (1985)
97. *P. R. Gribik and K. O. Kortanek*, Extremal Methods of Operations Research (1985)
98. *J.-A. Chao and W. A. Woyczynski, eds.*, Probability Theory and Harmonic Analysis (1986)
99. *G. D. Crown et al.*, Abstract Algebra (1986)
100. *J. H. Carruth et al.*, The Theory of Topological Semigroups, Volume 2 (1986)
101. *R. S. Doran and V. A. Belfi*, Characterizations of C*-Algebras (1986)
102. *M. W. Jeter*, Mathematical Programming (1986)
103. *M. Altman*, A Unified Theory of Nonlinear Operator and Evolution Equations with Applications (1986)
104. *A. Verschoren*, Relative Invariants of Sheaves (1987)
105. *R. A. Usmani*, Applied Linear Algebra (1987)
106. *P. Blass and J. Lang*, Zariski Surfaces and Differential Equations in Characteristic $p > 0$ (1987)
107. *J. A. Reneke et al.*, Structured Hereditary Systems (1987)

Additional Volumes in Preparation

MATHEMATICAL THEORY OF HEMIVARIATIONAL INEQUALITIES AND APPLICATIONS

Z. Naniewicz
University of Warsaw
Warsaw, Poland

P. D. Panagiotopoulos
Aristotle University
Thessaloniki, Greece
and
Rheinisch-Westfälische Technische Hochschule
Aachen, Germany

CRC Press
Taylor & Francis Group
Boca Raton London New York

CRC Press is an imprint of the
Taylor & Francis Group, an **informa** business
A TAYLOR & FRANCIS BOOK

CRC Press
Taylor & Francis Group
6000 Broken Sound Parkway NW, Suite 300
Boca Raton, FL 33487-2742

First issued in paperback 2019

© 1995 by Taylor Francis Group, LLC
CRC Press is an imprint of Taylor & Francis Group, an Informa business

No claim to original U.S. Government works

ISBN-13: 978-0-8247-9330-2 (hbk)
ISBN-13: 978-0-367-40198-6 (pbk)

Library of Congress Cataloging-in-Publication Data

Naniewicz, Z.
    Mathematical theory of hemivariational inequalities and
applications / Z. Naniewicz, P. D. Panagiotopoulos.
        p.   cm. — (Monographs and textbooks in pure and applied
mathematics ; 188)
    Includes bibliographical references and index.
    ISBN 0-8247-9330-7
    1. Hemivariational inequalities. 2. Engineering mathematics.
I. Panagiotopoulos, P. D. II. Title. III. Series.
QA316.N36   1995
515'.64—dc20   94-35419
                                                                                  CIP

Visit the Taylor & Francis Web site at
http://www.taylorandfrancis.com

and the CRC Press Web site at
http://www.crcpress.com

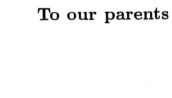

Preface

Inequality problems have within a very short time undergone a remarkable development in both pure and applied mathematics as well as in mechanics and the engineering sciences, largely because of new and efficient mathematical methods in this field or, more generally, in convex analysis and nonsmooth analysis. Moreover, the evolution of this area of mathematics has facilitated the solution of many open questions in mechanics and the engineering sciences and also allowed the formulation and the definitive mathematical study of new classes of interesting problems. These developments enabled a new branch of mechanics, called nonsmooth mechanics, to emerge during the last decade.

Inequality problems can be divided into two main classes: that of variational inequalities, which, with a research "life" of some thirty years now, is mainly concerned with convex energy functions, and that of hemivariational inequalities, which is more recent and is concerned with nonconvex energy functions. The area of hemivariational inequalities, stemming from the work of the second author in mechanics, was born only at the beginning of the last decade and has already produced an abundance of important results. The mathematical theory of hemivariational inequalities and some applications from mechanics, engineering, and economics are the subjects of this book.

The book is an outgrowth of twelve years of seminars and courses on the theory and applications of hemivariational inequalities we delivered to a variety of audiences at the University of Warsaw, the Technical University of Aachen, the Aristotle University of Thessaloniki, the University of Hamburg, and the Pontificia Universidade Católica in Rio de Janeiro. The book is primarily addressed to people working in pure and applied mathematics and, secondarily, to research workers in mechanics, engineering sciences, and economics: it points out important applications and their mathematical treatment, and introduces innovative mathematical methods which may be applied further to the study and solution of as yet unsolved or partially solved problems.

We would like to acknowledge the great assistance we received from Dr. P. A. Zervas, who diligently prepared the final text with the LaTeX program, and Dr. E. Koltsakis, Dr. E. Mistakidis, and Dipl. Ing. Th. Nikolaidis, for proofreading and the preparation of the final figures. We also wish to acknowledge the helpful comments received from Prof. D. Motreanu on some chapters of the book. The first author gratefully acknowledges the support received from the Aristotle University of Thessaloniki during his two-year stay there as Research Fellow, without which the realization of the present task would have been impossible. Many thanks are due also to our editors at Marcel Dekker, Inc., for their friendly assistance and for their cooperation during this project. We would like to apologize to those whose work was inadvertently neglected in compiling the literature for this book. We shall welcome all comments and corrections from readers.

Z. Naniewicz
P. D. Panagiotopoulos

Contents

Introduction

The scope of the present book is the mathematical study of hemivariational inequalities. This new type of inequality problem arises, e.g. in the variational formulation of mechanical problems whenever nonconvex energy functionals are involved. The basic form of the problem is the following: suppose that V is a Banach space, $a(\cdot,\cdot)$ is a bilinear form on $V \times V$, $J(\cdot)$ a locally Lipschitz functional on V and $f \in V^*$, the dual space of V. Denoting by $(\cdot,\cdot)$ the duality pairing we seek a $u \in V$ such as to satisfy

$$a(u, v - u) + J^0(u, v - u) \geq (f, v - u) \quad \forall v \in V.$$

This type of variational inequality is a generalization of the classical variational inequalities, because $J(\cdot)$ is nonconvex and $J^0(\cdot,\cdot)$ is the directional differential in the sense of F.H. Clarke, i.e. $J(\cdot)$ is in general nonsmooth. More general forms of hemivariational inequalities studied in this book contain multivalued pseudomonotone operators whose solution is sought in convex, or nonconvex but star-shaped with respect to a ball and closed, admissible sets. Thus the theory of hemivariational inequalities is a new field of nonsmooth analysis.

In Chapter 1 of this book notions and propositions of nonsmooth analysis that are used in the next chapters are presented. Also, some elements of the theory of maximal monotone operators are given and their relation to the theory of variational inequalites is illustrated. The chapter closes with a section concerning the formulation of hemivariational inequalities in mechanics. The material of this section is sufficient for the understanding of the applications contained in the other chapters.

In Chapter 2 we deal with the notions of pseudo-monotonicity and generalized pseudo-monotonicity for multivalued mappings according to Browder and Hess, and with the main results of this theory which will subsequently be applied to the study of hemivariational inequalities. Next we study some general properties of functions having pseudo-monotone and generalized pseudo-monotone generalized gradients. Two classes of functions will be considered: the first class includes locally Lipschitz functions, while the second one includes indicator functions of some nonconvex closed sets. Finally the notion of quasi-pseudo-monotonity is introduced and corresponding propositions are proved.

Chapter 3 contains a type of hemivariational inequality that was the first to be studied mathematically. They arise from one-dimensional nonconvex super-potential laws, i.e. in the case of one-dimensional, nonmonotone, multivalued nonlinearities. We begin with the coercive case and subsequently give an existence result for the more complicated semicoercive case. After the treatment of variational inequalities we deal with variational-hemivariational inequalities and we give some existence and approximation results. Next the relation between a hemivariational inequality and the corresponding substationarity problem is investigated. The theory of this chapter is illustrated by means of applications from mechanics and engineering.

In Chapter 4 a more general multidimensional type of hemivariational inequality is treated. Initially, two classes of locally Lipschitz functions are introduced and their properties are studied. The first class, denoted by $QPM(V)$, includes functions whose generalized gradient is quasi-pseudo-monotone, while the second, denoted as $PM(V)$, contains functions with pseudo-monotone generalized gradients. Next we show how the theory of pseudo-monotone operators permits the formulation of existence results for hemivariational inequalities involving functions from the classes $QPM(V)$ and $PM(V)$. Variational-hemivariational inequalities and quasi-hemivariational inequalities are then studied and certain existence results are obtained. The chapter closes with applications of the theory to mechanics.

Chapter 5 deals with hemivariational inequalities and variational-hemivariational inequalities for multidimensional superpotential laws, i.e. for hemivariational inequalities on vector-valued function spaces. The basic assumption is that certain new classes of directional growth conditions hold. The methods developed here are based on the Galerkin approximation combined with the fixed point theorems of Brouwer and Kakutani, the Dunford-Pettis criterion, and the well known topological result concerning a family of compact sets with the finite intersection property. This chapter closes with some applications from mechanics, engineering and economics.

Chapter 6 concerns noncoercive hemivariational inequalities related to free boundary problems. As a pilot problem a system consisting of a variational and a hemivariational inequality related to the continuous model of a delamination process in multilayered structures is studied.

In Chapter 7 the theory of constrained problems for nonconvex star-shaped admissible sets is developed. Making use of hemivariational inequalities we prove the existence of solutions to the following type of nonconvex constrained problems: find $u \in C$ such that

$$a(u,v) \geq (f,v) \quad \forall v \in T_C(u),$$

where the admissible set $C \subset V$ is a star-shaped set with respect to a certain ball, and $T_C(u)$ denotes Clarke's tangent cone of C at $u \in C$. It is worth noting that the approach here is based on a penalization method where the small parameter does not have to tend to zero.

The first author has written the mathematical part of Chapters 2, 4, 5, 6 and 7, as well as Section 7.5. The applications in Chapters 4,5 and Section 7.6 have been written by the second author, who has also written Chapters 1 and 3. The mathematical results of Chapters 2 (in part), 4, 5, 6 and 7 are due to the first author, whereas those of Chapter 3 are due to the second author. Of course, both authors share equally the responsibility for the content of the book.

Guidelines for the Reader, Abbreviations

The choice of the material of Chapter 1 is governed by the requirements of the subsequent chapters. All propositions of nonsmooth analysis (convex and nonconvex) needed in this book are given in Chapter 1. We expect the reader to have some knowledge of convex analysis and nonsmooth analysis, but this is not an absolute prerequisite to understanding the contents of the book because the definitions of all new mathematical notions are included in the text. Thus, only a basic knowledge of functional analysis is necessary.

All applications are given at the end of each chapter in order to illustrate the theory developed therein.

Certain important notations and abbreviations used throughout the text and not defined there, are listed here. All other notations are defined in the text. Throughout the book the summation convention with respect to a repeated index is employed, unless otherwise stated. In the first chapter we have distinguished between the subdifferential ∂f and generalized gradient $\bar{\partial} f$ for purely didactic reasons. In the next chapters we eliminate the overbar on $\bar{\partial}$ and we use the same symbol ∂f for both the subdifferential and the generalized gradient.

N	the set of natural numbers
R	the set of real numbers
R^+	the set of nonnegative numbers
$\bar{R}$	the set of real numbers including $\pm \infty$
R^n	Euclidean n-dimensional space
$\|x\| = (\sum_{i=1}^{n} x_i^2)^{1/2}$	length of $x \in R^n$
$C_0^\infty(\Omega)$	space of infinitely differentiable functions with compact support in Ω
a.e.	almost every (almost everywhere)
$\mu -$ a.e.	almost every (almost everywhere) with respect to a measure μ
$\ker T$	kernel of T
epi f	epigraph of a functional f
l.s.c., u.s.c.	lower semicontinuous, upper semicontinuous
$\partial f(x)$	subdifferential of f at x
$\bar{\partial} f(x)$	generalized gradient of f at x (only in the first chapter)

supp	support of a function
sgn	signum
$\Omega - \omega$ or Ω/ω	difference of two sets
$x = \{x_1, \ldots, x_n\} = \{x_i\},$	a point $x \in R^n$
$i = 1, \ldots, n$	
$\{a_1, \ldots a_n \ldots\}$	set with elements $a_1, \ldots a_n, \ldots$ also written as (a_i) or $\{a_i\}$
$\bar{A}$	closure of a set A
$\mathrm{co}A$	convex hull of a set A
$X, X^\star$	a topological space X and its dual $X^\star$
$u_n \xrightarrow{w} u,$	
$u_n \to u$ weakly:	two notations for the weak convergence
BVP (BVPs):	Boundary Value Problem (Problems)

MATHEMATICAL THEORY OF HEMIVARIATIONAL INEQUALITIES AND APPLICATIONS

1. Introductory Material

In Chapter 1 we give some notions and propositions of Nonsmooth Analysis that are used in the next Chapters for the formulation and study of hemivariational inequalities. More precisely certain general results from Convex Analysis as well as from Nonsmooth Nonconvex Analysis are given. Also, some elements of the theory of maximal monotone operators are presented and their relation to the theory of variational inequalities is illustrated. The Chapter closes with a section concerning the formulation of hemivariational inequalities. The propositions are given here without proofs. This Chapter is based on the monographs and works by Moreau [Mor67], Rockafellar [Rock60,68,70,79,80], Göpfert [Göp], Ekeland and Temam [Eke], Aubin [Aub77,79,79a,84,91], Aubin and Frankowska [Aub90], Clarke [Clar73,75,78], Panagiotopoulos [Pan85,93], Moreau, Panagiotopoulos, Strang [Mor88b], Moreau, Panagiotopoulos [Mor88a], Antes and Panagiotopoulos [Ant]. The reader is referred there for the proofs of the propositions and for more information concerning the mechanical background of the hemivariational inequalities. In this Chapter we use different notations for the subdifferential ∂ and the generalized gradient $\bar{\partial}$ for purely didactic reasons. In the next chapters we shall use the same notation ∂ for both notions, as it is usual in Nonsmooth Analysis.

1.1 Elements of Convex Analysis

Let X be a real Banach space and K a subset of X. The set K is said to be convex if

$$\lambda x_1 + (1 - \lambda)x_2 \in K \tag{1.1.1}$$

whenever $x_1 \in K, x_2 \in K$ and $0 < \lambda < 1$. All linear subspaces of X (including X) are convex. By convention, the empty set $\emptyset$ is convex. Of great interest are the convex cones. A set $K \subset X$ is a cone if, for $x \in K, \lambda x \in K$ for every $\lambda \geq 0$. Moreover K is a convex cone if $x_1 + x_2 \in K$ for $x_1 \in K$ and $x_2 \in K$. For any set $K_1 \subset X$, the set of all finite linear combinations $\sum_i \lambda_i x_i, x_i \in K_1$, with $\sum_i \lambda_i = 1$, $i = 1, 2, \ldots, n$, is called the affine hull of K_1. If, moreover, $\lambda_i \geq 0$, $i = 1, 2, \ldots, n$, then we have the convex hull of K_1 which is denoted by co K_1; it is the smallest convex subset of X which contains K_1.

A real-valued functional $f : K \to R$ is convex (resp. strictly convex) on K if for each $x_1 \in K, x_2 \in K$ and $0 < \lambda < 1$ (Fig. 1.1.1a)

$$f(\lambda x_1 + (1 - \lambda)x_2) \leq (\text{resp. } <)\lambda f(x_1) + (1 - \lambda)f(x_2). \tag{1.1.2}$$

1

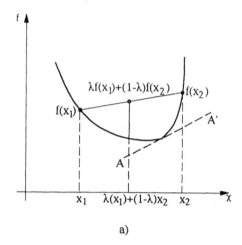

a)

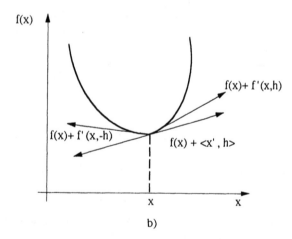

b)

Fig. 1.1.1. The definition of convexity. On the differentiability properties of convex functions

A functional f is concave (resp. strictly concave) if and only if $-f$ is convex (resp. strictly convex). A linear functional is at the same time convex and concave, but not strictly. We deal here with functionals taking values in the extended real line $\bar{R} = R \cup \{\pm\infty\} = [-\infty, +\infty]$. A functional $f : K \to \bar{R}$ is defined to be convex on K if for every $x_1 \in K$ and $x_2 \in K$ (1.1.2) holds whenever the right-hand side makes sense; this is not the case if $f(x_1) = -f(x_2) = \pm\infty$.

Using the fact that a convex functional may have infinite values, we can consider functionals defined on all of X. Indeed, if $f : K \to R$ is convex on K, we can define the extension $\bar{f}$ of f to all of X by setting $\bar{f}(x) = f(x)$ for $x \in K$ and $\bar{f}(x) = \infty$ for $x \notin K$.

The convex functional $I_K : X \to \bar{R}$, defined by the relation

$$I_K(x) = \begin{cases} 0 & \text{for } x \in K \\ \infty & \text{for } x \notin K \end{cases} \tag{1.1.3}$$

is called the indicator of K. This functional can be associated to every convex set K. With respect to a convex functional $f : X \to \bar{R}$ the set

$$\{(x, \lambda) : f(x) \leq \lambda, \ \lambda \in R, x \in X\} \tag{1.1.4}$$

is considered. It is called the epigraph of f, is denoted by epi f and it is a convex set in $X \times R$. An equivalent definition of convexity of a functional $f : X \to \bar{R}$ is the following: $f : X \to \bar{R}$ is defined to be convex, whenever epi f is a convex subset of $X \times R$. The effective domain $D(f)$ of a convex functional f on X is the set

$$D(f) = \{x : x \in X, f(x) < \infty\}. \tag{1.1.5}$$

A functional f is called proper if $f : X \to (-\infty, +\infty]$ and $f \not\equiv \infty$. If f is convex, $\lambda f (\lambda \geq 0)$ is convex. For f_1 and f_2 convex, $f_1 + f_2$ is convex as well (we assume that $(f_1 + f_2)(x) = +\infty$ for $f_1(x) = -f_2(x) = \pm\infty$, (cf. [Eke] p.67). A functional $f : X \to \bar{R}$ is called lower semicontinuous (l.s.c.) on X if for every $\lambda \in R$ the set

$$\{x : x \in X, f(x) \leq \lambda\} \tag{1.1.6}$$

is closed in X. For f l.s.c., $-f$ is upper semicontinuous (u.s.c.), and conversely. Similarly a functional $f : X \to \bar{R}$ is l.s.c. if and only if epi f is a closed subset of $X \times R$.

We shall give another definition of the lower semicontinuity. To this end let us recall now the notion of upper and lower limit.

Let $f : X_1 \to \bar{R}$ and $0 < \delta < 1$, where X_1 is a subset of X. We denote by $B_{X_1}(x, \delta)$ or simply $B(x, \delta)$ the ball $\{y \in X_1 : \|x - y\| \leq \delta\}$ for $x \in X_1$ and following Aubin [Aub77] we associate to δ the function

$$\alpha(\delta) = \inf\{f(y) : y \in B(x, \delta)\}, \tag{1.1.7}$$

which is a decreasing function of δ. Accordingly, the $\lim \alpha(\delta)$ as $\delta \to 0_+$ exists in $\bar{R}$ and is equal to $\sup\{\alpha(\delta) : \delta > 0\}$. We recall now the following definitions and relations

$$\liminf_{y \to x} f(y) = \lim_{\delta \to 0} \left(\inf_y \{f(y) : y \in B(x, \delta)\} \right) = \sup_{\delta > 0} \inf_y \{f(y) : y \in B(x, \delta)\} \tag{1.1.8}$$

$$\limsup_{y \to x} f(y) = \lim_{\delta \to 0} \left(\sup_y \{f(y) : y \in B(x, \delta)\} \right) = - \liminf_{y \to x} \left(-f(y) \right). \tag{1.1.9}$$

Analogously we may write that

$$\liminf_{n\to\infty} f(x_n) = \lim_{n\to\infty} \left(\inf_{p\geq 0} f(x_{n+p}) \right) = \sup_n \inf_{p\geq 0} f(x_{n+p}) \qquad (1.1.10)$$

$$\limsup_{n\to\infty} f(x_n) = \lim_{n\to\infty} \left(\sup_{p\geq 0} f(x_{n+p}) \right) = -\liminf_{n\to\infty} (-f(x_n)). \qquad (1.1.11)$$

From the above definition we can easily write the following inequalities for a function $f : X_1 \to \bar{R}$ and for every $x \in X_1$

$$\inf\{f(y) : y \in X_1\} \;\leq\; \liminf_{y\to x} f(y) \qquad\qquad (1.1.12)$$
$$\leq \; f(x) \leq \limsup_{y\to x} f(y) \leq \sup\{f(y) : y \in X_1\}.$$

Since we have defined the "liminf" and the "limsup" we can give now an equivalent definition of the lower semicontinuity.

Proposition 1.1 A functional $f : X \to \bar{R}$ is l.s.c if and only if

$$\forall x_0 \in X \quad \liminf_{x\to x_0} f(x) \geq f(x_0). \qquad (1.1.13)$$

Note that if (1.1.13) holds only at a point $x_0 \in X$ then f is called l.s.c. at x_0. If $f : X \to \bar{R}$ is a convex functional such that $f(x) = \{\tilde{f}(x)$ for $x \in D(\tilde{f})$, ∞ otherwise$\}$, then the lower semicontinuity of $\tilde{f}$ on $D(\tilde{f})$ does not necessarily imply the lower semicontinuity of f on X (cf. e.g. the function $f(x) = \{x$ for $x > 0$, ∞ for $x \leq 0\}$).

As we associate the epigraph epi f with a functional f, so we may associate with a set K the indicator I_K. It is easily shown that K is closed if and only if its indicator is l.s.c.

The following result on the continuity of convex functionals deserves mention. In a Banach space X, a convex, l.s.c. functional $f : X \to \bar{R}$ is continuous on int $D(f)$.

Let us denote further by X^* the dual space to the reflexive Banach space X and let $\langle \cdot, \cdot \rangle_X$ be the duality pairing between X and X^*. If no ambiguity occurs we will write simply $\langle \cdot, \cdot \rangle$.

Convex functionals $f : X \to \bar{R}$ are not necessarily everywhere differentiable. Then all the supporting hyperplanes (cf. AA' in Fig. 1.1.1) to the epi f "describe" in some sense the differentiability properties of f at a point where f is nondifferentiable in the classical sense. This leads to the notion of subdifferential. The vector $x^* \in X^*$, for which

$$f(x_1) - f(x) \geq \langle x^*, x_1 - x \rangle, \quad \forall x_1 \in X \qquad (1.1.14)$$

holds, where $f(x)$ is finite at $x \in X$, is called the subgradient of f at x. The set of all $x^* \in X^*$ satisfying (1.1.14) is called the subdifferential of f at x and is denoted by $\partial f(x)$. The usual notation is then

$$x^* \in \partial f(x). \qquad (1.1.15)$$

The set $\{x : \partial f(x) \neq \emptyset\}$ is denoted by $D(\partial f)$, and is called the domain of ∂f.

The mapping $\partial f : X \to X^\star$ is multivalued and is called the subdifferential of f. If $\partial f(x) \neq \emptyset$, f is subdifferentiable at x. Moreover $\partial f(x) = \emptyset$ for $x \notin D(f)$ and $f \not\equiv \infty$. From (1.1.14), it results that a necessary and sufficient condition for x_0 to minimize f on X is that

$$0 \in \partial f(x_0), \tag{1.1.16}$$

since $f(x_0) \leq f(x)$ $\forall x \in X$. This fact reveals the close relation of the subdifferentiability to the optimization theory.

The affine function $x_1 \to L(x_1) = f(x) + \langle x^*, x_1 - x \rangle$ is called the supporting hyperplane of epi f at $\{x, f(x)\}$. Thus (1.1.14) states that for $f(x)$ finite the supporting hyperplane to epi f at $\{x, f(x)\}$ is nonvertical. It can be proved that $\partial f(\cdot)$ is for every $x \in \text{int } D(f)$ a convex closed set of the dual space X^*. Actually if X is more generally a locally convex Hausdorff topological vector space and X^* its dual space then $\partial f(\cdot)$ is closed with respect to the weak topology $\Sigma(X^*, X)$ for the duality between X and X^*.

At this point we would like to recall the following important results.

Proposition 1.2 Let X be a locally convex Hausdorff topological vector space, X^* its dual space and let T be a topology on X compatible with the duality between X and X^*. A convex subset of X which is closed with respect to T, is closed with respect to every other topology compatible with the duality between X and X^*.

An analogous proposition holds also for every Banach space, where a convex set is norm-closed if and only if it is weakly closed. Applying this result to the epigraph we obtain the following proposition.

Proposition 1.3 Let X be a Banach space. A convex functional $f : X \to \bar{R}$ is l.s.c on X if and only if it is weakly l.s.c. on X.

In the case of convex functionals the existence of subdifferentials is ensured by means of the following result. We assume hereafter that X is a reflexive Banach space.

Proposition 1.4 Let $f : X \to \bar{R}$ be convex, and suppose that f is finite and continuous at $x_0 \in X$. Then $\partial f(x_0) \neq \emptyset$. Moreover $\partial f(x)$ is nonempty for every $x \in \text{int } D(f)$.

The case $f = I_K$, where K is a nonempty convex subset of X, is important. Then

$$\partial I_K(x) = \{x^* : I_K(x_1) - I_K(x) \geq \langle x^*, x_1 - x \rangle, \forall x_1 \in X\} \tag{1.1.17}$$

or, equivalently, for $x \in K$,

$$\partial I_K(x) = \{x^* : \langle x^*, x_1 - x \rangle \leq 0, \forall x_1 \in K\}. \tag{1.1.18}$$

The geometrical meaning of the variational inequality

$$\langle x^*, x_1 - x \rangle \leq 0, \quad \forall x_1 \in K, \quad x \in K \qquad (1.1.19)$$

is that x^* is an outward normal vector to K at x. In general, the set of all vectors x^* satisfying (1.1.19) forms an outward normal cone to K at x. This cone is empty for $x \notin K$, has at least the zero element for $x \in K$, and has only the zero element if $x \in \text{int } K$. We recall here that relint K (relative interior of K) is the interior of K, when K is regarded topologically as a subset of its affine hull. In several cases the set relint K plays an important role, e.g. if int $K = \emptyset$.

Subdifferentiability is closely related to the notion of "one-sided Gâteaux-differentiability." This provides a method for the "construction" of the subdifferential of a given functional.

A functional $f : X \rightarrow \bar{R}$, where X is a Banach space, is said to be one-sided directional Gâteaux-differentiable at x_0 if there exists $f'(x_0, h)$ such that

$$\lim_{\mu \to 0+} \frac{f(x_0 + \mu h) - f(x_0)}{\mu} = f'(x_0, h), \quad \forall h \in X. \qquad (1.1.20)$$

It should be noted that $+\infty$ and $-\infty$ are allowed as limits in (1.1.20). Functional $h \to f'(x_0, h)$ is the one-sided directional Gâteaux-differential of f at x_0 with respect to the direction h. It can be shown that $f'(x_0, \cdot)$ is a convex, positively homogeneous function of h.If $h \to f'(x_0, h)$ is continuous and linear, then f is said to be Gâteaux-differentiable at x_0. One important property of convex functionals is that they are one-sided directional Gâteaux-differentiable.

Proposition 1.5   Assume that $f : X \rightarrow \bar{R}$ is convex. Then f is one-sided directional Gâteaux-differentiable at every $x \in X$ with $f(x) \neq \pm\infty$. Moreover the following properties hold

$$f(x_1) - f(x) \geq f'(x, x_1 - x), \quad \forall x_1 \in X \qquad (1.1.21)$$

and

$$f'(x, x_1 - x) \geq -f'(x, -(x_1 - x)), \quad \forall x_1 \in X. \qquad (1.1.22)$$

If moreover f is bounded on a neighborhood of $x_0 \in X$, then

$$f'(x_0, h) = \max\{\langle x^*, h \rangle : x^* \in \partial f(x_0)\}, \quad \forall h \in X. \qquad (1.1.23)$$

Concerning the geometrical meaning of the inequalities (1.1.21)–(1.1.22) we refer to Fig. 1.1.1b on page 2.

This last proposition permits a simple construction (cf. Fig. 1.1.1b) of the set $\partial f(x)$: if f maps R into $\bar{R}$ then the subgradients x^* are the slopes of the non-vertical lines through $(x, f(x))$, which have no point in common with intepi f. From $f'(x, 1) = f'_+(x)$ and $f'(x, -1) = -f'_-(x)$ (right and left derivatives), it results that $f'_-(x) \leq x^* \leq f'_+(x)$. For f a convex, l.s.c., proper functional on R the right and left derivatives f'_+ and f'_- can be extended, when $x \notin D(f)$, by setting $f'_+ = f'_- = \infty$ (resp. $f'_+ = f'_- = -\infty$) for points lying to the right (resp. to the left) of $D(f)$. One can write then that

$$\partial f(x) = \{x^* \in R : f'_-(x) \leq x^* \leq f'_+(x)\}. \qquad (1.1.24)$$

Proposition 1.6 Suppose that $f : X \to \bar{R}$ is convex and that $\operatorname{grad} f(x)$ exists at x. Then $\partial f(x) = \{\operatorname{grad} f(x)\}$. Conversely, if f is finite and continuous at x and if $\partial f(x)$ has only one element, then $\operatorname{grad} f$ exists at x and $\partial f(x) = \{\operatorname{grad} f(x)\}$.

Examples illustrating the notion of the subdifferential are given in [Rock70, Pan85]. Here one only example (Fig. 1.1.2) is presented.

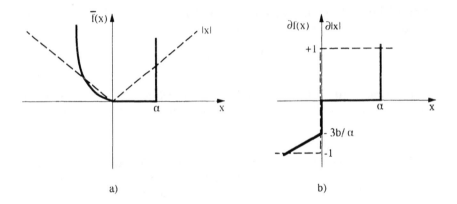

a)   b)

Fig. 1.1.2. The graphs of (1.1.25) and (1.1.26)

Let $f : R \to \bar{R}$, $f(x) = \bar{f}(x) + |x|$ where $\bar{f}(x)$ is given by $(b > 0)$

$$\bar{f}(x) = \begin{cases} \dfrac{3bx}{a}\left(\dfrac{x}{a} - 1\right) & \text{if } x \leq 0 \\ 0 & \text{if } 0 \leq x \leq a \\ \infty & \text{if } x > a. \end{cases} \qquad (1.1.25)$$

Then (cf.(1.1.24))

$$\partial f(x) = \begin{cases} \dfrac{3b}{a}\left(\dfrac{2x}{a} - 1\right) - 1 & \text{if } x < 0 \\ [-\dfrac{3b}{a} - 1, 1] & \text{if } x = 0 \\ 1 & \text{if } 0 < x < a \\ [1, \infty) & \text{if } x = a \\ \emptyset & \text{if } x > a. \end{cases} \qquad (1.1.26)$$

Now some propositions from the subdifferential calculus are given.

Proposition 1.7 Let $f : X \to \bar{R}$ and $\lambda > 0$. Then for every $x \in D(\partial f)$

$$\partial(\lambda f)(x) = \lambda \partial f(x). \tag{1.1.27}$$

Proposition 1.8 Let $f_1 : X \to \bar{R}$ and $f_2 : X \to \bar{R}$. Then for every $x \in D(\partial f_1) \cap D(\partial f_2)$

$$\partial f_1(x) + \partial f_2(x) \subset \partial(f_1 + f_2)(x). \tag{1.1.28}$$

This inclusion holds as an equality of sets, if certain additional assumptions hold. The next two Propositions concern this case.

Proposition 1.9 Assume that $f_1 : X \to \bar{R}$ and $f_2 : X \to \bar{R}$ are convex and that $\operatorname{grad} f_2$ exists at x and $x \in D(\partial f_1)$. Then

$$\partial(f_1 + f_2)(x) = \partial f_1(x) + \partial f_2(x). \tag{1.1.29}$$

Proposition 1.10 Let $f_1 : X \to (-\infty, +\infty]$ and $f_2 : X \to (-\infty, +\infty]$ be convex and l.s.c., and let $x_0 \in D(f_1) \cap D(f_2)$ exist at which f_1 is continuous. Then

$$\partial(f_1 + f_2)(x) = \partial f_1(x) + \partial f_2(x) \quad \forall x \in X. \tag{1.1.30}$$

Further we give a proposition concerning the subdifferential of a composite functional. Let X and Y be reflexive Banach spaces and let X^* and Y^* be their dual spaces. We denote by $L : X \to Y$ a linear continuous function and by $L^* : Y^* \to X^*$ its transpose.

Proposition 1.11 Let $f : Y \to (-\infty, +\infty]$ be convex and l.s.c. and let $y_0 = Lx_0$ exist at which f is finite and continuous. Then

$$\partial(f \circ L)(x) = L^* \partial f(Lx) \quad \forall x \in X. \tag{1.1.31}$$

Let us consider the minimization problem of a convex functional on a convex set $K \subset X$. We want to find a point x_0 which is a solution to the problem

$$f(x_0) = \inf\{f(x) : x \in K\} \text{ or } f(x_0) = \inf_K f(x). \tag{1.1.32a}$$

If f achieves the infimum on K at $x = x_0 \in K$, we write

$$f(x_0) = \min\{f(x) : x \in K\} \text{ or } f(x_0) = \min_K f(x). \tag{1.1.32b}$$

Let K be a nonempty convex subset of X and f a convex proper functional $f : K \to \bar{R}$. It is obvious that f can be extended to all of X, and hence the solution of (1.1.32b) is sought in X. Thus the minimization problem of f over K is equivalent to the minimization of $f + I_K$ over all of X. For x_0 to be a solution of this problem, it is necessary and sufficient that (cf. (1.1.16))

$$0 \in \partial(f(x_0) + I_K(x_0)). \tag{1.1.33}$$

Let us now assume that grad f exists at x_0. Then (1.1.33) is equivalent to the relation

$$-\text{grad } f(x_0) \in \partial I_K(x_0) \tag{1.1.34}$$

i.e $-\text{grad } f(x_0)$ is an element of the outward normal cone to K at x_0.

For any functional $f : X \to \bar{R}$ there arises the question as to whether an affine continuous function $x \to \langle x^*, x \rangle - \mu, \mu \in R$, can be determined which is a minorant of f, i.e

$$f(x) \geq \langle x^*, x \rangle - \mu \quad \forall x \in X. \tag{1.1.35}$$

A necessary and sufficient condition for it is that

$$\mu \geq \sup\{\langle x^*, x \rangle - f(x) : x \in X\}. \tag{1.1.36}$$

This relation introduces the conjugate functional f^c, which is defined on X^* by the relation

$$f^c(x^*) = \sup_{x \in X} (\langle x^*, x \rangle - f(x)). \tag{1.1.37}$$

Obviously the supremum in (1.1.37) may be taken only over $D(f)$. f^c can be regarded as the pointwise supremum of the family of affine continuous functionals $g(\cdot) = \langle \cdot, x \rangle - \mu$ with $(x, \mu) \in \text{epi } f$. We denote by $\Gamma(X)$ the set of functions $f : X \to \bar{R}$, which are the pointwise suprema of a family of affine continuous functionals $\langle x^*, \cdot \rangle + \alpha, \alpha \in R$ on X. The following propositions hold.

Proposition 1.12 The class $\Gamma(X)$ consists exactly of the convex, l.s.c., proper functionals $f : X \to \bar{R}$ and of the constants $+\infty$ and $-\infty$.

Further we denote by $\Gamma_0(X)$ the set of functionals $f \in \Gamma(X)$ such that $f \not\equiv \pm\infty$. It is obvious that $\Gamma_0(X)$ contains precisely the convex, l.s.c. and proper functionals on X.

Proposition 1.13 Suppose that f is a convex functional on X. Then f^c is a convex, l.s.c. functional on X^*. If, in addition f is proper, then f^c is proper as well, and conversely.

The conjugacy operation $f \to f^c$ can be seen as a one-to-one correspondence between $\Gamma_0(X)$ and $\Gamma_0(X^*)$ and is called Fenchel transformation (also Fenchel-Young, or Legendre-Fenchel, or polarity transformation). In the following proposition the relation between ∂f and ∂f^c is presented.

Proposition 1.14 Let f be a convex, proper functional on X. The following conditions are equivalent to one another

(i) $x^* \in \partial f(x)$; $\tag{1.1.38}$

(ii) $\sup_{z \in X} (\langle x^*, z \rangle - f(z))$ is achieved at $z = x$; $\tag{1.1.39}$

(iii) $f(x) + f^c(x^*) \leq \langle x^*, x \rangle$; and $\tag{1.1.40}$

(iv) $f(x) + f^c(x^*) = \langle x^*, x \rangle$. $\tag{1.1.41}$

If f is convex, proper and l.s.c, the foregoing conditions are equivalent to:

(v) $x \in \partial f^c(x^*)$; and   (1.1.42)

(vi) $\sup_{z^* \in X^*} (\langle z^*, x \rangle - f^c(z^*))$ is achieved at $z^* = x^*$ (1.1.43)

Suppose that $f = I_K$, where K is a nonempty, convex subset of X. Then I_K^c is given on X^* by

$$I_K^c(x^*) = \sup_{x \in K} \langle x^*, x \rangle.$$ (1.1.44)

I_K^c is called the support function of K. Now let K be a linear subspace M of R^n. Then the supremum is ∞, unless $\langle x^*, x \rangle = 0$, $\forall x \in M$. Accordingly, $I_K^c = I_{M^\perp}$, where $M^\perp$ is the orthogonal complement of M.

Let f be a nonconvex function and let epi f be its epigraph. We construct first the convex hull of epi f and then its closure, i.e. the closed convex hull of the epi f. This closed convex hull is the epigraph of a functional f_1 which is called the Γ-regularization of f. f_1 is the largest minorant of f in $\Gamma(X)$ and is the pointwise supremum of the affine continuous functions which are for every x less than f. Obviously if $f \in \Gamma(X)$ then $f_1 \equiv f$. Let us further define the conjugate functional f^{cc} of f^c given on X by the relation $f^{cc}(x) = (f^c)^c(x)$. f^{cc} is the Γ-regularization of f and for $f \in \Gamma(X)$, $f^{cc} = f$.

Note that for f nonconvex, f^c and f^{cc} are convex functionals . If $f_1 \leq f_2$ on X, then $f_1^c \geq f_2^c$ on X^* and since f^{cc} is the Γ-regularization of f we obtain that $f \geq f^{cc}$ on X, which implies that $f^c \leq f^{ccc}$. But

$$f^{ccc}(x^*) = \sup_{x \in X} \left\{ \langle x^*, x \rangle - f^{cc}(x) \right\} \leq f^c(x^*)$$ (1.1.45)

and therefore $f^{ccc} = f^c$ for every functional $f : X \to \bar{R}$.

In Sect. 1.3 we shall give some propositions from the duality theory of minimization problems according to Ekeland and Temam [Eke].

1.2 Elements of Nonconvex Nonsmooth Analysis

Suppose that X is a reflexive Banach space, X^* its dual space and let $\langle \cdot, \cdot \rangle$ be the duality pairing. Now let $A : X \to X^*$ be a multivalued operator (or a set-valued map, or a multivalued mapping, or a multifunction) and let

$$G(A) = \{(x, y) \in X \times X^* : y \in A(x)\}$$ (1.2.1)

be the corresponding graph. Then the inverse mapping A^{-1} is again a multivalued mapping $A^{-1} : X^* \to X$ defined by the relation

$$x \in A^{-1}(y) \Longleftrightarrow (x, y) \in G(A).$$ (1.2.2)

The set $D(A) = \{x : x \in X, A(x) \neq \emptyset\}$ is called the domain of the multifunction A. We say that a set-valued map satisfies a property (e.g closedness, measurability etc), if and only if this property is satisfied by its graph [Aub90].

A set-valued function F from X to Y (X, Y are metric spaces) is called upper semicontinuous (u.s.c.) at $x \in D(F)$, if and only if for any neighbourhood A of $F(x)$, there exists $\delta > 0$, such that for every $x' \in B(x, \delta)$, $F(x') \subset A$. If F is u.s.c. at any $x \in D(F)$, it is called upper semicontinuous.

In the case of nonsmooth functions, the upper and lower limits "replace" the limit in the classical definition of the derivative; there are many research efforts in this direction [Aub90]. In this Section we deal with the generalized gradient of F.H.Clarke, which leads to the theory of hemivariational inequalities. Closing we would like to remark that until now there is not an optimal generalization of the notion of differentiability for nonsmooth functions.

Let now C be a nonempty subset of the space X and let B denote the unit ball with center at zero. Then $x_0 + \delta B$ denotes the ball $B(x_0, \delta)$ defined in the previous Section. By writing $x_n \to_C x$ we mean that x_n converges to x in the set C. The subset K_C defined by

$$K_C(x) = \bigcap_{\delta > 0} \bigcap_{\alpha > 0} \bigcup_{0 < \mu \le \alpha} \left(\frac{1}{\mu}(C - x) + \delta B\right) \tag{1.2.3}$$

is called the contingent cone to C at the point x. The contingent cone is closed [Aub90, p.121]. From this definition it results [Aub84] that $y \in K_C(x)$ if and only if there exist a sequence of positive numbers μ_n and a sequence $\{u_n\} \in X$ such that as $n \to \infty$

$$u_n \to y, \mu_n \to 0 \quad \text{and} \quad \forall n \ge 0 \quad x + \mu_n u_n \in C. \tag{1.2.4}$$

Another equivalent definition of the contingent cone uses the distance function

$$d_C(x) = \inf\{||x - y|| : y \in C\}. \tag{1.2.5}$$

of x from the set C. Note that $d_C(\cdot)$ satisfies the relation

$$|d_C(x) - d_C(y)| \le ||x - y||. \tag{1.2.5a}$$

It can be shown that

$$y \in K_C(x) \iff \liminf_{\mu \to 0+} \frac{d_C(x + \mu y)}{\mu} = 0. \tag{1.2.6}$$

One can easily verify that for $x \in \text{int}\, C$, $K_C(x) = X$. Related to the contingent cone is the notion of the tangent cone. In order to give the definition of it, it is necessary to introduce the lower limit "liminf" of a multivalued mapping. Let X_1 be a subset of X and let F be a set-valued function from X_1 to Y, where Y is a Banach space. We define that for $x \in D(F)$

$$\liminf_{x^* \to x} F(x^*) = \{v \in Y : \lim d_{F(x^*)}(v) = 0 \text{ as } x^* \to x \; x^* \in D(F)\}. \tag{1.2.7}$$

The multifunction F is called by definition l.s.c. at x, if

$$\liminf_{x^* \to x} F(x^*) = F(x). \tag{1.2.8}$$

A useful result is that F is l.s.c. at $x \in D(F)$, if and only if for any $y \in F(x)$ and for any sequence $\{x_n\} \in D(F)$ with $x_n \to x$, a sequence $y_n \in F(x_n)$ can be determined which converges to y. For a set $C \in X$ and for $x \in X$, the tangent cone $T_C(x)$ to C at x is defined by the expression

$$T_C(x) = \liminf_{\substack{\tilde{x} \to_C x \\ \mu \to 0_+}} \frac{1}{\mu}(C - \{\tilde{x}\}); \tag{1.2.9}$$

Equivalently,

$$T_C(x) = \{y : y \in X, \text{ for } \mu_n \to 0_+, \text{ and } x_n \to_C x, \tag{1.2.10}$$
$$\text{there exists } y_n \to y \text{ with } x_n + \mu_n y_n \in C\}.$$

and

$$y \in T_C(x) \iff \lim_{\substack{\tilde{x} \to_C x \\ \mu \to 0_+}} \frac{d_C(\tilde{x} + \mu y)}{\mu} = 0. \tag{1.2.11}$$

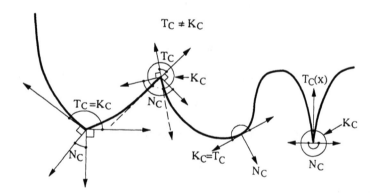

Fig. 1.2.1. A geometric representation of contingent, tangent and normal cones

$T_C(x)$ is a convex, closed cone which always contains 0. The cone $T_C(x)$ is always contained in $K_C(x)$, which is always contained in the closure of the set $\cup_{\mu > 0} \frac{1}{\mu}(C - \{x\})$. For $x \in \text{int } C$, $T_C(x) = X$, and for every $x \in X$, $T_X(x) = X$. By definition $T_\emptyset(x) = \emptyset$. The normal cone $N_C(x)$ to C at x is defined by the relation

$$N_C(x) = \{x^* : x^* \in X^*, (z, x^*) \le 0, \quad \forall z \in T_C(x)\}. \tag{1.2.12}$$

Obviously

$$T_C(x) = \{y : y \in X, (y, z^*) \le 0, \quad \forall z^* \in N_C(x)\}. \tag{1.2.13}$$

If C is convex, then $K_C(x) = T_C(x)$; if the boundary of C is continuously differentiable around a boundary point x, then $K_C(x) = T_C(x)$ and they are identified with the usual tangent vector space to C at this point. In Fig. 1.2.1 the geometical forms of the aforementioned cones with respect to certain types of set boundaries are depicted. We say that a set C is regular at x_0 if $T_C(x_0) = K_C(x_0)$.

Finally let us give the definition of the set $H_C(x)$ of the hypertangents to C at x: We say that $y \in H_C(x)$ if and only if there exists $\varepsilon > 0$ such that $v + tw \in C$ for all $\frac{v-x}{\varepsilon} \in B(0,1)$, $v \in C$, $\frac{w-y}{\varepsilon} \in B(0,1)$, $t \in (0,\varepsilon)$. If $H_C(x) \neq 0$ then $H_C(x) = \text{int} T_C(x)$ [Rock80].

We shall first define the generalized gradient only for Lipschitz functions according to the initially developed theory of Clarke [Clar75,81], [Aub79] and then we shall consider the more general case of $f : X \to \bar{R}$. We recall that $f : X \to \bar{R}$ is called locally Lipschitz at x, if a neighborhood U of x exists such that f is finite on U and

$$|f(x_1) - f(x_2)| \leq c||x_1 - x_2|| \quad \forall x_1, x_2 \in U, \tag{1.2.14}$$

where c is a positive constant depending on U. If f is locally Lipschitz at every $x \in X_1 \subset X$, then f is called locally Lipschitz on X_1. Function f is Lipschitz at x, for instance, if it is continuously differentiable on a neighborhood of x, or if it is convex (resp. concave) and bounded from above (resp. below) on a neighbourhood of x, or if it is a linear combination of Lipschitzian functions at x.

Let f be locally Lipschitz at $x \in X$ and let y be a vector in X. The directional differential in the sense of Clarke of f at x in the direction y, denoted by $f^0(x,y)$, is defined by the relation

$$f^0(x,y) = \limsup_{\substack{\mu \to 0_+ \\ h \to 0}} \frac{f(x + h + \mu y) - f(x + h)}{\mu}. \tag{1.2.15}$$

$f^0(x,y)$ is also called generalized directional differential and has the following properties .

Proposition 1.15 Let f be locally Lipschitz at x. Then i) $g : y \to f^0(x,y)$ is finite, convex, l.s.c., positively homogeneous and satisfies the inequality

$$|f^0(x,y)| \leq c||y||, \tag{1.2.16}$$

where c is the same constant as in (1.2.14).
ii) $(x,y) \to f^0(x,y)$ is u.s.c., and $g : y \to f^0(x,y)$ is locally Lipschitz at y.
iii) The following relations hold

$$\text{epi}\, g = T_{\text{epi}\, f}(x, f(x)) \tag{1.2.17}$$

$$f^0(x,-y) = (-f)^0(x,y). \tag{1.2.18}$$

By means of the directional differential $f^0(x,y)$ one can now define the generalized gradient $\bar{\partial} f(x)$.

Let $f : X \to R$ be a locally Lipschitz functional at $x \in X$. The following two equivalent definitions of the generalized gradient $\bar{\partial}f(x) : X \to X^*$ are given:

i) $\bar{\partial}f(x) = \{x^* : x^* \in X^*, f^0(x, x_1 - x) \geq \langle x^*, x_1 - x \rangle \quad \forall x_1 \in X\}$ (1.2.19)

and

ii) $\bar{\partial}f(x) = \{x^* : x^* \in X^*, (x^*, -1) \in N_{\text{epi}\,f}(x, f(x))\}.$ (1.2.20)

Note that $\bar{\partial}f(\cdot)$ is a multivalued mapping. The following propositions hold.

Proposition 1.16 Let $f : X \to R$ be a locally Lipschitz functional at $x \in X$. Then $\bar{\partial}f(x)$ is a nonempty convex, closed and bounded subset of X^*.

Proposition 1.17 Let $f : X \to R$ be a locally Lipschitz functional at $x \in X$. Then

$$f^0(x, y) = \max\{\langle y, x^* \rangle : x^* \in \bar{\partial}f(x)\}.$$ (1.2.21)

Until now we have defined the generalized gradient only for a Lipschitz function. Relation (1.2.20) can be used to define the generalized gradient $\bar{\partial}f(x)$ for any type of function $f : X \to \bar{R}$ which is finite at the point x: The set $\bar{\partial}f(x)$ is the set of all $x^* \in X^*$ such that (1.2.20) holds. Note that $\bar{\partial}f(x)$ may be empty. The above definition of $\bar{\partial}f(x)$ for any function $f : X \to \bar{R}$ makes sense, because the normal cone $N_C(x)$ can be defined with respect to any set epi f. Let us define now the generalized directional differential $f^\uparrow(x; y)$ at x in the direction y by the relation

$$f^\uparrow(x, y) = \sup\{\langle y, x^* \rangle : x^* \in \bar{\partial}f(x)\}.$$ (1.2.22)

Thus we can write that

$$\bar{\partial}f(x) = \{x^* : x^* \in X^*, f^\uparrow(x, x_1 - x) \geq \langle x^*, x_1 - x \rangle \quad \forall x_1 \in X\}.$$ (1.2.23)

The directional differential $f^\uparrow(x; y)$ is also called directional differential in the sense of Rockafellar, who has given for it another equivalent definition [Rock79,80]. Note that $\bar{\partial}f(x) = \emptyset$ if $f^\uparrow(x, 0) = -\infty$, and if $f^\uparrow(x, y)$ is finite for every y then $\bar{\partial}f(x) \neq \emptyset$. The following propositions hold [Rock80].

Proposition 1.18 Let $f : X \to \bar{R}$ and $f(x)$ finite. Then

i) $\bar{\partial}f(x)$ is a convex, closed subset of X^*.

ii) $g : y \to f^\uparrow(x; y)$ is convex, l.s.c. and positively homogeneous, if $f^\uparrow(x; y) > -\infty$ for all $y \in X$.

iii) relation (1.2.17) holds.

If X is a nonnecessarily reflexive Banach space then in Prop. 1.16 $\bar{\partial}f(x)$ is weakly⋆ compact and in Prop. 1.18 $\bar{\partial}f(x)$ is weakly⋆ closed [Clar83].

If f is convex, then

$$f^\uparrow(x,y) = \liminf_{\tilde{y} \to y} f'(x,\tilde{y}) \quad \forall y \in X, \qquad (1.2.24)$$

where $f'(\cdot,\cdot)$ denotes the one-sided directional Gâteaux differential (cf. (1.1.20)). If f is locally Lipschitz at x

$$f^\uparrow(x,y) = f^0(x,y) \quad \forall y \in X \qquad (1.2.25)$$

and if f is continuously differentiable at x

$$\bar{\partial} f(x) = \{\text{grad} f(x)\}. \qquad (1.2.26)$$

The indicator function I_C of a set C is defined as in the convex case, i.e. $I_C(x) = \{0 \text{ if } x \in C, \infty \text{ otherwise}\}$. It is proved [Rock79,80] that

$$\bar{\partial} I_C(x) = N_C(x) \qquad (1.2.27)$$

and

$$I_C^\uparrow(x,y) = I_{T_{C(x)}}(y). \qquad (1.2.28)$$

Note also the interesting property that $y \in T_C(x)$, if and only if $d_C^0(x,y) = 0$, with d_C given by (1.2.5). For f convex

$$\bar{\partial} f(x) = \partial f(x) \qquad (1.2.29)$$

and for f concave and bounded below on a neighborhood of x

$$\bar{\partial} f(x) = -\partial(-f)(x) \qquad (1.2.30)$$

at every x where f is finite. The following proposition holds.

Proposition 1.19 If f has at x_0 a finite local minimum, then

$$0 \in \bar{\partial} f(x_0). \qquad (1.2.31)$$

Moreover due to (1.2.29) and the convexity of $f^\uparrow(x,\cdot)$ we may write that

$$x^* \in \bar{\partial} f(x) \iff x^* \in \bar{\partial} f^\uparrow(x,0) = \partial f^\uparrow(x,0). \qquad (1.2.32)$$

Let us suppose now that $f, g : X \to R$ are Lipschitz functions at x. Then

$$\bar{\partial}(f+g)(x) \subset \bar{\partial} f(x) + \bar{\partial} g(x) \qquad (1.2.33)$$

and

$$\bar{\partial}(\lambda f)(x) = \lambda \bar{\partial} f(x) \text{ for } \lambda \in R. \qquad (1.2.34)$$

The finite dimensional case $X \equiv R^n$ is also important. Then for f locally Lipschitz at x a definition equivalent to the definition (1.2.19) is the following: $\bar{\partial} f(x)$ is the convex hull of all points $y \in R^n$ of the form

$$y = \lim_{i \to \infty} \operatorname{grad} f(x_i), \tag{1.2.35}$$

where x_i converges as $i \to \infty$ to x, avoiding the nondifferentiability points and any other points of a set of measure zero (in the sense of Lebesgue) and such that $\operatorname{grad} f(x_i)$ converges. Recall at this point Rademacher's theorem stating that a Lipschitz function f on an open subset of R^n is almost everywhere (a.e.) in the sense of Lebesgue measure differentiable.

An important notion is the notion of the substationarity [Rock79] of a functional $f : X \to \bar{R}$ at a point x_0. Point x_0 is a substationarity point of f if

$$0 \in \bar{\partial} f(x_0). \tag{1.2.36}$$

Equivalent to this definition is the statement that

$$f^{\uparrow}(x_0, y) \geq 0 \quad \forall y \in X. \tag{1.2.37}$$

Substationarity points are all the classical stationarity points, all the local minima, a large class of local maxima (e.g. if at a local maximum point x_0 there is y such that $\limsup\{[f(\tilde{x}+\mu\tilde{y}) - f(\tilde{x})]/\mu\} < \infty$, where $\tilde{x} \to x_0, f(\tilde{x}) \to f(x_0), \tilde{y} \to y, \mu \to 0_+$; then f is called locally Lipschitz at x_0 in the direction y), as well as all the saddle points. Point x is said to be a substationarity point of f with respect to a set C, if $f + I_C$ is substationary at x.

The notion of substationarity plays an important role in the theory of hemivariational inequalities because it permits the formulation of the propositions of substationary potential and complementary energy which generalize the corresponding classical minimum energy propositions in Mechanics [Pan81,82,83,84,85,93]. In the sequel we will calculate the generalized gradients of some important cases for the applications.

i) Suppose that f is a maximum-type function, i.e., $f = \max\{\varphi_i, \ldots, \varphi_m\}$ (cf. Fig. 1.2.2), where $\varphi_i = \varphi_i(x)$, $i = 1, \ldots, m$, $x \in R^n$ are continuously differentiable functions. Let us denote the sets $\{x : \varphi_i = f\}$ by A_i. It is easy to prove that f is a locally Lipschitz function and that

$$\bar{\partial} f(x) = \{\operatorname{grad} \varphi_i(x)\} \quad \text{if} \quad x \in A_i, \quad x \notin A_i \cap A_j \text{ etc.} \tag{1.2.38}$$

$$\bar{\partial} f(x) = \operatorname{co}\{\operatorname{grad} \varphi_i(x), \operatorname{grad} \varphi_j(x)\} \quad \text{if} \tag{1.2.39}$$

$$x \in A_i \cap A_j, x \notin (A_i \cap A_j) \cap A_k \text{ etc.}$$

If we introduce the notation $I(x) = \{i : \varphi_i(x) = f(x)\}$ then we may write that

$$\bar{\partial} f(x) = \operatorname{co}\{\operatorname{grad} \varphi_i(x) : \ i \in I(x)\}. \tag{1.2.40}$$

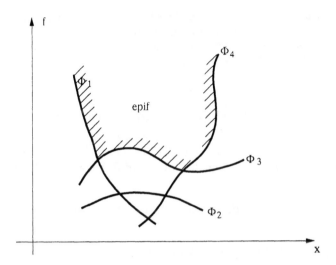

Fig. 1.2.2. A maximum type function

ii) Suppose also that $C = \{x \in R^n | f(x) \leq 0\}$. Then at a point x_0 with $f(x_0) = 0$ we have that

$$N_C(x_0) \subset \{\lambda x^* : x^* \in R^n, \lambda \geq 0, x^* \in \bar{\partial} f(x_0)\}, \qquad (1.2.41)$$

whenever f is Lipschitzian on a neighborhood of x_0 and $0 \notin \bar{\partial} f(x_0)$. If for a locally Lipschitz function at a point x

$$f^0(x,y) = f'(x,y) \quad \forall y \in X, \qquad (1.2.42)$$

f is called $\bar{\partial}$-regular at the point x. This definition is equivalent to the statement that epi f is regular at $(x, f(x))$. For instance, a convex function and a maximum type function as the one in i) are $\bar{\partial}$-regular at a point x where they are finite. If f and g are $\bar{\partial}$-regular at x then (1.2.33) holds as set equality. Similarly (1.2.41) holds as an equality if f is $\bar{\partial}$-regular at x_0. The combination of (1.2.40) with (1.2.41) leads for $f = \max\{\varphi_1, \ldots, \varphi_m\}$ to the relation

$$N_C(x_0) = \bar{\partial} I_C(x_0) = \qquad (1.2.43)$$
$$\{z : z = \sum_{i=1}^{m} \lambda_i \operatorname{grad} \varphi_i(x_0), \lambda_i \geq 0, \ \varphi_i(x_0) \leq 0, \ \lambda_i \varphi_i(x_0) = 0\},$$

if $0 \notin \bar{\partial} f(x_0)$. The above relation permits the extension of the Lagrange multiplier rule for optimization problems subjected to the nonconvex inequality constraints $\varphi_i(x) \leq 0$, $i = 1, \ldots, m$. This becomes obvious, e.g. if one considers the search for a local minimum problem of a continuously differentiable function $g : R^n \to R$ over $C = \{x \in R^n : \varphi_i(x) \leq 0 \ i = 1, \ldots, m\}$. A necessary condition is $0 \in \bar{\partial}(g + I_C)(x)$ which implies that

$$-\text{grad } g(x) \in \bar{\partial} I_C(x), \tag{1.2.44}$$

which together with (1.2.43) leads to the Lagrange multiplier rule.

iii) Further we calculate a generalized gradient which will be useful in Ch. 3. Suppose that $\beta : R \to R$ is a function such that $\beta \in L_{loc}^\infty(R)$, i.e. a function essentially bounded on any bounded interval of R. For any $\rho > 0$ and $\xi \in R$ let us define

$$\bar{\beta}_\rho(\xi) = \operatorname*{ess\,inf}_{|\xi_1 - \xi| \leq \rho} \beta(\xi_1) \quad \text{and} \quad \bar{\bar{\beta}}_\rho(\xi) = \operatorname*{ess\,sup}_{|\xi_1 - \xi| \leq \rho} \beta(\xi_1). \tag{1.2.45}$$

Obviously the monotonicity properties of $\rho \to \bar{\beta}_\rho(\xi)$ and $\rho \to \bar{\bar{\beta}}_\rho(\xi)$ imply that the limits as $\rho \to 0_+$ exist. Therefore one may write that

$$\bar{\beta}(\xi) = \lim_{\rho \to 0_+} \bar{\beta}_\rho(\xi) \quad \text{and} \quad \bar{\bar{\beta}}(\xi) = \lim_{\rho \to 0_+} \bar{\bar{\beta}}_\rho(\xi) \tag{1.2.46}$$

and define the multivalued function (cf. Fig. 1.2.3)

$$\tilde{\beta}(\xi) = [\bar{\beta}(\xi), \bar{\bar{\beta}}(\xi)] \tag{1.2.47}$$

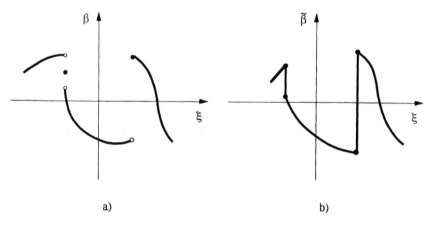

a)   b)

Fig. 1.2.3. On the definition of β and $\tilde{\beta}$

where $[\cdot, \cdot]$ denotes the interval. For instance, if at ξ, $\beta(\xi_+) > \beta(\xi_-)$ (resp. $\beta(\xi_+) < \beta(\xi_-)$) then $\tilde{\beta}(\xi) = [\beta(\xi_-), \beta(\xi_+)]$ (resp. $\tilde{\beta}(\xi) = [\beta(\xi_+), \beta(\xi_-)]$). Now we can apply a result proved by Chang [Ch]: a locally Lipschitz function j can be determined up to an additive constant by the relation

$$j(\xi) = \int_0^\xi \beta(\xi_1)d\xi_1 \qquad (1.2.48)$$

such that $\bar{\partial}j(\xi) \subset \tilde{\beta}(\xi)$. If moreover $\beta(\xi_\pm)$ exist for each $\xi \in R$ then

$$\bar{\partial}j(\xi) = \tilde{\beta}(\xi). \qquad (1.2.49)$$

1.3 Maximal Monotone Operators and Variational Inequalities

Let X and X^* be two vector spaces and A a multivalued mapping from X into X^*, i.e. precisely into the power set $\mathcal{P}(X^*)$ or 2^{X^*} of X^*. Considering A as a subset of $X \times X^*$, we can write $A(x) = \{y \in X^* : (x,y) \in A\}$. The set $D(A) = \{x : x \in X, A(x) \neq \emptyset\}$ is called the domain of A and the set $R(A) = \bigcup_x A(x), x \in X$, the range of A. Because A is multivalued, we will write $y \in A(x)$, where $x \in D(A)$ and $y \in X^*$. If A and B are two multivalued operators on X, then $\lambda A + \mu B, \lambda, \mu \in R$, is a multivalued operator mapping x into $\lambda A(x) + \mu B(x) = \{\lambda y + \mu z : y \in A(x), z \in B(x)\}$. Moreover, $D(\lambda A + \mu B) = D(A) \cap D(B)$. Suppose further that X is a reflexive Banach space and X^* is its dual space with the duality pairing $\langle x^*, x \rangle$ for $x \in X, x^* \in X^*$. The multivalued mapping $A : X \to 2^{X^*}$ is said to be monotone if

$$\langle y_1 - y_2, x_1 - x_2 \rangle \geq 0, \qquad (1.3.1)$$

$$\forall x_1, x_2 \in D(A), \quad \forall y_1 \in A(x_1), \quad \forall y_2 \in A(x_2).$$

If $\geq$ is replaced by $>$ for $x_1 \neq x_2$, then A is said to be strictly monotone.

Let f be a convex proper functional on X. Then it can be shown that ∂f is a monotone multivalued function from X into 2^{X^*}. The graph of the multivalued operator $A : X \to 2^{X^*}$ is a set $G(A) = \{(x,y) : (x,y) \in D(A) \times X^*, y \in A(x)\}$. Then $G(A_1) \subset G(A_2)$, if and only if $A_1(x) \subset A_2(x) \ \forall x \in D(A_1) \cap D(A_2)$. The set $\mathcal{A}$ of the monotone operators from X into 2^{X^*} can be partially ordered by graph inclusion. It can also be shown that every totally ordered subset of $\mathcal{A}$ has an upper bound. Then the Lemma of Zorn implies that $\mathcal{A}$ contains at least one maximal element, which is called a maximal monotone operator. Accordingly, a monotone operator $A : X \to 2^{X^*}$ is called maximal monotone if and only if $G(A) \subset G(B)$ implies that $A = B$, where $B : X \to 2^{X^*}$ is an arbitrary monotone operator, i.e., if and only if $G(A)$ is not properly contained in any other monotone subset of $X \times X^*$. From the above we obtain equivalently that an operator $A : X \to 2^{X^*}$ is called maximal monotone if and only if i) A is monotone and ii) for every $x \in X$ and $y \in X^*$ such that

$$\langle y - y_1, x - x_1 \rangle \geq 0, \quad \forall x_1 \in D(A), \quad \forall y_1 \in A(x_1) \qquad (1.3.2)$$

the relation

$$y \in A(x) \tag{1.3.3}$$

holds.

The following proposition relates the theory of maximal monotone operators to subdifferentiation.

Proposition 1.20 The subdifferential ∂f of a convex, l.s.c., proper functional f on X, where X is a reflexive Banach space, is a maximal monotone operator.

The class of the monotone operators $\beta : R \to 2^R$ is subsequently considered. A complete nondecreasing curve in R^2 is the graph $G(\beta)$ of a maximal monotone mapping $\beta : R \to 2^R$. In a Cartesian coordinate system such a graph is the graph of a continuous nondecreasing function, with the difference that it may contain vertical segments as well. The maximal monotone graphs in R^2 are applied for the formulation of monotone unilateral boundary conditions in Mechanics [Pan76,85]. A proposition now follows relating the complete nondecreasing curves in R^2 and the subdifferentials ∂f of convex, l.s.c and proper functionals on R.

Proposition 1.21 Let $\beta : R \to 2^R$ be a maximal monotone mapping. A convex, l.s.c., proper functional $f : R \to \bar{R}$ can be determined up to an additive constant such that

$$\beta = \partial f. \tag{1.3.4}$$

Thus the graphs of the subdifferentials ∂f, where $f \in \Gamma_0(R)$, are precisely the complete nondecreasing curves of R^2.

In the previous two Sections we have given some results connecting the search for a local or global minimum with the solution of a multivalued equation or inclusion of the type $0 \in \bar{\partial} f(x)$. In this Section some additional results are presented concerning the convex minimization problems, the corresponding variational inequalities and the duality of convex minimum problems. Let us consider the convex miminum problem (1.1.32b):
Find $x_0 \in K$ such that

$$f(x_0) = \min\{f(x) : x \in K\}, \tag{1.3.5}$$

where K is a nonempty convex closed subset of a reflexive Banach space X and $f : X \to (-\infty, +\infty], f(x) \not\equiv \infty$, is a convex, l.s.c. functional. The following proposition concerns the existence of a minimum over K.

Proposition 1.22 Let $||.||$ be the norm of the space X and let

$$\lim f(x) = \infty \text{ when } ||x|| \to \infty, x \in K \subset X, \tag{1.3.6}$$

or let

$$K \text{ be bounded.} \tag{1.3.7}$$

The problem (1.3.5) admits at least one solution. If f is strictly convex, the solution is unique.

The solutions of problem (1.3.5) constitute a convex closed subset of X. Some variational inequalities equivalent to problem (1.3.5) will now be presented.

Proposition 1.23 Let $f = f_1 + f_2$ be a proper functional, where f_1 and f_2 are convex, l.s.c. functionals on K and suppose, that $\operatorname{grad} f_1$ exists on X. For $x_0 \in K$, the following conditions are equivalent to each other:

$$f(x_0) = \inf_K f(x); \tag{1.3.8}$$

$$\langle \operatorname{grad} f_1(x_0), x - x_0 \rangle + f_2(x) - f_2(x_0) \geq 0, \quad \forall x \in K; \tag{1.3.9}$$

and

$$\langle \operatorname{grad} f_1(x), x - x_0 \rangle + f_2(x) - f_2(x_0) \geq 0, \quad \forall x \in K. \tag{1.3.10}$$

Further some elements of the duality theory of convex minimum problems, as it is developed by Ekeland and Temam [Eke], are given.

Let again f be a convex l.s.c. and proper functional on a reflexive Banach space X and let X^* be the dual space of X with the duality pairing $\langle \cdot, \cdot \rangle$. We consider the problem

$$\min\{f(x) : x \in X\} \tag{1.3.11}$$

which includes the problem (1.3.5) as a special case. (Set $f(x) = \infty$ for $x \notin K$). Problem (1.3.11) is called primal problem or problem I.

Let us further introduce a reflexive Banach space Y, and let Y^* be its dual space. For $p \in Y$ and $p^* \in Y^*$ we denote by $\langle\langle p, p^* \rangle\rangle$ the duality pairing between Y and Y^*. Then a functional $F : X \times Y \to \bar{R}$ is introduced such that

$$F(x', 0) = f(x'). \tag{1.3.12}$$

Then the problem

$$\inf_{x' \in X} F(x', p), \tag{1.3.13}$$

also called problem I_p, is considered. For $p = 0$ problem I_p coincides with problem I. Further, let F^c be the conjugate functional of F defined on $X^* \times Y^*$. We consider the problem

$$\sup_{p^{*'} \in Y^*} \{-F^c(0, p^{*'})\}, \tag{1.3.14}$$

which is called problem I^c. Problem I_p is a "perturbed" form of problem I, and I^c is the dual problem of I. Let us further denote by $\inf I$ and $\sup I^c$ the infimum and the supremum of problems I and I^c respectively. Inf I and $\sup I^c$ are the reals $f(x)$ and $\sup \{-F^c(0, p^{*'}) : p^{*'} \in Y^*\}$. One may easily verify that

$$-F^c(0, p^{*'}) \leq F(x', 0), \quad \forall p^{*'} \in Y^*, \quad \forall x' \in X \tag{1.3.15}$$

and, therefore, that

$$-\infty \leq \sup I^c \leq \inf I < \infty. \tag{1.3.16}$$

We denote now by g the function

$$g(p) = \inf I_p = \inf_{x' \in X} F(x', p) \tag{1.3.17}$$

and we assume that

$$F(x', p) \in \Gamma_0(X \times Y). \tag{1.3.18}$$

It can be proved that if (1.3.18) holds, $g : Y \to \bar{R}$ is convex. Further we consider the conjugate functional g^c of g. It holds that

$$g^c(p^*) = F^c(0, p^*), \quad \forall p^* \in Y^* \tag{1.3.19}$$

and

$$\sup I^c = \sup_{p^{*'} \in Y^*} \{-g^c(p^{*'})\} = g^{cc}(0). \tag{1.3.20}$$

Indeed (1.3.14) implies that for every p'

$$g^c(p^*) = \sup_{p \in Y} \left[\langle \langle p^*, p \rangle \rangle - g(p) \right] = \sup_{p \in Y} \sup_{x' \in X} \left[\langle \langle p^*, p \rangle \rangle - F(x', p) \right] = F^c(0, p^*). \tag{1.3.21}$$

Now we associate with I^c the "perturbed" problem $\sup\{-F^c(x^*, p^{*'}) : p^{*'} \in Y^*\}$. Hence the dual problem of I^c with respect to the perturbation x^* reads

$$\inf_{x^* \in X} \{F^{cc}(x^*, 0)\} \tag{1.3.22}$$

and is denoted by I^{cc}. On the assumption (1.3.18), $F^{cc}(x', 0) = F(x', 0)$ for every $x' \in X$, and thus problem I^{cc} coincides with the primal problem I. Moreover, since $P^{ccc} = P^c$, problem I^{ccc}, which results by continuing the dualization procedure, is identified with I^c. The following propositions concerns the important result $\inf I = \sup I^c$.

Proposition 1.24 On the assumption (1.3.18), the following three conditions are equivalent to each other:

(i) $-\infty < \inf I = \sup I^c < \infty$;

(ii) $g(0)$ is finite and g is l.s.c. at $p = 0$ (this is called the normality property of problem I).

(iii) Problem I^c has the normality property.

Proposition 1.25 Suppose that solutions to problems I and I^c exist and that

$$-\infty < \inf I = \sup I^c < \infty. \tag{1.3.23}$$

Then any solution x of I and any solution p^* of I^c satisfy the relation

$$(0, p^*) \in \partial F(x, 0). \tag{1.3.24}$$

Conversely, if x and p^* satisfy (1.3.24), then x is a solution of I, p^* is a solution of I^c and (1.3.23) holds.

Condition (1.3.24) is called the extremality condition of the problem and may equivalently be written as

$$F(x,0) + F^c(0,p') = 0. \tag{1.3.25}$$

We shall close this Section by giving certain propositions concerning the existence of solution of variational inequalities

i) Let $a(u,v)$ be for $u,v \in V$, where V is a real Hilbert space, a symmetric, continuous bilinear form and let (f,v) be a linear form. Then there exists a symmetric, bounded linear operator $A \colon V \to V^*$ such that

$$a(u,v) = (Au,v), \quad \forall u,v \in V. \tag{1.3.26}$$

Assume that $f \in V^*$, and let H be a pivot Hilbert space (i.e. $H \equiv H^*$) such that

$$V \subset H \subset V^* \tag{1.3.27}$$

holds where the injections are continuous and dense. The duality pairing is denoted by $(\cdot, \cdot)$, the norm on H by $|\cdot|$ and on V by $\|\cdot\|$. The following problem is now considered: find $u \in K$ such that

$$a(u, v-u) - (f, v-u) \geq 0, \quad \forall v \in K, \tag{1.3.28}$$

where K is a convex closed subset of V.

By Prop. 1.23 the solution $u \in K$ of (1.3.28), if any exists, is a solution of the minimization problem

$$\Pi(u) = \min\{\Pi(v) : v \in K\}, \tag{1.3.29}$$

where $\Pi(v) = \frac{1}{2}(Av, v) - (f, v)$, and conversely, on the assumption that $a(v,v) \geq 0 \ \forall v \in V$. Here first, we assume that $a(u,v)$ is coercive on V, i.e.,

$$a(u,u) \geq c\|u\|^2, \quad \forall u \in V, \quad c = \text{const} > 0. \tag{1.3.30}$$

Proposition 1.26 Suppose that (1.3.30) holds. Then the variational inequality (1.3.28) admits a unique solution.

The proof is based on (1.3.30) which guarantees that (1.3.6) holds. Next we deal with the semicoercive case. Let

$$\ker A = \{v : v \in V, \ Av = 0\}, \tag{1.3.31}$$

and let Q (resp. $\tilde{Q}$) be the orthogonal projector of V onto $\ker A$ in the topology of H (resp. of V). Let $P = I - Q$ (resp. $\tilde{P} = I - \tilde{Q}$), where I denotes the identity mapping, and assume that

$$a(v,v) \geq c|Pv|^2, \quad \forall v \in V, \quad c = \text{const} > 0. \tag{1.3.32}$$

Obviously, ker $A = \{v : v \in V,\ a(v,v) = 0\}$. Moreover we assume that

$$\text{ker } A \text{ is finite-dimensional} \tag{1.3.33}$$

and that

$$c_1(a(u,u)^{1/2} + |u|) \le \|u\| \le c_2(a(u,u)^{1/2} + |u|)\quad c_1, c_2 = \text{ const} > 0. \tag{1.3.34}$$

Accordingly $a(u,u)^{1/2}$ is a seminorm on V and $\||u\|| = a(u,u)^{1/2} + |u|$ is a norm on V equivalent to $\|u\|$. Let U be a subset of V which contains nonzero elements and let for $u \in U,\ u \ne 0$

$$p(u, U) = \sup_t \left\{ t : t \ge 0,\ \frac{tu}{\|u\|} \in U \right\}. \tag{1.3.35}$$

From (1.3.32) and (1.3.34) we obtain that

$$a(v,v) \ge c\|Pv\|^2 \quad \forall v \in V,\quad c \text{ const} > 0. \tag{1.3.36}$$

The kernel of f, ker $f = \{v : v \in V, (f,v) = 0\}$, is now considered. We symbolize by L the intersection of ker A and ker f and let L_1 be a subspace of ker A such that

$$\text{ker } A = L \oplus L_1. \tag{1.3.37}$$

Now we denote by $\bar{Q}$ and $\bar{\bar{Q}}$ the orthogonal projectors of V onto L and L_1, respectively in the topology of V, and by $\bar{P}$ the operator $I - \bar{Q}$. Easily we can show that $\bar{\bar{Q}} = \bar{P} - \tilde{P}$ and $\bar{\bar{Q}}v$ is orthogonal to $\tilde{P}v$. With respect to $u_0 \in K$, we denote by K_{u_0} the set $\{v : v \in V, v + u_0 \in K\}$. The following propositions hold:

Proposition 1.27 Assume that (1.3.32)–(1.3.34) hold and that a $u_0 \in K$ exists such that

(i) $(f, \rho) < 0$ for $\rho \in \text{ker } A \cap K_{u_0}$, $p(\bar{\bar{Q}} \rho, \bar{\bar{Q}}\, (\text{ker } A \cap K_{u_0})) = \infty$, and (1.3.38)

(ii) the set $\bar{P}(K_{u_0})$ is closed in V. (1.3.39)

Then (1.3.28) has a solution.

Proposition 1.28 Assume that (1.3.28) admits a solution. Then for any $u_0 \in K$ and any $\rho \in \text{ker } A \cap K_{u_0}$ such that $p(\bar{\bar{Q}} \rho, \bar{\bar{Q}}\, (\text{ker } A \cap K_{u_0})) = \infty$, the condition

$$(f, \rho) \le 0 \tag{1.3.40}$$

must hold.

   Let us now denote by $V/\text{ker } A$ the quotient space and by $[u]$ an element of it.

Proposition 1.29 If u is a solution of (1.3.28), then $[u] \in V/\text{ker } A$ is uniquely determined. Every other solution u' of (1.3.28) can be written as $u' = u + \rho$ where

$$(f, \rho) = 0 \quad \text{and} \quad \rho \in \ker A \cap K_{u_0}. \tag{1.3.41}$$

For the proofs of these propositions, we refer to [Pan85], where the initial proofs given by Fichera [Fich72], for $H \equiv V$ have been appropriately modified. For nonsymmetric bilinear forms see [Fich72].

ii) Further let us study a more general type of variational inequalities. Let V be a real Hilbert space and V^* its dual and let the functional framework of (1.3.28) hold. For $f \in V^*$ we want to find a $u \in V$ such that

$$(T(u), v - u) + \Phi(v) - \Phi(u) \geq (f, v - u) \quad \forall v \in V. \tag{1.3.42}$$

Here Φ is a convex l.s.c and proper functional on V and $T : V \to V^*$ is a pseudomonotone (generally singlevalued nonlinear) operator. We recall that an operator $T : X \to X^*$, where X is of a reflexive B-space is called pseudomonotone if

(i) T is bounded; and

(ii) for any sequence $\{x_n\} \in X$ such that $x_n \to x_0$ weakly in X and $\lim \sup \langle T(x_n), x_n - x_0 \rangle \leq 0$, the inequality

$$\lim \inf \langle T(x_n), x_n - x \rangle \geq \langle T(x_0), x_0 - x \rangle, \quad \forall x \in X \tag{1.3.43}$$

holds.

Note that if Φ is the indicator then (1.3.42) reduces to (1.3.28). In the following proposition we denote by $R(\partial \Phi)$ the range of the multivalued operator $\partial \Phi$ and by $\overline{R(\partial \Phi)}$ the closure of it.

Proposition 1.30 Assume that

(i)   the norm $\|v\|$ on V is equivalent to $p(v) + |v|$, where $p(v)$ and $|v|$ are a seminorm and a norm on V and on H respectively;

(ii) $Q = \{q : q \in V, p(q) = 0\}$ is a finite-dimensional subspace of V, Q is the orthogonal projection of V onto Q with respect to $|\cdot|$, $P = I - Q$, and

$$|Pv| \leq cp(v), \quad \forall v \in V, \quad c = \text{ const} > 0; \tag{1.3.44}$$

(iii) $(Tv, q) = 0, \quad \forall v \in V$ and $\forall q \in Q$;

and

(iv) there exists $b > 1$ such that

$$T(v, v) \geq c(p(v))^b, \quad c = \text{ const} > 0 \tag{1.3.45}$$

holds for every v.

Let us denote by Φ_{u_0} a convex, l.s.c., proper functional on Q defined by

$$\Phi_{u_0}(q) = \Phi(u_0 + q), \quad \text{for } u_0 \in D(\Phi) \tag{1.3.46}$$

and by $f|_Q \in Q^*$ the restriction of f to Q, i.e.

$$(f|_Q, v) = (f, v) \quad \forall v \in Q. \tag{1.3.47}$$

Then

(a) If assumptions (ii) and (iii) hold, the relation

$$f|_Q \in \overline{R(\partial\Phi_{u_0})}, \quad \forall u_0 \in D(\Phi) \tag{1.3.48}$$

is a necessary condition for the existence of a solution of (1.3.42);

(b) If assumptions (i) through (iv) are valid, then a sufficient condition for the existence of a solution of (1.3.42) is that there exists at least one $u_0 \in D(\Phi)$ such that

$$f|_Q \in \text{ relint } R(\partial\Phi_{u_0}). \tag{1.3.49}$$

For the proof of this proposition we refer to [Pot, Pan85]. Other results of coercive, semicoercive or noncoercive variational inequalities may be found in [Lio67,71, Fré71, Glow, Ba84,86,88, Gas88a,b, Boi, Kin, Scha].

1.4 On the Formulation of Hemivariational Inequalities and Related Topics

The theory of hemivariational inequalities constitutes a direct generalization of the theory of variational inequalities. Hemivariational inequalities arise in Mechanics, Engineering Sciences, and Economics in connection with nonconvex energy functionals or equivalently in connection with nonmonotone possibly multivalued laws, e.g. between stresses and strains or reactions and displacements in deformable bodies, between heat flux and temperature in thermal problems or between differentials and flow intensities (cf. e.g. [Prag, Oet, Pan82a]) in economic network problems. This Section is oriented towards applications and therefore can be omitted by a reader interested only in the mathematical theory of hemivariational inequalities. In order to illustrate the method for the formulation of hemivariational inequalities we shall consider, as a pilot problem, a linear elastic body subjected to certain nonmonotone multivalued boundary conditions which lead to hemivariational inequalities. We recall here that monotone multivalued boundary conditions give rise to variational inequalities [Duv72, Pan85].

We denote by Ω an open bounded subset of R^3 which is occupied by the body. The boundary of Ω is denoted by Γ and is assumed to be appropriately regular ($C^{0,1}$, i.e. a Lipschitzian boundary, is sufficient). The points $x \in \Omega, x = \{x_i\}, i = 1, 2, 3$, are referred to a Cartesian orthogonal coordinate system. We denote by $S = \{S_i\}$ the stress vector on Γ. We recall that $S_i = \sigma_{ij}n_j$, where $\sigma = \{\sigma_{ij}\}$ is an appropriately defined stress tensor and $n = \{n_i\}$ is the outward unit normal vector on Γ. The vector S may be decomposed into a normal component S_N and a tangential component S_T with respect to Γ, i.e.,

$$S_N = \sigma_{ij} n_j n_i \quad \text{and} \quad S_{T_i} = \sigma_{ij} n_j - (\sigma_{ij} n_i n_j) n_i. \tag{1.4.1}$$

Analogously to S_N and S_T, u_N and u_T denote the normal and the tangential components of the displacement vector u with respect to Γ. S_N and u_N are considered as positive if they are parallel to n. Let us first consider some monotone boundary conditions which are needed as a preliminary knowledge for the introduction of the nonmonotone boundary conditions.

A maximal monotone operator $\beta_N : R \to 2^R$ is introduced and a boundary condition of the form

$$-S_N \in \beta_N(u_N) \tag{1.4.2}$$

is considered in the normal direction. Then (Prop. 1.21) a convex, l.s.c and proper functional j_N on R may be determined up to an additive constant such that

$$\beta_N = \partial j_N. \tag{1.4.3}$$

Then (1.4.2) is written as

$$-S_N \in \partial j_N(u_N). \tag{1.4.4}$$

This relation is a monotone subdifferential boundary condition and obviously it may also be written in the inverse form (cf. Prop. 1.14)

$$u_N \in \partial j_N^c(-S_N) \tag{1.4.5}$$

and

$$u_N \in \beta_N^c(-S_N), \tag{1.4.6}$$

where $\beta_N^c = \partial j_N^c$ is again a maximal monotone operator on R and is the inverse operator of β_N. The graph of β_N, referred to a Cartesian system Oxy, is a complete nondecreasing curve in R^2 which is generally multivalued; thus the graph may include segments parallel to both coordinate axes. "Superpotential" j_N (resp. j_N^c) is a local superpotential (resp. conjugate superpotential) and expresses the potential (resp. complementary energy) of the contact constraint [Mor68,70, Pan76,85]. The notion of convex superpotentials has been introduced and first studied by Moreau [Mor68]. For the study of convex superpotentials we refer also to the monograph of the second author [Pan85].

Assume now that j is a convex, l.s.c., proper functional on R^3. Then a contact relation of the form

$$-S \in \partial j(u) \tag{1.4.7}$$

may be defined pointwise on Γ, i.e., as a monotone relation between $S(x)$ and $u(x)$. Equivalently to (1.4.7), we may write (cf. Prop. 1.14)

$$u \in \partial j^c(-S) \tag{1.4.8}$$

and

$$j(u) + j^c(-S) = -u_i S_i. \tag{1.4.9}$$

Similarly to (1.4.7), a subdifferential law

$$-S_T \in \partial j_T(u_T) \tag{1.4.10}$$

may be considered in the tangential direction to Γ.

All the above laws are multivalued and monotone laws between $-S_N$ and u_N, $-S$ and u etc. They include as special cases all the classical boundary conditions, e.g. $u_N = 0$ or $S_N = 0$ and several interesting classes of the so-called unilateral or inequality boundary conditions. Let us give an important unilateral condition, the Signorini boundary condition, which holds when an elastic body is in contact with a rigid support [Fich63,64,72, Duv72].

It reads (Fig. 1.4.1a)

$$\begin{aligned} \text{if } u_N < 0, \quad & \text{then } S_N = 0; \\ \text{if } u_N = 0, \quad & \text{then } S_N \leq 0, \end{aligned} \qquad (1.4.11)$$

or equivalently

$$S_N \leq 0, \quad u_N \leq 0, \quad \text{and} \quad S_N u_N = 0. \qquad (1.4.12)$$

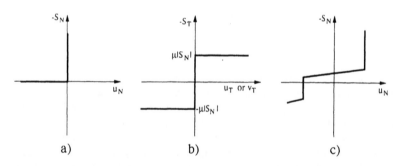

Fig. 1.4.1. a) The Signorini boundary condition b) The friction boundary condition for $\Omega \subset R^2$ c) A general subdifferential boundary condition

The respective operator β_N is

$$\beta_N(u_N) = \begin{cases} 0 & \text{if } u_N < 0 \\ [0, +\infty) & \text{if } u_N = 0 \\ \emptyset & \text{if } u_N > 0, \end{cases} \qquad (1.4.13)$$

and the corresponding convex superpotential

$$j_N(u_N) = \begin{cases} 0 & \text{if } u_N \leq 0 \\ \infty & \text{if } u_N > 0. \end{cases} \qquad (1.4.14)$$

In the tangential direction we assume that S_T or u_T are given.

The next example concerns friction conditions [Duv71,80, Mor70,86,88c]. We consider the following boundary conditions

$$\text{if } |S_T| < \mu|S_N|, \quad \text{then} \quad u_{T_i} = 0, \quad i = 1, 2, 3, \tag{1.4.15a}$$

if $|S_T| = \mu|S_N|$, then there exists $\lambda \geq 0$ such that $u_{T_i} = -\lambda S_{T_i}, \quad i = 1, 2, 3.$
$$\tag{1.4.15b}$$

Here $\mu = \mu(x) > 0$ denotes the coefficient of friction and $|\cdot|$ the usual R^3-norm. If Ω is a two-dimensional body, Γ is a curve, and thus S_T, u_T may be referred to a local right-handed coordinate system (n, τ) on Γ where τ denotes the unit vector tangential to Γ. Then (1.4.15) can be put in the form

$$-S_T \in \beta_T(u_T), \tag{1.4.16}$$

where (Fig. 1.4.1b)

$$\beta_T(u_T) = \begin{cases} [-\mu|S_N|, +\mu|S_N|] & \text{if } u_T = 0 \\ \mu|S_N| & \text{if } u_T > 0 \\ -\mu|S_N| & \text{if } u_T < 0. \end{cases} \tag{1.4.17}$$

Assume further that $S_N = C_N$, where C_N is given, and denote $\mu|C_N|$ by S_{T_0}. Then

$$\beta_T(u_T) = \partial(S_{T_0}|u_T|). \tag{1.4.18}$$

If Ω is a three-dimensional body, then (1.4.15) can be put only in the subdifferential form (1.4.10) with

$$j_T(u_T) = S_{T_0}|u_T|. \tag{1.4.19}$$

We can verify that

$$j_T^c(-S_T) = \begin{cases} 0 & \text{if } |S_T| \leq S_{T_0} \\ \infty & \text{otherwise}, \end{cases} \tag{1.4.20}$$

and thus (1.4.15) is equivalently written in the form

$$u_T \in \partial j_T^c(-S_T). \tag{1.4.21}$$

In dynamic problems, a friction law of the form

$$-S_T \in \partial j_T(v_T) = \partial(S_{T_0}|v_T|) \tag{1.4.22}$$

can be considered (Coulomb's law of friction). Here v_T denotes the tangential velocity which is equal to $\partial u_T/\partial t$ if the displacements are sufficiently small (e.g. in quasistatic problems).

In the normal direction the friction boundary conditions can be combined with any type of boundary conditions. For instance, one may consider the classical boundary conditions that S_N is given, or that a general monotone multivalued boundary condition of the type (1.4.2) (cf. e.g. Fig. 1.4.1c). In this context we refer to [Ant, Pan75,85].

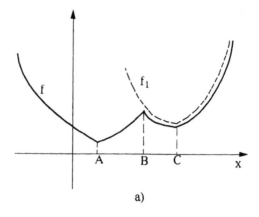

a)

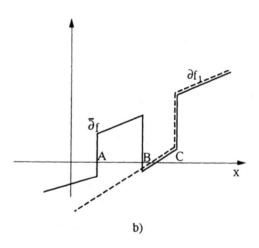

b)

Fig. 1.4.2. Convex and nonconvex superpotential laws

Further we consider nonmonotone boundary conditions expressed through the generalized gradient. They read

$$-S_N \in \bar{\partial} j_N(u_N) \tag{1.4.23}$$

$$-S_T \in \bar{\partial} j_T(u_T), \tag{1.4.24}$$

or

$$-S \in \bar{\partial} j(u), \tag{1.4.25}$$

where j is a locally Lipschitz functional and $\bar{\partial}$ is the generalized gradient of Clarke (Sect. 1.2). For instance, (1.4.25) is by definition equivalent to the inequality

$$j^0(u, v - u) \geq -S_i(v_i - u_i) \quad \forall v = \{v_i\} \in R^3, \tag{1.4.26}$$

where $j^0(\cdot, \cdot)$ is the directional differential in the sense of Clarke (cf. (1.2.15)). If j is not locally Lipschitz but any functional taking values in the extended real line $\bar{R}$, then in (1.4.26) $j^0(\cdot, \cdot)$ is replaced by the directional differential in the sense of Rockafellar $j^\uparrow(\cdot, \cdot)$. From the relation (1.2.29) we obtain that, for j, j_N and j_T convex, the relations (1.4.23), (1.4.24), (1.4.25) are identified with the relations (1.4.4), (1.4.10) and (1.4.7) respectively (Fig. 1.4.2). The generally nonconvex energy functionals in (1.4.23), (1.4.24) and (1.4.25) are called nonconvex superpotentials [Pan81,82,85]. The relations (1.4.23), (1.4.24), (1.4.25) describe nonmonotone, possibly multivalued relations, e.g. between $-S_N$ and u_N etc. Such nonmonotone multivalued boundary conditions do not lead to any type of variational formulations for the corresponding boundary value problems (B.V.Ps), unless they are expressed as nonconvex superpotential relations using the notion of the generalized gradient. Then they lead to a new type of variational expressions, the hemivariational inequalities.

The second author of the present book, who first introduced [Pan81,82, 83] and studied this new type of variational forms, has given to them the name "hemivariational inequalities" in order to point out the clear difference between them and the "variational inequalities." Note that if the superpotentials involved in a hemivariational inequality are convex, then the latter reduces to a variational inequality.

There exist several contact problems of elastic bodies which must be expressed through nonmonotone multivalued boundary conditions between $-S_N$ and u_N, between $-S_T$ and u_T, or generally between $-S$ and u.

The diagram of Fig. 1.4.3a concerns the adhesive contact problem. The adhesive material between body and support may sustain small tensile force. Then a debonding takes place which may obey the brittle type diagram ABCOD or the semibrittle diagram ABC'OD. Note that the vertical branch (i.e. the multivaluedness) is complete, i.e. for an appropriate loading the reaction and the normal boundary displacement u_N can define a point on the vertical branch.

In Fig. 1.4.3b,c certain nonmonotone friction laws for $\Omega \subset R^2$ are depicted. The first can be applied in geomechanics and rock interface analysis, whereas the second appears between reinforcement and concrete in a concrete structure. Finally, the law of Fig. 1.4.3d appears in the tangential direction in adhesively bonded parts and describes the partial cracking of the adhesive interface material due to slippage.

Using the relations (1.2.48) and (1.2.49) we obtain from the reaction displacement diagrams the nonconvex locally Lipschitz superpotentials j_N and j_T.

In Fig. 1.4.4 reaction-displacement laws with infinite branches are depicted. For instance, the law of Fig. 1.4.4a describes the adhesive contact with a rubber support which presents in compression ideal locking effects (the infinite branch

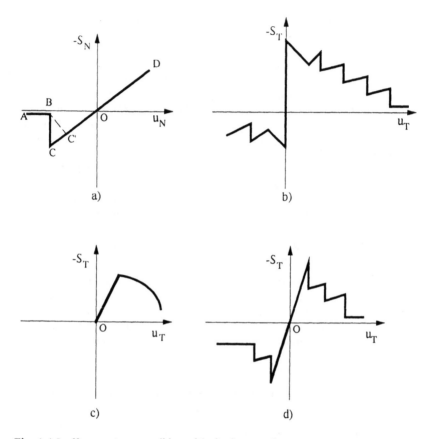

Fig. 1.4.3. Nonmonotone, possibly multivalued contact laws

EF), i.e. always $u_N \leq a$, whereas $u_N > a$ is impossible. Thus for $u_N = a$ the reaction may become infinite. In the case of adhesive contact with a rigid support, $a = 0$ and we have to consider the diagram ABCDF′. The diagram of Fig. 1.4.4b describes a nonmonotone friction law, where a mechanism does not permit a slip $u_T \notin [-b, b]$. Due to the infinite branches of the present laws the corresponding nonconvex superpotentials (resulting by "integration" along the horizontal axis) are no longer locally Lipschitz. Let us consider the law of Fig. 1.4.4a. It takes the form

$$
\begin{array}{lll}
\text{if} & u_N < a & \text{then} \quad -S_N \in \tilde{\beta}(u_N) \\
\text{if} & u_N = a & \text{then} \quad -\infty < -S_N \leq \tilde{\beta}(a) \\
\text{if} & u_N > a & \text{then} \quad S_N = \emptyset,
\end{array}
\qquad (1.4.27)
$$

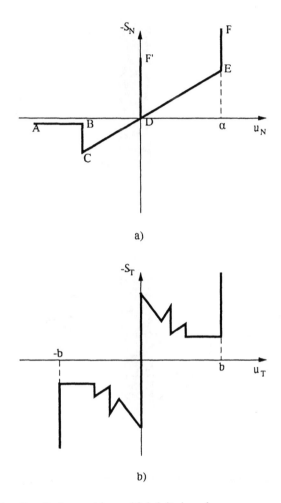

Fig. 1.4.4. Reaction-displacement laws with infinite branches

where $\tilde{\beta}$ is a multivalued function defined by (1.2.47). Relation (1.4.27) may be written as

$$-S_N \in \tilde{\beta}(u_N) + N_K(u_N) = \bar{\partial}j_N(u_N) + \partial I_K(u_N) \qquad (1.4.28)$$

where $K = \{u_N | u_N \le a\}$, $\bar{\partial}j_N$ results from $\tilde{\beta}$ as in (1.2.48) (here we assume that $\tilde{\beta}$ results from $\beta \in L^\infty_{\mathrm{loc}}(R)$ as in eqs. (1.2.45)–(1.2.49)), N_K is the normal cone to K at u_N and I_K is the indicator of the set K. Recall that $N_K(u_N) = \{0$ if $u_N < \alpha,\ [0,\infty)$ for $u_N = \alpha,\ \emptyset$ for $u_N > \alpha\}$. Relation (1.4.28) implies that

$$u_N \in R, \qquad j_N^o(u_N, u_N^\star - u_N) + I_K(u_N^\star) - I_K(u_N) \qquad (1.4.29)$$
$$\ge -\langle S_N, u_N^\star - u_N\rangle \quad \forall u_N^\star \in R$$

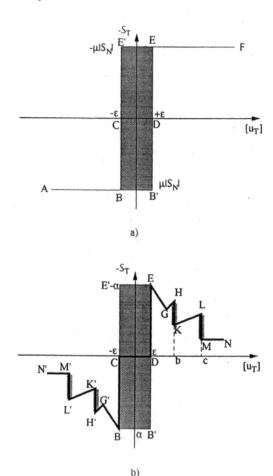

a)

b)

Fig. 1.4.5. Nonmonotone laws with nonfully determined values (fuzzy laws)

and gives rise to a variational-hemivariational inequality. Equivalently (1.4.29) can be written as

$$u_N \in K, \quad j_N^0(u_N, u_N^\star - u_N) \geq -\langle S_N, u_N^\star - u_N \rangle \quad \forall u_N^\star \in K. \quad (1.4.30)$$

This case can be extended to three-dimensional contact problems with an ideally locking support having a nonconvex locking criterion. We assume that $K = \{u \in R^3 | f_1(u) \leq 0, \ldots, f_n(u) \leq 0\}$, where $f_i, i = 1, \ldots, n$, are continuously differentiable functions such that K is a closed, but not necessarily convex subset of R^3. Thus, we may consider a generalization of (1.4.28) of the type

$$-S \in \bar{\partial}j(u) + \bar{\partial}I_K(u), \quad (1.4.31)$$

where $\bar{\partial}I_K(u)$ is given by (1.2.43) for $u \in K$ and $\bar{\partial}I_K(u) = \emptyset$ if $u \notin K$.

In Fig. 1.4.5 reaction displacement laws with nonfully determined values (fuzzy laws) in some regions are depicted. For instance, in Fig. 1.4.5a we give a friction law where $|S_T| \leq \mu|S_N|$ is verified as a strict inequality, not only for $u_T = 0$ but for $-\varepsilon \leq u_T < \varepsilon$, where ε is a small number, i.e. the reaction S_T may take for $-\varepsilon \leq u_T \leq \varepsilon$ any value in the interval $[-\mu|S_N|, \mu|S_N|]$. We say that the limit between adhesive frictional behaviour and sliding frictional behaviour is not clear. Therefore we have called this type of laws "fuzzy laws" in [Pan93] where they have for the first time been introduced. In order to describe such a law it is necessary to introduce the following nonconvex superpotential (see [Rock79]):

Let l be an open subset of the real line R and let M be a measurable subset of l such that for every open and nonempty subset I of l, mes $(I \cap M)$ and mes $(I \cap (l - M))$ are positive. Let

$$g(u) = \begin{cases} +b_1 & \text{if } u \in M \\ -b_2 & \text{if } u \notin M \end{cases} \tag{1.4.32}$$

and

$$f(u) = \int_0^u g(u^\star)du^\star. \tag{1.4.33}$$

Then f is Lipschitzian and it can be verified that

$$\bar{\partial}f(u) = [-b_2, b_1] \tag{1.4.34}$$

for every $u \in l$, i.e., we obtain an infinite number of jumps in l (Fig. 1.4.6).

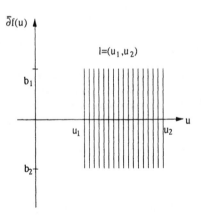

Fig. 1.4.6. The graph of (1.4.34)

Let us show how a superpotential of the type (1.4.34) may be applied to the description of fuzzy laws. We consider the law of Fig. 1.4.5. Indeed the law ABB′EE′F may be written in the form (fuzzy law)

$$-S_T \in \partial j([u_T]) + \bar{\partial} f([u_T]), \tag{1.4.35}$$

where j is convex and results from the maximal monotone graph ABCDEF according to Prop. 1.21, and f is given by the relations (1.4.32)–(1.4.34) with $b_1 = b_2 = \mu|S_N|$ and $l = (-\varepsilon, \varepsilon)$. Analogously is written the law N'M'L'K'H'G'BB'E'EGHKLMN. Now $\partial j([u_T])$ is replaced in (1.4.35) by $\bar{\partial} j([u_T])$, where j is nonconvex and results from the graph N'M'L'K'H'G'BCDEGHKLMN by means of the relation (1.2.48), (1.2.49). These diagrams describe monotone and nonmonotone interface friction laws with a nonfully determined behaviour around the adhesive friction region (fuzzy adhesive behaviour). Such nonfully determined regions may be considered around the complete vertical segments of the nonmonotone law (see the shaded areas around HK and LM in Fig. 1.4.5b). The expression of such nonfully determined laws in terms of nonconvex superpotentials permits for the first time the formulation of a variational theory for this class of problems.

After introducing the nonconvex superpotentials and the corresponding nonmonotone possibly multivalued boundary conditions we can proceed to the formulation of the corresponding hemivariational inequalities. We denote further by $H^1(\Omega)$ the classical Sobolev space and by $H^{1/2}(\Gamma)$ the Sobolev space of the traces on Γ of the functions $v_i \in H^1(\Omega)$. Let us denote further by $\langle \cdot, \cdot \rangle$ the duality pairing between $[H^{1/2}(\Gamma)]^3$ and its dual space.

We denote by H_T the space

$$H_T = \{ v | v \in [H^{1/2}(\Gamma)]^3, \ v_i n_i = 0 \text{ a.e. on } \Gamma \} \tag{1.4.36}$$

and we recall ([Pan85] p.32) that if $a = \{a_i\} \in [H^{1/2}(\Gamma)]^3$, and $a_N = a_i n_i$, $a_T = \{a_{T_i}\}$ where $a_{T_i} = a_i - a_N n_i$, then the mapping $a \to \{a_N, a_T\}$ is an isomorphism from $[H^{1/2}(\Gamma)]^3$ onto $H^{1/2}(\Gamma) \times H_T$. In the dual spaces, a'_N and a'_T are uniquely determined by the relation

$$\langle a', a \rangle = \langle a'_N, a_N \rangle_{1/2} + \langle a'_T, a_T \rangle_{H_T} \quad \forall a \in [H^{1/2}(\Gamma)]^3, \tag{1.4.37}$$

where $\langle \cdot, \cdot \rangle_{1/2}$ and $\langle \cdot, \cdot \rangle_{H_T}$ denote the duality pairings on $H^{1/2}(\Gamma) \times H^{-1/2}(\Gamma)$ and $H'_T \times H_T$. Obviously $a' \to \{a'_N, a'_T\}$ is again an isomorphism from $([H^{1/2}(\Gamma)]^3)'$ onto $H^{-1/2}(\Gamma) \times H'_T$. For all the above it is sufficient that Γ be $C^{1,1}$-regular ($C^{0,1}$-regularity, i.e. a Lipschitz boundary is also possible [Has82] with minor modifications).

Let $\Omega \subset R^3$ be an open bounded subset occupied by a deformable body in its undeformed state. On the assumption of small strains $\varepsilon = \{\varepsilon_{ij}\}$ we can write the relation

$$\int_\Omega \sigma_{ij}(u)\varepsilon_{ij}(v - u)d\Omega = (f, v - u) + \langle \sigma_{ij}n_j, (v_i - u_i) \rangle, \quad \forall v \in [H^1(\Omega)]^3, \tag{1.4.38}$$

where $u \in [H^1(\Omega)]^3$, $\sigma_{ij} \in L^2(\Omega)$, $f_i \in L^2(\Omega)$, and $(f, v) = \int_\Omega f_i v_i d\Omega$ $i, j = 1, 2, 3$. Relation (1.4.38), which is obtained from the "operator equations of the problem" by applying the Green-Gauss theorem, is the expression of the principle of virtual work for the body when it is considered to be free, i.e., with no constraint on its boundary Γ. Note that for the derivation of (1.4.38) we have multiplied the equilibrium equation

$$\sigma_{ij,j} + f_i = 0, \tag{1.4.39}$$

where the $f_i \in L^2(\Omega)$ is the volume force vector, by $v_i - u_i$ and then we have integrated over Ω. On the assumption of "appropriately smooth" functions, we have applied the Green-Gauss theorem by taking into account the strain-displacement relationship

$$\varepsilon_{ij} = \frac{1}{2}(u_{i,j} + u_{j,i}). \tag{1.4.40}$$

An extension by density of the arising functionals leads to the variational equality (1.4.38).

Let us assume further that the body is linear elastic, i.e. that

$$\sigma_{ij} = C_{ijhk}\varepsilon_{hk}, \tag{1.4.41}$$

where $C = \{C_{ijhk}\}$, $i, j, h, k = 1, 2, 3$, is the elasticity tensor which has the well-known symmetry and ellipticity properties

$$C_{ijhk} = C_{jihk} = C_{khij} \tag{1.4.42a}$$

$$C_{ijhk}\varepsilon_{ij}\varepsilon_{hk} \geq c\varepsilon_{ij}\varepsilon_{hk} \quad \forall \varepsilon = \{\varepsilon_{ij}\} \in R^6, \quad c = \text{const} > 0. \tag{1.4.42b}$$

We denote the bilinear form of linear elasticity by $a(\cdot, \cdot)$, i.e.

$$a(u, v) = \int_\Omega C_{ijhk}\varepsilon_{ij}(u)\varepsilon_{hk}(v)d\Omega, \tag{1.4.43}$$

where $C_{ijhk} \in L^\infty(\Omega)$. Note also that instead of (1.4.38) we can write the relation (cf. eq.(1.4.37))

$$\int_\Omega \sigma_{ij}\varepsilon_{ij}(v - u)d\Omega = (f, v - u) + \langle S_N, v_N - u_N \rangle_{1/2} + \langle S_T, v_T - u_T \rangle_{H_T}$$

$$\forall v \in [H^1(\Omega))]^3. \tag{1.4.44}$$

Let us assume now that on Γ the classical boundary conditions $S_{T_i} = 0$, $i = 1, 2, 3$, and $u_N = 0$ hold. Then (1.4.44) with (1.4.43) leads to the following variational equality: Find $u \in V_0 = \{v : v \in [H^1(\Omega)]^3, v_N = 0 \text{ on } \Gamma\}$ such that

$$a(u, v) = (f, v) \quad \forall v \in V_0. \tag{1.4.45}$$

Let us assume that on Γ the general nonmonotone possibly multivalued boundary condition

$$-S_N \in \bar{\partial}j_N(u_N), \quad S_{T_i} = 0 \quad i = 1, 2, 3 \tag{1.4.46}$$

holds, where j_N is generally a function taking values in the extended real line $\bar{R}$. It is then by definition equivalent to the inequality

$$j_N^\uparrow(u_N, v_N - u_N) \geq -S_N(v_N - u_N) \quad \forall v_N \in R, \qquad (1.4.47)$$

which combined with (1.4.44) and with the boundary condition with respect to S_T leads to the following variational expression, called hemivariational inequality: Find $u \in [H^1(\Omega)]^3$ such as to satisfy the inequality

$$a(u, v - u) + \int_\Gamma j_N^\uparrow(u_N, v_N - u_N)d\Gamma \geq (f, v - u) \quad \forall v \in [H^1(\Omega)]^3. \quad (1.4.48)$$

Here the second term is set equal to ∞ if the integral is not defined.

Obviously, if j_N is locally Lipschitz, (1.4.48) will be replaced by the hemivariational inequality: Find $u \in [H^1(\Omega)]^3$ such that

$$a(u, v - u) + \int_\Gamma j_N^0(u_N, v_N - u_N)d\Gamma \geq (f, v - u)d\Omega \quad \forall v \in [H^1(\Omega)]^3. \quad (1.4.49)$$

Let us assume now that j_N in (1.4.46) is a convex l.s.c and proper functional. We introduce the new functional on $[H^1(\Omega)]^3$

$$\Phi(v) = \begin{cases} \int_\Gamma j_N(v_N)d\Gamma & \text{if } j_N(\cdot) \in L^1(\Gamma) \\ \infty & \text{otherwise.} \end{cases} \qquad (1.4.50)$$

We know ([Pan85] p. 104 or [Bréz72]) that Φ is convex, l.s.c and proper on $[H^1(\Omega)]^3$. Then (1.4.48) reduces to the following variational inequality: Find $u \in [H^1(\Omega)]^3$ such that

$$a(u, v - u) + \Phi(v) - \Phi(u) \geq (f, v - u) \quad \forall v \in [H^1(\Omega)]^3. \qquad (1.4.51)$$

Note that if j_N is defined by (1.4.14), i.e. if the Signorini boundary condition holds, then (1.4.51) takes the form: Find $u \in K$, where

$$K = \{v : v \in [H^1(\Omega)]^3, \ v_N \leq 0 \text{ on } \Gamma\}, \qquad (1.4.52)$$

such that

$$a(u, v - u) \geq (f, v - u) \quad \forall v \in K. \qquad (1.4.53)$$

Indeed, if (1.4.14) holds, (1.4.51) holds by replacing $\Phi(v)$ by $I_K(v)$, the indicator of the convex closed set K, and thus it is equivalent to the variational inequality (1.4.53).

A combined variational expression having the characteristics of both the variational inequalities and the hemivariational inequalities results, when monotone boundary conditions hold on a part of the boundary and nonmonotone boundary conditions hold on the rest of it, or if the solution of a hemivariational inequality of the type (1.4.49) is constrained to belong to a convex closed subset of the solution space. This is the case, for instance, if the boundary conditions of Fig. 1.4.4 hold (cf. also (1.4.30)). Indeed the hemivariational inequality (1.4.49) for $v \in K \subset [H^1(\Omega)]^3$ reduces to the problem: Find $u \in [H^1(\Omega)]^3$ such as to satisfy

$$a(u, v - u) + I_K(v) - I_K(u) + \int_\Gamma j_N^0(u_N, v_N - u_N)d\Omega \geq (f, v - u) \quad \forall v \in [H^1(\Omega)]^3.$$

$$(1.4.54)$$

This variational hemivariational inequality is a special case of the problem: Find $u \in [H^1(\Omega)]^3$ such that

$$a(u, v - u) + \Phi(v) - \Phi(u) + \int_\Gamma j_N^0(u_N, v_N - u_N)d\Omega \geq (f, v - u) \quad \forall v \in [H^1(\Omega)]^3.$$

$$(1.4.55)$$

Here Φ is a convex, l.s.c, proper functional on $[H^1(\Omega)]^3$. Obviously the most general type of variational-hemivariational inequalities reads: Find $u \in V$ (e.g. a Hilbert space) such that

$$a(u, v - u) + \Phi(v) - \Phi(u) + \Phi_1^\uparrow(u, v - u) \geq (f, v - u) \quad \forall v \in V, \qquad (1.4.56)$$

where $\Phi_1 : V \to [-\infty, +\infty]$.

In the case of nonlinear elasticity the linear stress-strain law (1.4.41) may be replaced by a monotone, possibly multivalued law between the (symmetric) stress tensor $\sigma = \{\sigma_{ij}\}$ and the (symmetric) strain tensor $\varepsilon = \{\varepsilon_{ij}\}$ $i, j = 1, 2, 3$,

$$\sigma \in \partial w(\varepsilon), \qquad (1.4.57)$$

where w is a convex, l.s.c., proper functional on R^6, or by a nonmonotone possibly multivalued law

$$\sigma \in \bar{\partial} w(\varepsilon), \qquad (1.4.58)$$

where w is a locally Lipschitz functional on R^6 and $\bar{\partial}$ is the generalized gradient, defined by (1.2.19), or w is any functional on R^6 taking values on $[-\infty, +\infty]$ and $\bar{\partial}$ is generalized gradient defined by (1.2.23). Relation (1.4.57) describes the constitutive laws of the deformation theory of plasticity, of Hencky plasticity with convex yield function, of locking materials with convex locking function etc. (cf. [Pan85]) and (1.4.58) describes the sawtooth laws in composite materials, locking materials with nonconvex locking function etc (cf. [Pan85,93]). If (1.4.57) holds, then in the above variational formulations $a(u, v - u)$ must be replaced by $W(v) - W(u)$, where [Pana88c]

$$W(v) = \begin{cases} \int_\Omega w(\varepsilon(v))d\Omega & \text{if } w(\cdot) \in L^1(\Omega) \\ \infty & \text{otherwise.} \end{cases} \qquad (1.4.59)$$

If (1.4.58) holds and w is locally Lipschitz (resp. is not locally Lipschitz) then in the aforementioned variational formulations $a(u, v - u)$ must be replaced by $\int_\Omega w^0(\varepsilon(u), \varepsilon(v - u))d\Omega$ (resp. by $W(u, v - u) = \{\int_\Omega w^\uparrow(\varepsilon(u), \varepsilon(v - u))d\Omega$ if $w^\uparrow(\cdot) \in L^1(\Omega)$, ∞ otherwise$\}$.

Concerning the exact relationship of the above variational expressions, called also in Mechanics and Engineering Science variational "principles" for historical reasons, to the classical formulations of the B.V.Ps we refer to [Duv72, Pan85, Mor88a,b]. Here we briefly note that, usually, a solution of a variational formulation satisfies the operator equations and the boundary conditions of the problem in a generalized sense (e.g. as equalities in the sense of distributions over Ω, or in the sense of $H^{-1/2}(\Gamma)$ etc.). A more elaborate mathematical study may lead to additional regularity results for the variational solution of a B.V.P.

In Mechanics the aforementioned variational formulations express the principle of virtual work or power in equality or in inequality form. Following Hamel [Ham] we accept that this principle is one of the basic principles of Mechanics. Analogous variational formulations, which include the variations of the stresses may be derived. Then we speak about the "principle" of complementary virtual work ([Duv72, Pan85]). A B.V.P. is called bilateral (resp. unilateral) if it leads to a variational equality (resp. to a variational, or to a hemivariational, or to a variational-hemivariational inequality).

As Fourier has noticed [Lan] the inequality form of the principle of virtual work is due to the fact that the variations of certain variables involved into the problem are "irreversible." For instance, if (1.4.53) held for $u, v \in V$, where V is a vector space, then the substitution $v - u = \pm w$ would lead to a variational equality. But since $u, v \in K$, where K is a closed convex set, we cannot set $v - u = \pm w$, i.e., the variation $v - u$ is irreversible. Irreversible variations are called "unilateral" variations. The variations in (1.4.48) and (1.4.49) are also unilateral, unless $\operatorname{grad} j_N(\cdot)$ exists everywhere.

Concerning the mechanical aspects of the theory of hemivariational inequalities, the mathematical study of certain mechanical problems giving rise to hemivariational inequalities and their numerical treatment we refer the reader to the books and monographs [Pan85, Mor88a,b, Ant, Pan93] and to the papers [Pan87a,b,c,88,88a,89a,b,90,91,92a,b,c, Pana85a,b,88,89,91, Panag83,84,88,89, 90,91,92, Nan88, Nan89a,b, Nan92a,b, Nan93a,b, Nan94, Mot93, Miet92,93, Hasli89].

Closing this Chapter let us note that the variational inequalities and the hemivariational inequalities involve nonsmooth energy functionals and therefore they belong to Nonsmooth Mechanics [Mor88a,b], as this class of mechanical problems has been called by the second author in [Pan85 p.374]. Other topics of Nonsmooth Mechanics result after the introduction into Mechanics of Warga's derivate containers, Ioffe's fans, and Demyanov's quasidifferentials. In this context we refer to [Pan85,87b, Mor88a, Stav91,93a,b, Pan92].

2. Pseudo-Monotonicity and Generalized Pseudo-Monotonicity

In the first Section of this Chapter we present the main results of the theory of pseudo-monotone and generalized pseudo-monotone multivalued mappings developed by Browder and Hess. We confine ourselves to those parts of this theory that will be used in the study of hemivariational inequalities. The notion of pseudo-monotone operators first given in [Bréz68] has been applied to the treatment of multivalued mappings in [Bro72] and [Bro75]. For applications of the theory of pseudo-monotonicity in relation to elliptic problems we refer to [Lio69, Bro70b, Bro75, Zeid]. A self-contained exposition of the degree theory for nonlinear mappings of monotone type in reflexive Banach spaces can be found in [Bro83]. We refer also to [Avr, Com, Crou, Kara, Brü90,92] for other generalizations of the notions of convexity and monotonicity.

The second Section is devoted to the study of some general properties of functions having pseudo-monotone and generalized pseudo-monotone generalized gradients. Two classes of functions will be considered: the first class includes locally Lipschitz functions while the second one includes indicator functions of some nonconvex closed sets. We examine the relation between the pseudo-monotonicity of the generalized gradient of a locally Lipschitz function and the generalized pseudo-monotonicity of the normal cone mapping corresponding to the epigraph of the same function. Finally the notion of quasi-pseudo-monotonity is introduced and corresponding propositions are proved. Some results of this Section were first presented in [Nan92a] and [Nan89a].

2.1 Pseudo-Monotone and Generalized Pseudo-Monotone Mappings. Basic Properties

We start with the definition of pseudo-monotonicity.

Let T be a mapping from a real reflexive Banach space V into 2^{V^*}. Then T is said to be pseudo-monotone if the following conditions hold:

(a) The set Tu is nonempty, bounded, closed and convex for all $u \in V$;

(b) T is upper semicontinuous from each finite dimensional subspace F of V to the weak topology on V^*;

(c) If $\{u_i\}$ is a sequence in V converging weakly to u, and if $u_i^* \in Tu_i$ is such that $\limsup\langle u_i^*, u_i - u\rangle_V \leq 0$, then to each element $v \in V$ there exists $u^*(v) \in Tu$ with the property that

$$\liminf\langle u_i^*, u_i - v\rangle_V \geq \langle u^*(v), u - v\rangle_V. \tag{2.1.1}$$

In the next definition we introduce the class of generalized pseudo-monotone mappings.

A mapping T from V into 2^{V^*} is said to be generalized pseudo-monotone if the following holds: Let us consider a sequence $\{u_i\}$ in V and the corresponding sequence $\{u_i^*\}$ in V^* with $u_i^* \in Tu_i$. If $\{u_i\}$ converges weakly to u, $\{u_i^*\}$ converges weakly to u^* and

$$\limsup\langle u_i^*, u_i - u\rangle_V \leq 0, \tag{2.1.2}$$

then the element u^* lies in Tu and

$$\langle u_i^*, u_i\rangle_V \rightarrow \langle u^*, u\rangle_V. \tag{2.1.3}$$

Proposition 2.1 Let V be a reflexive Banach space, and $T : V \rightarrow 2^{V^*}$ a pseudo-monotone mapping from V into 2^{V^*}. Then T is generalized pseudo-monotone.

Proof. Let $\{(u_i, u_i^*)\}$ be a sequence of points of the graph $G(T)$ of T, i.e. of $G(T) = \{(v, v^*) \in V \times V^* : v^* \in Tv\}$, converging weakly to (u, u^*) in $V \times V^*$ while (2.1.2) holds. Since T is pseudo-monotone, for each $v \in V$ there exists $u^*(v) \in Tu$ such that

$$\liminf\langle u_i^*, u_i - v\rangle_V \geq \langle u^*(v), u - v\rangle_V. \tag{2.1.4}$$

By passing to a subsequence we may assume that $\langle u_i^*, u_i\rangle_V \rightarrow d$, where d is some real number. Then

$$\limsup\langle u_i^*, u_i - u\rangle_V = d - \langle u^*, u\rangle_V, \tag{2.1.5}$$

i.e., $d \leq \langle u^*, u\rangle_V$. Furthermore,

$$d - \langle u^*, v\rangle_V \geq \liminf\langle u_i^*, u_i - v\rangle_V \geq \langle u^*(v), u - v\rangle_V, \tag{2.1.6}$$

and thus

$$\langle u^*, u - v\rangle_V \geq \langle u^*(v), u - v\rangle_V \tag{2.1.7}$$

for all $v \in V$. Further we will show that $u^* \in Tu$. By condition (a) in the definition of pseudo-monotonicity, Tu is closed convex subset of V^*. If u^* does not lie in Tu, there will exist a $w \in V$ such that (due to Hahn-Banach theorem)

$$\langle u^*, w\rangle_V < \inf\{\langle z^*, w\rangle_V : z^* \in Tu\}; \tag{2.1.8}$$

setting $v = u - w$ in (2.1.7) leads to a contradiction.

Finally we note that

$$\liminf\langle u_i^*, u_i - u\rangle_V \geq \langle u^*(u), u - u\rangle_V = 0, \qquad (2.1.9)$$

i.e., $\liminf\langle u_i^*, u_i\rangle_V \geq \langle u^*, u\rangle_V$. Since we already know that $\limsup\langle u_i^*, u_i\rangle_V \leq \langle u^*, u\rangle_V$, we may obtain that $\langle u_i^*, u_i\rangle_V \to \langle u^*, u\rangle_V$.   q.e.d.

The following proposition is the converse of Proposition 2.1, on the additional assumption that T is bounded.

Proposition 2.2 Let V be a reflexive Banach space and let T be a bounded generalized pseudo-monotone mapping from V into 2^{V^*}. Assume that for each $u \in V$, Tu is a nonempty closed convex subset of V^*. Then T is pseudo-monotone.

Proof. It suffices to prove two results: firstly that T is upper semicontinuous from V into 2^{V^*} (with V^* equiped with the weak topology), and secondly that for any sequence $\{u_i\} \subset V$ converging weakly to some $u \in V$, for which $\limsup\langle u_i^*, u_i - u\rangle_V \leq 0$ with $u_i^* \in Tu_i$, we have that

$$\liminf\langle u_i^*, u_i - v\rangle_V \geq \langle u^*(v), u - v\rangle_V \qquad (2.1.10)$$

for each $v \in V$ and some $u^*(v) \in Tu$. Let us begin with the proof of the second result.

We note that the sequence $\{u_i^*\}$ is bounded. Suppose now that the pseudo-monotonicity condition (2.1.10) is not valid. Then there exists $v \in V$ such that

$$\liminf\langle u_i^*, u_i - v\rangle_V < \inf\{\langle z^*, u - v\rangle_V : z^* \in Tu\}. \qquad (2.1.11)$$

By passing to a subsequence, if necessary, we may assume that

$$\lim\langle u_i^*, u_i - v\rangle_V < \inf\{\langle z^*, u - v\rangle_V : z^* \in Tu\}. \qquad (2.1.12)$$

Using the fact that the sequence $\{u_i^*\}$ is bounded, we obtain by choosing a further subsequence, that $\{u_i^*\}$ converges weakly to some element $u^* \in V^*$. The generalized pseudo-monotonicity implies that $u^* \in Tu$, and that

$$\langle u_i^*, u_i\rangle_V \to \langle u^*, u\rangle_V. \qquad (2.1.13)$$

Hence

$$\lim\langle u_i^*, u_i - v\rangle_V = \langle u^*, u - v\rangle_V < \inf\{\langle z^*, u - v\rangle_V : z^* \in Tu\}, \qquad (2.1.14)$$

which is a contradiction.

To show the first result, we suppose that $\{u_i\}$ is a sequence in V converging strongly to $u \in V$, and such that $u_i^* \in Tu_i$ for each i. Let X be a weakly open neighborhood of Tu in V^*. We must show that $u_i^* \in X$ for i sufficiently large. Suppose that this is not true. Then by choosing a subsequence, we can assume that $u_i^* \in V^* \setminus X$ for all i. We can further assume that the bounded sequence

$\{u_i^*\}$ converges weakly to an element u^* of the weakly closed set $V^* \setminus X$. On the other hand,

$$\langle u_i^*, u_i - u \rangle_V \to 0 \qquad (2.1.15)$$

due to the strong convergence of $\{u_i\}$ to u and the boundedness of $\{u_i^*\}$. Hence, to each $v \in V$ there exists because of (2.1.15) and (2.1.10) an element $u^*(v) \in Tu$ such that

$$\langle u^*, u - v \rangle_V = \lim \langle u_i^*, u_i - v \rangle_V \geq \langle u^*(v), u - v \rangle_V. \qquad (2.1.16)$$

Since Tu is closed and convex, and since u^* does not belong to Tu, we can find an element $v_0 \in V$ such that

$$\langle u^*, u - v_0 \rangle_V < \inf\{\langle z^*, u - v_0 \rangle_V : z^* \in Tu\}, \qquad (2.1.17)$$

which yields, with (2.1.16), a contradiction. Hence T is upper semicontinuous from V into 2^{V^*}, where V^* has its weak topology, and V its strong topology. q.e.d.

Proposition 2.3 A maximal monotone mapping $T : V \to 2^{V^*}$ from the reflexive Banach space V into 2^{V^*} with effective domain $D(T) = V$ is pseudo-monotone.

Proof. We have to show that a maximal monotone mapping T with $D(T) = V$ satisfies the pseudo-monotonicity conditions (a)–(c).

(a) It is known that the image Tu of a maximal monotone mapping is closed and convex for each $u \in V$. The fact that a monotone operator is locally bounded at interior points of its domain implies that the set Tu is bounded. Obviously $Tu \neq \emptyset$.

(b) We prove that T is upper semicontinuous from the strong topology on V to the weak topology on V^*. Assume that for a given weakly open neighborhood X of Tu there exists a sequence $\{u_i\} \subset V$ with $u_i \to u$ and $u_i^* \in Tu_i$ such that $u_i^* \notin X$ for all i. The local boundedness of T at u implies, that the sequence $\{u_i^*\}$ is bounded in V^*, and we can find a subsequence converging weakly to an element u^*. As a consequence of the maximal monotonicity of T, $u^* \in Tu$. But on the other hand the weak closedness of $V^* \setminus X$ implies that $u^* \notin X$, which is a contradiction.

(c) Let $\{u_i\}$ be a sequence in V with $u_i \to u$ weakly, and let $u_i^* \in Tu_i$ be such that $\limsup \langle u_i^*, u_i - u \rangle_V \leq 0$. If u^* denotes an arbitrary element of Tu,

$$\langle u^*, u_i - u \rangle_V \leq \langle u_i^*, u_i - u \rangle_V \qquad (2.1.18)$$

by the monotonicity of T, where the left expression tends to 0 as $i \to \infty$. Hence $\langle u_i^*, u_i - u \rangle_V \to 0$. Let now $(z, z^*) \in G(T)$ be arbitrary. Since

$$\langle u_i^*, u_i - z \rangle_V = \langle u_i^*, u_i - u \rangle_V + \langle u_i^*, u - z \rangle_V, \qquad (2.1.19)$$

it follows that

$$\liminf \langle u_i^*, u_i - z \rangle_V = \liminf \langle u_i^*, u - z \rangle_V. \tag{2.1.20}$$

But $\langle z^*, u_i - z \rangle_V \leq \langle u_i^*, u_i - z \rangle_V$, with the left term converging to $\langle z^*, u - z \rangle_V$. Consequently,

$$\langle z^*, u - z \rangle_V \leq \liminf \langle u_i^*, u - z \rangle_V. \tag{2.1.21}$$

For a given $v \in V$ and $t > 0$, set $z_t = u + t(v - u)$. Let $z_t^* \in T z_t$. If we replace (z, z^*) in (2.1.21) by (z_t, z_t^*), we get

$$\langle z_t^*, u - v \rangle_V \leq \liminf \langle u_i^*, u - v \rangle_V. \tag{2.1.22}$$

By the local boundedness of T at u, we can assume the existence of sequences $t_k \to 0+$, $z_{t_k} \to u$ strongly and $z_{t_k}^* \to z^*(v)$ weakly. Again the maximal monotonicity of T implies that $z^*(v) \in Tu$. We further may write that

$$\langle z^*(v), u - v \rangle_V \leq \liminf \langle u_i^*, u - v \rangle_V = \liminf \langle u_i^*, u_i - v \rangle_V. \tag{2.1.23}$$

The proof is complete.   q.e.d.

The class of pseudo-monotone mappings is invariant under addition of operators.

Proposition 2.4 Let V be a reflexive Banach space, T_1 and T_2 two pseudo-monotone mappings from V into 2^{V^*}. Then $T_1 + T_2$ is pseudo-monotone.

Proof. For each $u \in V$, $(T_1 + T_2)u = T_1 u + T_2 u$ is a nonempty convex subset of V^* which is closed and bounded. Since T_1 and T_2 are upper semicontinuous from each finite dimensional subspace F of V into 2^{V^*}, with V^* having the weak topology, the same happens also for $T_1 + T_2$ as a consequence of the weak compactness of the sets $T_1 u$ and $T_2 u$ for each $u \in V$. It remains to verify that if $\{u_i\}$ is a sequence in V which converges weakly to $u \in V$, $u_i^* \in (T_1 + T_2)(u_i)$, and if

$$\limsup \langle u_i^*, u_i - u \rangle_V \leq 0, \tag{2.1.24}$$

then to each $v \in V$ an element $u^*(v) \in (T_1 + T_2)(u)$ can be determined such that

$$\liminf \langle u_i^*, u_i - v \rangle_V \geq \langle u^*(v), u - v \rangle_V. \tag{2.1.25}$$

By the definition of $T_1 + T_2$, for each integer i there exist elements w_i^* and z_i^* such that

$$u_i^* = w_i^* + z_i^*, \quad w_i^* \in T_1(u_i), \quad z_i^* \in T_2(u_i). \tag{2.1.26}$$

Hence, by (2.1.24), relation

$$\limsup [\langle w_i^*, u_i - u \rangle_V + \langle z_i^*, u_i - u \rangle_V] \leq 0 \tag{2.1.27}$$

holds. Now we will prove that (2.1.27) implies that

$$\limsup \langle w_i^*, u_i - u \rangle_V \leq 0, \quad \limsup \langle z_i^*, u_i - u \rangle_V \leq 0. \tag{2.1.28}$$

Indeed, suppose that (2.1.28) does not hold and let us assume e.g. that $\lim \sup \langle z_i^*, u_i - u \rangle_V > 0$. We can find $d > 0$ and a subsequence of $\{u_i\}$, which we identify for simplicity of notation with $\{u_i\}$, such that

$$\lim \langle z_i^*, u_i - u \rangle_V = d > 0. \tag{2.1.29}$$

Then from (2.1.29) and (2.1.27), we have that

$$\lim \sup \langle w_i^*, u_i - u \rangle_V \leq -d < 0. \tag{2.1.30}$$

It follows from the pseudo-monotonicity of T_1, that for each $v \in V$ there exists $w^*(v) \in T_1 u$ such that

$$\lim \inf \langle w_i^*, u_i - v \rangle_V \geq \langle w^*(v), u - v \rangle_V. \tag{2.1.31}$$

In particular, setting $v = u$, we obtain

$$\lim \inf \langle w_i^*, u_i - u \rangle_V \geq 0, \tag{2.1.32}$$

which with the preceding inequality (2.1.30) leads to a contradiction. This contradiction establishes the above assertions.

Since each T_1 and T_2 is pseudo-monotone, it follows that there exist elements $w^*(v)$ and $z^*(v)$ in $T_1(u)$ and $T_2(u)$ such that

$$\lim \inf \langle w_i^*, u_i - v \rangle_V \geq \langle w^*(v), u - v \rangle_V \tag{2.1.33}$$

and

$$\lim \inf \langle z_i^*, u_i - v \rangle_V \geq \langle z^*(v), u - v \rangle_V \tag{2.1.34}$$

for each $v \in V$. If we combine these inequalities, we obtain

$$
\begin{aligned}
\lim \inf \langle u_i^*, u_i - v \rangle_V &\geq \lim \inf \langle w_i^*, u_i - v \rangle_V + \lim \inf \langle z_i^*, u_i - v \rangle_V \\
&\geq \langle w^*(v), u - v \rangle_V + \langle z^*(v), u - v \rangle_V \\
&\geq \langle u^*(v), u - v \rangle_V,
\end{aligned}
\tag{2.1.35}
$$

where $u^*(v) = w^*(v) + z^*(v)$ lies in $(T_1 + T_2)u$. Thus the pseudo-monotonicity of $T_1 + T_2$ has been proved.   q.e.d.

We give now the following result concerning the range of pseudo-monotone mappings. To prove it we apply the following result which is a consequence of the Kakutani fixed point theorem (see e.g. [Bro72]).

Theorem 2.5 Let F be a finite-dimensional Banach space, T_0 a mapping from F into 2^{F^*} such that for each $u \in F$, $T_0 u$ is a nonempty bounded closed convex subset of F^*. Suppose that T_0 is coercive and upper semicontinuous from F to 2^{F^*}. Then range $R(T_0) = F^*$.

Now we will prove the following result.

Theorem 2.6 Let V be a reflexive Banach space and T a pseudo-monotone mapping from V into 2^{V^*}. Suppose that T is coercive, i.e., that there exists a real-valued function c on R^+ with

$$\lim_{r \to \infty} c(r) = \infty, \tag{2.1.36}$$

such that for all $(u, u^*) \in G(T)$, $\langle u^*, u \rangle_V \geq c(||u||_V)||u||_V$. Then T has as its range the whole space V^*, i.e., $R(T) = V^*$.

Proof. Let Λ be the family of all finite-dimensional subspaces F of V, ordered by inclusion. For $F \in \Lambda$, let $j_F : F \to V$ denote the inclusion mapping of F into V, and let $j_F^* : V^* \to F^*$ be its transpose operator. The mapping

$$T_F = j_F^* T j_F \tag{2.1.37}$$

then maps F into 2^{F^*}. For each $u \in F$, Tu is a weakly compact convex subset of V^* and j_F^* is continuous from the weak topology on V^* to the (unique) topology on F^*. Thus $T_F u$ is a nonempty closed convex subset of F^*. Since T is upper semicontinuous from V into 2^{V^*} with V^* having its weak topology, T_F is upper semicontinuous from F to 2^{F^*}.

Let (u, u_F^*) be an element of the graph of T_F. Then $u_F^* = j_F^* u^*$ where u^* is some element of Tu. It follows that

$$\langle u_F^*, u \rangle_F = \langle j_F^* u^*, u \rangle_F = \langle u^*, u \rangle_V \geq c(||u||_V)||u||_V, \tag{2.1.38}$$

where $c = c(r)$ is the coercivity function of T. Here the symbol $\langle \cdot, \cdot \rangle_F$ denotes the duality pairing over $F^* \times F$. Thus, each mapping T_F is coercive, having the same coercivity function as the original mapping T.

To complete the proof of our theorem it suffices to show, without loss of generality, that $0 \in R(T)$ (each variable is subjected to a translation by a constant quantity). By Theorem 2.5 to each $F \in \Lambda$ there exists an element $u_F \in F$ such that $0 \in T_F u_F$, i.e., $0 = j_F^* u_F^*$ for some $u_F^* \in Tu_F$. The coerciveness of T implies that

$$0 = \langle j_F^* u_F^*, u_F \rangle_F = \langle u_F^*, u_F \rangle_V \geq c(||u_F||_V)||u_F||_V. \tag{2.1.39}$$

Consequently, the elements $\{u_F\}$ are uniformly bounded in V by a constant M not depending on $F \in \Lambda$. For $F \in \Lambda$, let

$$W_F = \cup_{\substack{F' \in \Lambda \\ F' \supset F}} \{u_{F'}\}. \tag{2.1.40}$$

Then the set W_F is contained in the closed ball $B_V(0, M)$ in V with center 0 and radius M. Since $B_V(0, M)$ is weakly compact, and since the family $\{\text{weakcl}(W_F)\}$ has the finite intersection property (for any $F_1, \ldots, F_n \in \Lambda$, $W_{F_1} \cap \ldots \cap W_{F_n} \supset W_F$ with $F = F_1 \cup \ldots \cup F_n$), the intersection $\cap_{F \in \Lambda}$ $\{\text{weakcl}(W_F)\}$ is not empty. Here "weakcl(W_F)" denotes the closure of W_F in the weak topology of V. Let $\hat{u}$ be an element contained in this intersection. The proof will be completed if we show that $0 \in T\hat{u}$. Let $v \in V$ be an arbitrary element. We choose $F \in \Lambda$ such that it contains $\hat{u}$ and v. Let $\{u_{F_k}\}$ denote a sequence in W_F converging weakly to $\hat{u}$. Since $0 = j_{F_k}^* u_{F_k}^*$, we obtain that

$$\langle u_{F_k}^*, u_{F_k} - \hat{u}\rangle_V = 0 \qquad (2.1.41)$$

for all k. Consequently, due to the pseudo-monotonicity of T, to the given $v \in V$ a $u^*(v) \in T\hat{u}$ corresponds with

$$0 = \liminf \langle u_{F_k}^*, u_{F_k} - v\rangle_V \geq \langle u^*(v), \hat{u} - v\rangle_V. \qquad (2.1.42)$$

Suppose now that $0 \notin T\hat{u}$. Then 0 can be separated from the nonempty closed convex set $T\hat{u}$, i.e., there exists an element $w = \hat{u} - v \in V$ such that

$$0 < \inf\{\langle z^*, \hat{u} - v\rangle_V : z^* \in T\hat{u}\}. \qquad (2.1.43)$$

But this is a contradiction to (2.1.42). The proof is complete.   q.e.d.

Remark 2.7 The conclusion of Theorem 2.6 remains true when the coercivity condition, formulated there, is replaced by the following one: There exists a function $c : R^+ \to R$ with $c(r) \to \infty$ as $r \to \infty$, such that for all $(u, u^*) \in G(T), \langle u^*, u - u_0\rangle_V \geq c(\|u\|_V)\|u\|_V$ for some $u_0 \in V$. Indeed, the mapping $T_{u_0}(v) = T(v + u_0)$, $v \in V$, has the same pseudo-monotone properties as the original mapping T. Moreover, for all $(u, u^*) \in G(T_{u_0})$, $\langle u^*, u\rangle_V \geq c(\|u + u_0\|_V)\|u + u_0\|_V$. In this way there follows the existence of a $\hat{c} : R^+ \to R$, with $\hat{c}(r) \to \infty$ as $r \to \infty$, such that $\langle u^*, u\rangle_V \geq \hat{c}(\|u\|_V)\|u\|_V$ for any $(u, u^*) \in G(T_{u_0})$. But this implies that $R(T_{u_0}) = V^*$ and consequently, $R(T) = V^*$.

In the last part of this section we consider mappings of the form $T + \bar{T}$, with T pseudo-monotone from V into 2^{V^*} and $\bar{T}$ maximal monotone from V into 2^{V^*}. To study such mappings an approximation procedure constructed by Brézis-Crandall-Pazy [Bréz70] will be applied.

Let us recall that $J : V \to 2^{V^*}$ is said to be the normalized duality mapping of V into 2^{V^*} if

$$Ju = \{u^* \in V^* : \langle u^*, u\rangle_V = \|u^*\|_{V*}\|u\|_V = \|u\|_V^2\}. \qquad (2.1.44)$$

Proposition 2.8 Let V be a reflexive Banach space, $\bar{T}$ a maximal monotone mapping from V into 2^{V^*} with $(0, 0) \in G(\bar{T})$, and J the normalized duality mapping from V into 2^{V^*}. Let $\lambda > 0$. If we set

$$\bar{T}_\lambda u = \{w^* \in V^* : \text{ there exists } w \in D(\bar{T}) \text{ such that } w^* \in \bar{T}w, \lambda w^* \in J(u - w)\}, \qquad (2.1.45)$$

then

(i) Each $\bar{T}_\lambda$ is bounded maximal monotone mapping from V into 2^{V^*} with effective domain $D(\bar{T}_\lambda) = V$.

(ii) For each $v^* \in \bar{T}_\lambda u$ and $u^* \in \bar{T}u$,

$$||v^\star||_{V^*} \leq ||u^\star||_{V^*}. \tag{2.1.46}$$

Proof. By definition, $w^\star$ lies in $\bar{T}_\lambda u$ if and only if there exists $w \in V$ such that

$$w^\star \in \bar{T}w, \qquad \lambda w^\star \in J(u - w), \tag{2.1.47}$$

or equivalently

$$w \in \bar{T}^{-1}w^\star, \qquad u - w \in J^{-1}(\lambda w^\star) = \lambda J^{-1}(w^\star). \tag{2.1.48}$$

This last pair of conditions is equivalent to

$$u \in (\bar{T}^{-1} + \lambda J^{-1})w^\star. \tag{2.1.49}$$

Hence

$$\bar{T}_\lambda = (\bar{T}^{-1} + \lambda J^{-1})^{-1}. \tag{2.1.50}$$

Since $\bar{T}$ is maximal monotone from V into 2^{V^*}, $\bar{T}^{-1}$ is maximal monotone from V^* into 2^V. J^{-1} is the normalized duality mapping from V^* into 2^V and is bounded, coercive and maximal monotone. Hence, by the known results concerning coercive maximal monotone operators, the range of $\bar{T}^{-1} + \lambda J^{-1}$ coincides with the whole space V (cf. [Bro75]). Consequently $\bar{T}_\lambda$, its inverse, is a maximal monotone mapping from V into 2^{V^*} which is bounded and defined onto V. The proof of assertion (i) is thus complete.

Let $v^\star \in \bar{T}_\lambda u, u^\star \in \bar{T}u$, and let w denote an element corresponding by definition to $v^\star$, such that $v^\star \in \bar{T}w$, $\lambda v^\star \in J(u-w)$. Since $\langle u^\star - v^\star, u - w \rangle_V \geq 0$, by the monotonicity of $\bar{T}$ we have

$$||u - w||_V^2 = \lambda \langle v^\star, u - w \rangle_V \leq \lambda \langle u^\star, u - w \rangle_V. \tag{2.1.51}$$

Therefore

$$||u - w||_V^2 \leq \lambda ||u^\star||_{V^*} ||u - w||_V, \tag{2.1.52}$$

and thus

$$||u - w||_V \leq \lambda ||u^\star||_{V^*}. \tag{2.1.53}$$

Since $||v^\star||_{V^*} = \lambda^{-1}||u - w||_V$, we obtain $||v^\star||_{V^*} \leq ||u^\star||_{V^*}$, which proves (ii). q.e.d.

The following lemma will be applied further.

Lemma 2.9 Let T be a generalized pseudo-monotone mapping from V into 2^{V^*}, and $\{(u_i, u_i^\star)\}$ a sequence in $G(T)$ converging weakly to some $(u, u^\star) \in V \times V^*$. Suppose that

$$\limsup_{i,k \to \infty} \langle u_i^\star - u_k^\star, u_i - u_k \rangle_V \leq 0. \tag{2.1.54}$$

Then $(u^\star, u) \in G(T)$ and

$$\langle u_i^\star, u_i \rangle_V \to \langle u^\star, u \rangle_V. \tag{2.1.55}$$

Proof. Since T is assumed to be generalized pseudo-monotone, it suffices to prove that $\limsup\langle u_i^*, u_i\rangle_V \leq \langle u^*, u\rangle_V$. By passing to a subsequence, we may assume that $\langle u_i^*, u_i\rangle_V \to p$ for some real number p. We prove that $p \leq \langle u^*, u\rangle_V$. Due to (2.1.54) for a given $\varepsilon > 0$ there exists $K(\varepsilon)$ such that for $i, k \geq K(\varepsilon)$,

$$\langle u_i^* - u_k^*, u_i - u_k\rangle_V \leq \varepsilon. \tag{2.1.56}$$

We write this inequality in the form

$$\langle u_i^*, u_i\rangle_V + \langle u_k^*, u_k\rangle_V \leq \langle u_i^*, u_k\rangle_V + \langle u_k^*, u_i\rangle_V + \varepsilon. \tag{2.1.57}$$

Now we keep i fixed with $i \geq K(\varepsilon)$, and let $k \to \infty$. We obtain

$$\langle u_i^*, u_i\rangle_V + p \leq \langle u^*, u_i\rangle_V + \langle u_i^*, u\rangle_V + \varepsilon. \tag{2.1.58}$$

Now let $i \to \infty$. Then $2p \leq 2\langle u^*, u\rangle_V + \varepsilon$. Since this is true for all $\varepsilon > 0$, the desired estimate $p \leq \langle u^*, u\rangle_V$ follows.   q.e.d.

Lemma 2.10 Let V be a reflexive Banach space, $\bar{T}$ a maximal monotone mapping from V into 2^{V^*} with $(0,0) \in G(\bar{T})$, and T a pseudo-monotone operator from V into 2^{V^*}. Suppose that for a given $g_0 \in V^*$, the following condition holds: For each λ with $0 < \lambda \leq \lambda_0$,

$$g_0 = w_\lambda^* + z_\lambda^*, \tag{2.1.59}$$

where

$$w_\lambda^* \in \bar{T}_\lambda u_\lambda, \quad z_\lambda^* \in T u_\lambda, \tag{2.1.60}$$

with

$$||w_\lambda^*||_{V^*} \leq M, \quad ||u_\lambda||_V \leq M, \quad (0 < \lambda < \lambda_0), \quad M = \text{const} > 0. \tag{2.1.61}$$

Then $g_0 \in R(T + \bar{T})$.

Proof. Since $w_\lambda^* \in \bar{T}_\lambda u_\lambda$, the definition of $\bar{T}_\lambda$ implies that $\lambda w_\lambda^* \in J(u_\lambda - w_\lambda)$ and $w_\lambda^* \in \bar{T} w_\lambda$ for a suitable element w_λ of V. It follows from the hypotheses that

$$||u_\lambda - w_\lambda||_V = \lambda ||w_\lambda^*||_{V^*} \leq \lambda M \leq \lambda_0 M, \tag{2.1.62}$$

and hence

$$||w_\lambda||_V \leq (\lambda_0 + 1)M. \tag{2.1.63}$$

If we consider the sequential weak compactness of closed balls in V and V^*, we may find a sequence $\lambda_i \to 0_+$ such that with the notation

$$w_i^* = w_{\lambda_i}^*, z_i^* = z_{\lambda_i}^*, u_i = u_{\lambda_i}, w_i = w_{\lambda_i}, \tag{2.1.64}$$

the sequences $\{u_i\}$ and $\{w_i\}$ converge weakly to a common limit u_0 as $i \to \infty$, while $\{w_i^*\}$ converges weakly to w_0^* in V^*; $\{z_i^*\}$ converges weakly to z_0^*, with $w_0^* + z_0^* = g_0$. It suffices to prove that $w_0^* \in \bar{T} u_0$ and $z_0^* \in T u_0$.

Let i and k be two positive integers. Then

$$\langle w_i^* - w_k^*, w_i - w_k \rangle_V + \langle z_i^* - z_k^*, w_i - w_k \rangle_V = 0. \qquad (2.1.65)$$

By the monotonicity of $\bar{T}$, $\langle w_i^* - w_k^*, w_i - w_k \rangle_V \geq 0$, so that

$$\limsup_{i,k\to\infty} \langle z_i^* - z_k^*, w_i - w_k \rangle_V \leq 0. \qquad (2.1.66)$$

We observe that

$$\limsup_{i,k\to\infty} \langle z_i^* - z_k^*, u_i - u_k \rangle_V = \limsup_{i,k\to\infty} \langle z_i^* - z_k^*, w_i - w_k \rangle_V$$
$$+ \lim_{i,k\to\infty} \langle z_i^* - z_k^*, (u_i - w_i) - (u_k - w_k) \rangle_V \leq 0 + 0 = 0. \qquad (2.1.67)$$

The pseudo-monotonicity of T and Lemma 2.9 imply that $z_0^* \in Tu_0$ and $\lim \langle z_i^*, u_i - u_0 \rangle_V = 0$. Consequently,

$$\lim \langle z_i^*, w_i - u_0 \rangle_V = \lim \langle z_i^*, u_i - u_0 \rangle_V + \lim \langle z_i^*, w_i - u_i \rangle_V = 0 + 0 = 0. \qquad (2.1.68)$$

For each i,

$$\langle w_i^* - w_0^*, w_i - u_0 \rangle_V + \langle z_i^* - z_0^*, w_i - u_0 \rangle_V = 0. \qquad (2.1.69)$$

We conclude that

$$\lim \langle w_i^*, w_i - u_0 \rangle_V = 0. \qquad (2.1.70)$$

By the generalized pseudo-monotonicity of the maximal monotone mapping $\bar{T}$, we have that $w_0^* \in \bar{T}u_0$, which proves the assertion.   q.e.d.

To formulate the main result for the range of the sum of maximal monotone and pseudo-monotone mappings let us recall the following definitions [Bro72].

Let T be a mapping from V into 2^{V^*}. Then T is said to be quasi-bounded, if for each $M > 0$ there exists $K(M) > 0$ such that, whenever (u, u^*) lies in the graph $G(T)$ of T and

$$\langle u^*, u \rangle_V \leq M \|u\|_V, \quad \|u\|_V \leq M, \qquad (2.1.71)$$

then

$$\|u^*\|_{V^*} \leq K(M). \qquad (2.1.72)$$

T is strongly quasi-bounded if for each $M > 0$ there exists $K(M) > 0$ such that for all $(u, u^*) \in G(T)$ with

$$\langle u^*, u \rangle_V \leq M, \quad \|u\|_V \leq M, \qquad (2.1.73)$$

we have

$$\|u^*\|_{V^*} \leq K(M). \qquad (2.1.74)$$

Theorem 2.11 Let V be a reflexive Banach space and $\bar{T}$ a maximal monotone mapping from V into 2^{V^*} with $0 \in D(\bar{T})$. Let T be a pseudo-monotone mapping from V into 2^{V^*} and let c be a real-valued function on R^+ with $c(r) \to \infty$ as $r \to \infty$, such that for all $(u, u^*) \in G(T)$, $\langle u^*, u \rangle_V \geq c(||u||_V)||u||_V$. Suppose further that either T is quasi-bounded, or $\bar{T}$ is strongly quasi-bounded. Then $R(T + \bar{T}) = V^*$.

Proof. We may assume without loss of generality (the variable is subjected to a translation by a constant element) that $(0, 0) \in G(\bar{T})$. We form the mappings $\bar{T}_\lambda$, $\lambda > 0$, of Proposition 2.8 and apply Proposition 2.4 and Theorem 2.6 to assure that

$$R(T + \bar{T}_\lambda) = V^* \quad \text{for each } \lambda > 0. \tag{2.1.75}$$

Let g_0 be an arbitrary element of V^*. It suffices to prove that g_0 lies in $R(T + \bar{T})$. By the foregoing condition we have the following representation of g_0:

$$g_0 = w_\lambda^* + z_\lambda^*, \quad w_\lambda^* \in \bar{T}_\lambda u_\lambda, \quad z_\lambda^* \in T u_\lambda, \tag{2.1.76}$$

for some $u_\lambda \in V$. Since $(0, 0) \in G(\bar{T}_\lambda)$ for each $\lambda > 0$, it follows that $\langle w^*, u \rangle_V \geq 0$ for $(u, w^*) \in G(\bar{T}_\lambda)$. Therefore the family $\{T + \bar{T}_\lambda\}$ is uniformly coercive for $\lambda > 0$, and consequently $\{u_\lambda\}$ is uniformly bounded by a constant $M > 0$. Suppose that T is quasi-bounded. Since

$$\langle w_\lambda^*, u_\lambda \rangle_V + \langle z_\lambda^*, u_\lambda \rangle_V = \langle g_0, u_\lambda \rangle_V \tag{2.1.77}$$

and $\langle w_\lambda^*, u_\lambda \rangle_V \geq 0$, it follows that

$$\langle z_\lambda^*, u_\lambda \rangle_V \leq \langle g_0, u_\lambda \rangle_V \leq ||g_0||_{V^*}||u_\lambda||_V. \tag{2.1.78}$$

Hence $\{z_\lambda^*\}$ is uniformly bounded by the quasi-boundness of T, and the same is true for $\{w_\lambda^*\}$. Our conclusion then follows from Lemma 2.10.

Suppose on the other hand that $\bar{T}$ is strongly quasi-bounded. The coercivity of T implies that there exists a function $c : R^+ \to R$ such that for $(u, u^*) \in G(T)$,

$$\langle u^*, u \rangle_V \geq c(||u||_V)||u||_V. \tag{2.1.79}$$

Hence

$$\langle z_\lambda^*, u_\lambda \rangle_V \geq c(||u_\lambda||_V)||u_\lambda||_V \geq -M_1 \tag{2.1.80}$$

and

$$\langle w_\lambda^*, u_\lambda \rangle_V = \langle g_0 - z_\lambda^*, u_\lambda \rangle_V \leq ||g_0||_{V^*}M + M_1 = M_2. \tag{2.1.81}$$

Since

$$\langle \lambda w_\lambda^*, u_\lambda - w_\lambda \rangle_V = ||u_\lambda - w_\lambda||_V^2 \geq 0, \quad w_\lambda^* \in \bar{T}w_\lambda, \tag{2.1.82}$$

it results that

$$\langle w_\lambda^*, w_\lambda \rangle_V \leq \langle w_\lambda^*, u_\lambda \rangle_V \leq M_2. \tag{2.1.83}$$

Since $\bar{T}$ is strongly quasi-bounded, we obtain that $\{w_\lambda^*\}$ is uniformly bounded for $0 < \lambda \leq \lambda_0$, and the conclusion follows again from Lemma 2.10.   q.e.d.

It is useful to adapt the last result to the coercivity condition given in Remark 2.7.

Theorem 2.12 Let V be a reflexive Banach space and $\bar{T}$ a maximal monotone mapping from V into 2^{V^*} with $u_0 \in D(T)$. Let T be a pseudo-monotone mapping from V into 2^{V^*}. Suppose that there exists a function $c : R^+ \to R$ with $c(r) \to \infty$ as $r \to \infty$, such that for all $(u, u^*) \in G(T)$, $\langle u^*, u - u_0 \rangle_V \geq c(\|u\|_V)\|u\|_V$ for some $u_0 \in V$, and that either T_{u_0} is quasi-bounded, or $\bar{T}_{u_0}$ is strongly quasi-bounded. Then $R(T + \bar{T}) = V^*$.

2.2 Nonconvex Functions with Generalized Gradient of Pseudo- or Generalized Pseudo-Monotone Type

In this Section the class of locally Lipschitz and indicator functions on a reflexive Banach space will be studied by taking into account the pseudo-monotone and generalized pseudo-monotone properties of their generalized gradient. Throughout this Section the notion of a locally Lipschitz function on a Banach space V will be understood in the sense that the function is Lipschitz continuous on each bounded subset of V.

We begin with the formulation of sufficient conditions for the generalized gradient of a locally Lipschitz function to be pseudo-monotone.

Proposition 2.13 Let $f : V \to R$ be a locally Lipschitz function from a reflexive Banach space V into R. Suppose that the generalized gradient $\partial f : V \to 2^{V^*}$ fulfills the condition: for any sequence $\{u_n\}$ in V converging weakly to u we have that

$$[\liminf f^0(u_n, u - u_n) \geq 0] \Longrightarrow [\limsup f^0(u_n, w - u_n) \leq f^0(u, w - u) \; \forall w \in V]. \tag{2.2.1}$$

Then ∂f is pseudo-monotone.

Proof. To begin with let us observe that ∂f fulfills all the hypotheses of Proposition 2.2. Indeed, $\partial f(u)$ is closed, convex and nonempty for each $u \in V$, moreover, it is bounded from V into 2^{V^*}. Thus to show its pseudo-monotonicity it is enough to check that ∂f is generalized pseudo-monotone. For this purpose let us suppose that (2.2.1) holds and choose sequences $\{u_n\}$ in V and $\{u_n^*\}$ in V^* with $u_n^* \in \partial f(u_n)$ converging weakly to u and u^*, respectively, such that $\limsup \langle u_n^*, u_n - u \rangle_V \leq 0$. Since

$$f^0(u_n, u - u_n) \geq \langle u_n^*, u - u_n \rangle_V, \tag{2.2.2}$$

we are led to

$$\liminf f^0(u_n, u - u_n) \geq 0. \tag{2.2.3}$$

Thus, due to (2.2.1) we may write that

$$\limsup f^0(u_n, w - u_n) \leq f^0(u, w - u) \quad \forall w \in V. \tag{2.2.4}$$

Hence, by virtue of (2.2.2) we obtain

$$\limsup \langle u_n^*, u - u_n \rangle_V \leq \limsup f^0(u_n, u - u_n) \leq f^0(u, u - u) = 0 \qquad (2.2.5)$$

which implies that $\liminf \langle u_n^*, u_n - u \rangle_V \geq 0$. Consequently,

$$\lim \langle u_n^*, u_n - u \rangle_V = 0. \qquad (2.2.6)$$

Thus the first condition of the generalized pseudo-monotonicity has been established. What we need now is to show that u^* lies in $\partial f(u)$. Indeed, taking into account (2.2.6) and

$$f^0(u_n, w - u_n) \geq \langle u_n^*, w - u_n \rangle_V \quad \forall w \in V, \qquad (2.2.7)$$

we obtain by means of (2.2.1) that

$$f^0(u, w - u) \geq \limsup f^0(u_n, w - u_n) \geq \langle u^*, w - u \rangle_V \quad \forall w \in V. \qquad (2.2.8)$$

Hence, by definition, $u^* \in \partial f(u)$.   q.e.d.

A locally Lipschitz function with pseudo-monotone generalized gradient enjoys the weak lower semicontinuity property.

Proposition 2.14 Let $f : V \to R$ be a locally Lipschitz function from a reflexive Banach space into R. Suppose that ∂f is pseudo-monotone. Then f is weakly lower semicontinuous, i.e., for any sequence $\{u_n\}$ in V converging weakly to u,

$$\liminf f(u_n) \geq f(u). \qquad (2.2.9)$$

Proof. Suppose that the contrary is true. Then we can find a sequence $\{u_n\} \subset V$ converging weakly to u and such that

$$f(u_n) - f(u) < -\varepsilon \qquad (2.2.10)$$

for some positive ε. Due to the continuity of f at u we can find $\delta > 0$ small enough so that the following inequality holds

$$f(u_n) - f(u + \delta(u_n - u)) < -\frac{\varepsilon}{2} \qquad (2.2.11)$$

($\{u_n\}$ is bounded). By means of Lebourg theorem (cf. [Clar83]) there exist $\{u_n^*\} \subset V^*$ and $\{t_n\} \subset R$ with $\delta < t_n < 1$, such that

$$u_n^* \in \partial f(y_n) \text{ for some } y_n = u + t_n(u_n - u) \qquad (2.2.12)$$

and

$$f(u_n) - f(u + \delta(u_n - u)) = \langle u_n^*, (1 - \delta)(u_n - u) \rangle_V. \qquad (2.2.13)$$

Then by means of (2.2.11) it results that

$$\langle u_n^*, u_n - u \rangle_V < -\frac{\varepsilon}{2(1 - \delta)}. \qquad (2.2.14)$$

Noting that $\{y_n\}$ converges weakly to u and that

$$\langle u_n^\star, y_n - u \rangle_V = t_n \langle u_n^\star, u_n - u \rangle_V < -\frac{\varepsilon t_n}{2(1-\delta)} < -\frac{\varepsilon \delta}{2(1-\delta)} \qquad (2.2.15)$$

implies that

$$\limsup \langle u_n^\star, y_n - u \rangle_V \leq 0. \qquad (2.2.16)$$

Thus taking into account the pseudo-monotonicity of ∂f it results that

$$\langle u_n^\star, y_n - u \rangle_V \to 0. \qquad (2.2.17)$$

But, on the other hand, due to (2.2.15) we may write that

$$\limsup \langle u_n^\star, y_n - u \rangle_V \leq -\frac{\varepsilon \delta}{2(1-\delta)}, \qquad (2.2.18)$$

which leads to a contradiction. Thus the proof is complete.   q.e.d.

Now let us consider a closed subset $C \subset V$ of a reflexive Banach space V. Denote by

$$I_C(v) = \begin{cases} 0 & \text{if } v \in C \\ +\infty & \text{if } v \notin C \end{cases} \qquad (2.2.19)$$

the indicator function of C. Let us recall that (cf. Ch. 1)

$$\partial I_C = N_C \qquad (2.2.20)$$

where $N_C(v)$ is the normal cone to C at v, i.e.

$$v^\star \in N_C(v) \Longleftrightarrow \langle v^\star, k \rangle_V \leq 0 \text{ for any } k \in T_C(v). \qquad (2.2.21)$$

Here $T_C(v)$ denotes the Clarke's tangent cone defined by (1.2.10) i.e.

$$T_C(v) = \{ k \in V : \forall v_n \to_C v, \forall \lambda_n \to 0_+, \text{ there exists } k_n \to k$$
$$\text{with } v_n + \lambda_n k_n \in C \}. \qquad (2.2.22)$$

We recall the property

$$k \in T_C(v) \Longleftrightarrow d_C^0(v, k) = 0, \qquad (2.2.23)$$

where $d_C : V \to R$ is the distance function corresponding to the set C.

We can obtain the following result concerning the generalized pseudo-monotonicity of N_C.

Proposition 2.15 Let $C \subset V$ be a closed subset of a reflexive Banach space V. Suppose that N_C is generalized pseudo-monotone. Then C is weakly closed.

Proof. Let $\{u_n\}$ be a sequence in C converging weakly to u. It is obvious that $0 \in N_C(u_n)$. Thus $\langle 0, u_n - u \rangle_V \to 0$ and by the generalized pseudo-monotonicity of N_C it follows that $0 \in N_C(u)$ which means that $u \in C$.   q.e.d.

Let us consider a bounded operator $A : V \to V^*$ with the domain $D(A) = V$, and a locally Lipschitz function $f : V \to R$ from V into R. Define $\bar{A} : V \times R \to V^* \times R$ by

$$\bar{A}(v, a) = (Av, a), \quad (v, a) \in V \times R. \tag{2.2.24}$$

Recalling that the epigraph of f is given by the formula

$$\text{epi} \, f = \{(v, a) \in V \times R : f(v) \leq a\}. \tag{2.2.25}$$

and that for a locally Lipschitz function f the relation

$$v^* \in \partial f(v) \iff (v^*, -1) \in N_{\text{epi} \, f}(v, f(v)), \tag{2.2.26}$$

holds, we obtain the following formula for the normal cone to epi f

$$N_{\text{epi} \, f}(v, a) = \begin{cases} \{t(v^*, -1) : t \geq 0, v^* \in \partial f(v)\} & \text{if } a = f(v) \\ \{(0, 0)\} & \text{if } a > f(v) \\ \emptyset & \text{if } a < f(v). \end{cases} \tag{2.2.27}$$

The following proposition deals with the relation between the generalized pseudo-monotonicity of the mapping $\bar{A} + N_{\text{epi} \, f}$ and the pseudo-monotonicity of $A + \partial f$.

Proposition 2.16 Let $f : V \to R$ be a locally Lipschitz function from a reflexive Banach space V into R and let $A : V \to V^*$ be a bounded operator from V into V^* with the domain $D(A) = V$. Suppose that $\bar{A} + N_{\text{epi} \, f}$ is generalized pseudo-monotone. Then $A + \partial f : V \to V^*$ is pseudo-monotone.

Proof. Note that the multivalued mapping $A + \partial f$ is defined at every $v \in V$ and that the corresponding set $Av + \partial f(v)$ is bounded, closed, nonempty and convex. Due to Proposition 2.2 in order to prove the assertion it suffices to show that $A + \partial f$ is generalized pseudo-monotone. Let us choose sequences $\{u_n\}$ in V and $\{\bar{u}_n^*\}$ in V^* with the properties that

$$\begin{aligned} &u_n \xrightarrow{w} u \text{ in } V, \\ &Au_n + \bar{u}_n^* \xrightarrow{w} u^* \text{ in } V^*, \quad \bar{u}_n^* \in \partial f(u_n), \\ &\limsup \langle Au_n + \bar{u}_n^*, u_n - u \rangle_V \leq 0. \end{aligned} \tag{2.2.28}$$

What we need now is to deduce that the following two conditions hold

$$\begin{aligned} &\langle Au_n + \bar{u}_n^*, u_n - u \rangle_V \to 0, \\ &u^* \in Au + \partial f(u). \end{aligned} \tag{2.2.29}$$

To do this let us observe that the boundedness of $\{u_n\}$ implies the boundedness of $\{f(u_n)\}$. Thus without loss of generality we may assume that $f(u_n) \to \alpha$ for some $\alpha \in R$. Moreover, from (2.2.28) we obtain immediately that

(i) $(u_n, f(u_n)) \xrightarrow{w} (u, \alpha) \quad \text{in } V \times R, \quad \alpha \in R,$ $\tag{2.2.30}$

(ii) $\bar{A}(u_n, f(u_n)) + N_{\text{epi}\,f}(u_n, f(u_n)) \ni (Au_n + u_n^\star, f(u_n) - 1)$ (2.2.31)

and

$$(Au_n + u_n^\star, f(u_n) - 1) \overset{w}{\to} (u^\star, \alpha - 1) \text{ in } V^\star \times R,$$ (2.2.32)

(iii) $\lim \sup \langle (Au_n, f(u_n)) + (\bar{u}_n^\star, -1), (u_n, f(u_n)) - (u, \alpha) \rangle_V \leq 0.$ (2.2.33)

From here on we retain the same symbol "$\langle \cdot, \cdot \rangle_V$" for the pairing over $(V^\star \times R) \times (V \times R)$, if no ambiguity occurs. Accordingly, in view of the generalized pseudo-monotonicity of $\bar{A} + N_{\text{epi}\,f}$ it follows

$$\lim \langle (Au_n, f(u_n)) + (\bar{u}_n^\star, -1), (u_n, f(u_n)) - (u, \alpha) \rangle_V = 0$$ (2.2.34)

and

$$(u^\star, \alpha - 1) \in \bar{A}(u, \alpha) + N_{\text{epi}\,f}(u, \alpha).$$ (2.2.35)

This leads directly to

$$\lim \langle Au_n + \bar{u}_n^\star, u_n - u \rangle_V = 0,$$ (2.2.36)

and to

$$(u^\star - Au, -1) \in N_{\text{epi}\,f}(u, \alpha).$$ (2.2.37)

The last inclusion means that $\alpha = f(u)$. Indeed, since the element $(u^\star - Au, -1)$ is different from $(0, 0)$ the definition of $N_{\text{epi}\,f}$ implies that the only possibility is that α is equal to $f(u)$. Thus we obtain immediately that $u^\star - Au \in \partial f(u)$ and we have shown the two relations in (2.2.29).   q.e.d.

As an immediate consequence of Proposition 2.16 we obtain the following Corollary by setting $A = 0$.

Corollary 2.17 Let $f : V \to R$ be a locally Lipschitz function from a reflexive Banach space V into R. Suppose that $N_{\text{epi}\,f}$ is generalized pseudo-monotone. Then $\partial f : V \to V^\star$ is pseudo-monotone.

Note that the converse is not true i.e., it is not possible to deduce from the pseudo-monotonicity of ∂f the generalized pseudo-monotonicity of $N_{\text{epi}\,f}$. An appropriate example showing this fact will be presented in Chapter 4.

However, if we confine ourselves to the class of functions satisfying the (C) condition defined below, the equivalence between pseudo-monotonicity of ∂f and generalized pseudo-monotonicity of $N_{\text{epi}\,f}$ is guaranteed.

Let $f : V \to R$ be a locally Lipschitz function from a reflexive Banach space V into R. A function f is said to satisfy the (C) condition, or to have the (C) property if for any sequence $\{u_n\}$ in V converging weakly to $u \in V$, with $u_n^\star \in \partial f(u_n)$ such that $\lim \sup \langle u_n^\star, u_n - u \rangle_V \leq 0$, the relation $f(u_n) \to f(u)$ holds.

It is worth noting that a convex lower semicontinuous proper function $f : V \to \bar{R}$ fulfills the (C) property. Indeed, suppose that $\{u_n\}$ converges weakly to u, and the sequence $\{u_n^*\}$ with $u_n^* \in \partial f(u_n)$ is such that $\limsup \langle u_n^*, u_n - u \rangle_V \leq 0$. Then by the convexity of f, $f(u) - f(u_n) \geq \langle u_n^*, u - u_n \rangle_V$. Hence, $\limsup f(u_n) \leq f(u)$. On the other hand, the weak lower semicontinuity of f implies that $\liminf f(u_n) \geq f(u)$. Thus the assertion results.

The following proposition holds true for functions having the (C) property.

Proposition 2.18 Let $f : V \to R$ be a locally Lipschitz function from a reflexive Banach space V into R, the generalized gradient of which is pseudo-monotone. If f has the (C) property, then $N_{\mathrm{epi}\, f}$ is generalized pseudo-monotone.

Proof. Let $\{(u_n, a_n)\}$, with $f(u_n) \leq a_n$, be a sequence in $V \times R$ converging weakly to (u, a), and $(u_n^*, a_n^*) \in N_{\mathrm{epi}\, f}(u_n, a_n)$ a sequence in $V^* \times R$ converging weakly to (u^*, a^*), such that

$$\limsup \langle (u_n^*, a_n^*), (u_n, a_n) - (u, a) \rangle_V \leq 0. \tag{2.2.38}$$

We have to show that the foregoing conditions imply that

$$\lim \langle (u_n^*, a_n^*), (u_n, a_n) - (u, a) \rangle_V = 0. \tag{2.2.39}$$

and

$$(u^*, a^*) \in N_{\mathrm{epi}\, f}(u, a). \tag{2.2.40}$$

At the beginning let us notice that from the weak lower semicontinuity of f it follows that

$$a = \lim a_n \geq \liminf f(u_n) \geq f(u). \tag{2.2.41}$$

Hence $(u, a) \in \mathrm{epi}\, f$. For the case $f(u_n) < a_n$ the assertion follows immediately. Indeed, in such a case $(u_n^*, a_n^*) = (0, 0)$ and $(0, 0) \in N_{\mathrm{epi}\, f}(u, a)$. Consequently, (2.2.39) and (2.2.40) result. Thus it remains to consider the case $f(u_n) = a_n$. Then we have

$$(u_n^*, a_n^*) = t_n(\bar{u}_n^*, -1) \text{ for some } t_n \geq 0 \text{ and } \bar{u}_n^* \in \partial f(u_n). \tag{2.2.42}$$

The boundedness of the sequences $\{u_n^*\}$ and $\{\bar{u}_n^*\}$ implies the boundedness of $\{t_n\}$. Thus, without loss of generality we can assume that $\lim t_n = t$ for some $t \in R$. If $t = 0$ then

$$(u_n^*, a_n^*) = t_n(\bar{u}_n^*, -1) \to (0, 0) \tag{2.2.43}$$

and (2.2.39)–(2.2.40) follow immediately. For the case $t > 0$ we argue as follows: from (2.2.38) we easily deduce that

$$\limsup \langle t_n \bar{u}_n^*, u_n - u \rangle_V \leq 0, \tag{2.2.44}$$

which by virtue of $\lim t_n = t > 0$ implies that

$$\limsup \langle \bar{u}_n^*, u_n - u \rangle_V \leq 0. \tag{2.2.45}$$

Using the pseudo-monotonicity of ∂f we obtain the relation

$$\lim\langle \bar{u}_n^\star, u_n - u\rangle_V = 0. \tag{2.2.46}$$

and we are led directly to (2.2.39). To establish (2.2.40) let us notice that the weak convergence of the sequence $\{t_n \bar{u}_n^\star\}$ to $u^\star$ together with $\lim t_n = t > 0$ implies the weak convergence of $\{\bar{u}_n^\star\}$ to the element $u^\star/t$. Hence, by applying the pseudo-monotonicity of ∂f we can deduce again that $u^\star/t \in \partial f(u)$. But this is equivalent to $(u^\star, -t) \in N_{\mathrm{epi}\,f}(u, f(u))$. On the other hand, the (C) property of f implies that $a = \lim a_n = \lim f(u_n) = f(u)$. Therefore, $N_{\mathrm{epi}\,f}(u, a) = N_{\mathrm{epi}\,f}(u, f(u))$ and the desired condition holds true. The proof is complete. q.e.d.

From Proposition 2.18 and Corollary 2.17 we have the following result immediately.

Proposition 2.19 Let $f : V \to R$ be a locally Lipschitz function from a reflexive Banach space V into R. Suppose that f has the (C) property. Then ∂f is pseudo-monotone, if and only if $N_{\mathrm{epi}\,f}$ is generalized pseudo-monotone.

In many practical problems we deal with multivalued mappings of the form $A + \partial f$, where $A : V \to V^\star$ satisfies the following $(S)_+$ condition.

An operator $A : V \to V^\star$ is said to satisfy the $(S)_+$ condition if the following condition is fulfilled: the weak convergence of a sequence $\{u_n\}$ in V to u together with $\limsup\langle Au_n, u_n - u\rangle_V \leq 0$, implies the strong convergence of $\{u_n\}$ to u.

The class of operators satisfying the $(S)_+$ condition was first introduced and studied in [Bro68] (see also [Bro70a,b,75,83, Lio69]). Note that e.g. the strongly monotone operators belong to this class.

It is well known that the sum of two pseudo-monotone operators is again pseudo-monotone (see Proposition 2.4). It turns out that if A is not only pseudo-monotone but also satisfies the $(S)_+$ condition, ∂f does not need to be pseudo-monotone in order that the sum $A + \partial f$ be so. It is enough to impose on ∂f the following condition, which is the relaxation of property (c) in the definition of pseudo-monotonicity.

Let $f : V \to R$ be a locally Lipschitz function from V into R. Then ∂f is said to be "quasi-pseudo-monotone" if the following condition is satisfied: if $\{u_n\}$ is a sequence in V converging weakly to u and $u_n^\star \in \partial f(u_n)$ is such that $\limsup\langle u_n^\star, u_n - u\rangle_V \leq 0$, then $\lim\langle u_n^\star, u_n - u\rangle_V = 0$.

It is easily seen that this definition does not require all the cluster points of $\{u_n^\star\}$ (in the weak topology of $V^\star$) to lie in $\partial f(u)$. However, as we shall see below, if A has the $(S)_+$ property then the sufficient condition for the sum $A + \partial f$ to be pseudo-monotone is the quasi-pseudo-monotonicity of ∂f.

Proposition 2.20 Let $A : V \to V^\star$ be a pseudo-monotone operator from a reflexive Banach space V into $V^\star$ and $f : V \to R$ be a locally Lipschitz function from V into R. Suppose that A satisfies the $(S)_+$ condition and ∂f is quasi-pseudo-monotone. Then the sum $A + \partial f$ is pseudo-monotone.

Proof. To prove the pseudo-monotonicity of $A + \partial f$ it is sufficient to consider a sequence $\{u_n\}$ converging weakly to u in V, the corresponding sequence $\{Au_n + \bar{u}_n^*\}$ with $\bar{u}_n^* \in \partial f(u_n)$, and to check whether the condition $\limsup \langle Au_n + \bar{u}_n^*, u_n - u \rangle_V \le 0$ implies that for any $v \in V$ there exists $u^*(v) \in Au + \partial f(u)$ such that

$$\liminf \langle Au_n + \bar{u}_n^*, u_n - v \rangle_V \ge \langle u^*(v), u - v \rangle_V. \tag{2.2.47}$$

From the facts that a sequence $\{u_n\}$ converges weakly to u in V, and that the corresponding sequence $\{Au_n + \bar{u}_n^*\}$ with $\bar{u}_n^* \in \partial f(u_n)$ is such that

$$\limsup \langle Au_n + \bar{u}_n^*, u_n - u \rangle_V \le 0, \tag{2.2.48}$$

we first obtain that

$$\limsup \langle Au_n, u_n - u \rangle_V \le 0 \text{ and } \limsup \langle \bar{u}_n^*, u_n - u \rangle_V \le 0. \tag{2.2.49}$$

Indeed, suppose that this is not true. Assume that

$$\limsup \langle Au_n, u_n - u \rangle_V > 0. \tag{2.2.50}$$

Then we can find $d > 0$ and a subsequence of $\{u_n\}$, which we denote again by $\{u_n\}$, such that

$$\lim \langle Au_n, u_n - u \rangle_V = d > 0. \tag{2.2.51}$$

From (2.2.48) and (2.2.51) we have for the other summand that

$$\limsup \langle \bar{u}_n^*, u_n - u \rangle_V \le -d < 0. \tag{2.2.52}$$

But then, the quasi-pseudo-monotonicity of ∂f implies that

$$\lim \langle \bar{u}_n^*, u_n - u \rangle_V = 0, \tag{2.2.53}$$

which leads to a contradiction. If we consider the case

$$\limsup \langle \bar{u}_n^*, u_n - u \rangle_V > 0, \tag{2.2.54}$$

then following the lines of the previous case we arrive again at a contradiction by taking into account the pseudo-monotonicity of A. These contradictions establish the inequalities of (2.2.49).

Now from the pseudo-monotonicity of A one obtains

$$\liminf \langle Au_n, u_n - v \rangle_V \ge \langle Au, u - v \rangle_V \quad \forall v \in V. \tag{2.2.55}$$

The boundedness of $\{\bar{u}_n^*\}$ allows us to pass to a subsequence and suppose that $\bar{u}_n^* \xrightarrow{w} \bar{u}^*$ for an element $\bar{u}^*$ from V^*. It remains to show that $\bar{u}^* \in \partial f(u)$. For this purpose we make use of the strong convergence of $\{u_n\}$ to u which is guaranteed by the $(S)_+$ property of A. Accordingly, due to the upper semicontinuity of ∂f from V into the weak topology of V^* we conclude that $\bar{u}^* \in \partial f(u)$. Moreover, we have that

$$\lim \langle \bar{u}_n^*, u_n - v \rangle_V = \langle \bar{u}^*, u - v \rangle_V \quad \forall v \in V \tag{2.2.56}$$

which combined with (2.2.55) leads to (2.2.47) with $u^*(v) = Au + \bar{u}^*$, as desired.
q.e.d.

In the next proposition we establish the generalized pseudo-monotonicity of the mapping $\bar{A} + N_{\text{epi} f}$ under the hypotheses of the previous proposition strengthened by the boundedness of A.

Proposition 2.21 Let $A : V \to V^*$ be a bounded pseudo-monotone operator satisfying the $(S)_+$ condition and let $f : V \to R$ be a locally Lipschitz function having a quasi-pseudo-monotone generalized gradient. Then $\bar{A} + N_{\text{epi} f}$ is generalized pseudo-monotone.

Proof. Let $\{(u_n, a_n)\}$ be a sequence belonging to epi f, i.e. $f(u_n) \leq a_n$, with

$$(u_n, f(u_n)) \xrightarrow{w} (u, \alpha) \text{ in } V \times R, \quad \alpha \in R. \tag{2.2.57}$$

It is easily seen that without loss of generality one may consider only the boundary of epi f, i.e., those (u_n, a_n) for which $f(u_n) = a_n$. Thus let us suppose that

$$(Au_n, f(u_n)) + t_n(\bar{u}_n^*, -1) \in \bar{A}(u_n, f(u_n)) + N_{\text{epi} f}(u_n, f(u_n)), \tag{2.2.58}$$

where $\bar{u}_n \in \partial f(u_n)$, and

$$(Au_n, f(u_n)) + t_n(\bar{u}_n^*, -1) \xrightarrow{w} (u^*, \alpha - t) \quad \text{in } V^* \times R, \tag{2.2.59}$$

$$\limsup \langle (Au_n, f(u_n)) + t_n(\bar{u}_n^*, -1), (u_n, f(u_n)) - (u, \alpha) \rangle_V \leq 0. \tag{2.2.60}$$

We have to show that the foregoing conditions imply that

$$\lim \langle (Au_n, f(u_n)) + (\bar{u}_n^*, -1), (u_n, f(u_n)) - (u, \alpha) \rangle = 0 \tag{2.2.61}$$

$$(u^*, \alpha - t) \in \bar{A}(u, \alpha) + N_{\text{epi} f}(u, \alpha). \tag{2.2.62}$$

From (2.2.60) it follows that

$$\limsup \langle (Au_n + t_n \bar{u}_n^*, u_n - u) \rangle_V \leq 0. \tag{2.2.63}$$

We deduce that

$$\limsup \langle Au_n, u_n - u \rangle_V \leq 0 \text{ and } \limsup \langle t_n \bar{u}_n^*, u_n - u \rangle_V \leq 0. \tag{2.2.64}$$

Indeed, suppose that this is not true. Then we may assume that either

$$\limsup \langle Au_n, u_n - u \rangle_V > 0 \text{ or } \limsup \langle t_n \bar{u}_n^*, u_n - u \rangle_V > 0. \tag{2.2.65}$$

Consider the first case, i.e., $\limsup \langle Au_n, u_n - u \rangle_V > 0$. Passing to a subsequence, if necessary, we can assume that $\lim \langle Au_n, u_n - u \rangle_V = d > 0$. Then we easily obtain for the other summand, that

$$\limsup \langle t_n \bar{u}_n^*, u_n - u \rangle_V \leq -d < 0. \tag{2.2.66}$$

Moreover, the boundedness of $\{\bar{u}_n^*\}$ and $\{Au_n\}$ implies that the sequence $\{t_n\}$ is bounded and, without loss of generality, we may assume that it converges to some real number $t \geq 0$. It should be noticed that the limit t cannot be 0. In fact, if $t = 0$, then from the boundedness of the sequence $\{\bar{u}_n^*\}$ in V^* we would obtain $\lim \langle t_n \bar{u}_n^*, u_n - u \rangle_V = 0$ which contradicts (2.2.66). Thus $t > 0$. But in such a case (2.2.66) yields

$$\limsup \langle \bar{u}_n^*, u_n - u \rangle_V \leq -\frac{d}{t} < 0. \qquad (2.2.67)$$

However, in view of the quasi-pseudo-monotonicity of ∂f we are led again to the conclusion

$$\lim \langle \bar{u}_n^*, u_n - u \rangle_V = 0 \qquad (2.2.68)$$

which contradicts (2.2.67).

Assuming that $\lim \langle t_n \bar{u}_n^*, u_n - u \rangle_V > 0$ one obtains, analogously, the inequality

$$\limsup \langle Au_n, u_n - u \rangle_V \leq -d < 0 \qquad (2.2.69)$$

for an appropriate subsequence of $\{u_n\}$. Applying again the pseudo-monotonicity of A we obtain $\lim \langle Au_n, u_n - u \rangle_V = 0$ which contradicts (2.2.69). Thus the assertions (2.2.64) have been proved.

Consequently, the condition $\limsup \langle Au_n, u_n - u \rangle_V \leq 0$, implies, due to the $(S)_+$ property of A, the strong convergence of $\{u_n\}$ to u in V. It allows us to conclude that (2.2.61) holds. Moreover, the pseudo-monotonicity and the boundedness of A lead easily to the weak convergence of $\{Au_n\}$ to Au. To derive (2.2.62) we can apply the semicontinuity of ∂f from V to V^* with the weak topology. It implies that each cluster point of $\{\bar{u}_n^*\}$ in the weak topology of V^* lies in $\partial f(u)$. Thus without loss of generality we may assume that $\bar{u}_n^* \xrightarrow{w} \bar{u}^*$ for some $\bar{u}^* \in \partial f(u)$. The obtained results together with the strong convergence of $\{u_n\}$ to u permit us to write that

$$Au_n + t_n \bar{u}_n^* \to Au + t\bar{u}^* \quad \text{weakly in } V^*, \qquad (2.2.70)$$

and

$$f(u_n) \to f(u) = \alpha. \qquad (2.2.71)$$

Therefore

$$(Au, f(u)) + t(\bar{u}^*, -1) = (u^*, f(u) - t) \in \bar{A}(u, f(u)) + N_{\text{epi} f}(u, f(u)) \qquad (2.2.72)$$

and the proof is complete.   q.e.d.

The last part of this Section is devoted to the study of some properties of closed subsets of V on the assumption that the corresponding normal cone mapping satisfies a relaxed form of the generalized pseudo-monotonicity. Let us define the relaxed form.

Let T be a mapping from V into 2^{V^*}. Then T is said to be quasi generalized pseudo-monotone if the following condition is satisfied: for any sequence $\{u_n\}$ in

V converging weakly to u and the corresponding sequence $\{u_n^*\}$ with $u_n^* \in Tu_n$, converging weakly to u^*, the condition

$$\limsup \langle u_n^*, u_n - u \rangle_V \leq 0 \tag{2.2.73}$$

implies

$$\lim \langle u_n^*, u_n - u \rangle_V = 0. \tag{2.2.74}$$

As it is easily seen in this definition, we do not require that any cluster points of $\{u_n^*\}$ in the weak topology of V^* lie in Tu.

To formulate the next result let us recall the notion of hypertangent vectors to a set C at the point $u \in C$ (cf. Sect. 1.2).

Proposition 2.22 Let $A : V \to V^*$ be a pseudo-monotone bounded operator from a reflexive Banach space V into V^* and $C \subset V$ be a closed subset in V. Suppose that A satisfies the $(S)_+$ condition, $H_C(u) \neq \emptyset$ for all $u \in C$, and that $N_C : V \to 2^{V^*}$ is quasi generalized pseudo-monotone. Then $A + N_C$ is generalized pseudo-monotone.

Proof. Let $\{u_n\}$ be a sequence in C converging weakly to u. Suppose that $u_n^* \in N_C(u_n)$ is such that the sequence $\{Au_n + u_n^*\}$ converges weakly to $u^* \in V^*$ and

$$\limsup \langle Au_n + u_n^*, u_n - u \rangle_V \leq 0. \tag{2.2.75}$$

We have to show that

$$\lim \langle Au_n + u_n^*, u_n - u \rangle_V = 0 \tag{2.2.76}$$

and

$$u^* \in Au + N_C(u). \tag{2.2.77}$$

Taking into account the fact that N_C is assumed to be quasi generalized pseudo-monotone and A is bounded and pseudo-monotone we prove analogously to Proposition 2.21 that (2.2.75) leads to

$$\limsup \langle Au_n, u_n - u \rangle_V \leq 0 \tag{2.2.78}$$

and

$$\limsup \langle u_n^*, u_n - u \rangle_V \leq 0. \tag{2.2.79}$$

Hence, due to the $(S)_+$ property of A, the strong convergence $u_n \to u$ follows. It implies that $u \in C$ and, by the pseudo-monotonicity of A, we deduce the weak convergence of $\{Au_n\}$ to Au. Consequently, we may assume that $\{u_n^*\}$ converges weakly to $u^* - Au$. But in such a case the existence of a hypertangent to C at u implies that $u^* - Au \in N_C(u)$ (cf. [Clar75], Corollary, p.58), as desired. q.e.d.

3. Hemivariational Inequalities for Static One-dimensional Nonconvex Superpotential Laws

The present Chapter deals with a very common type of hemivariational inequalities. They arise in the case of one-dimensional, nonmonotone, multivalued nonlinearities or, equivalently, in the case of one-dimensional nonconvex superpotentials. This is the simplest type of hemivariational inequalities and therefore it was quite natural that this type of hemivariational inequalities was the first to be studied concerning the existence of its solution [Pan85,91]. Moreover this type of hemivariational inequality arises in many problems in Mechanics and Engineering. We begin first with the coercive case and then, we give an existence result for the more complicated semicoercive case. After the treatment of hemivariational inequalities we deal with variational-hemivariational inequalities arising from one-dimensional nonconvex superpotentials and we give some existence and approximation results. Then the relation between a hemivariational inequality and the corresponding substationarity problem is investigated. This Chapter closes by giving certain applications of the presented theory to problems from Mechanics and Engineering.

3.1 Coercive Hemivariational Inequalities

We recall that the theory of the existence of solution of variational inequalities is a well-developed theory in mathematics which is closely connected with the convexity of the energy functionals involved. Indeed the existence theory of variational inequalities is based on monotonicity arguments. On the contrary the study of hemivariational inequalities, due to the absence of convexity is based on compactness arguments as we shall see further.

Several types of hemivariational inequalities have already been studied (see, for example, [Pana83,85a,b,88a,b,c, Pan85,88a,89a,90,92c], and the references given there) with respect to certain engineering problems, e.g. in nonmonotone semipermeability problems, in the theory of multilayered plates (delamination), in the theory of composite structures, in the theory of partial debonding of adhesive joints etc. Let

$$V \text{ be a real Hilbert space} \qquad (3.1.1)$$

with the property that

$$V \subset [L^2(\Omega)]^n \subset V^*, \tag{3.1.2}$$

where V^* denotes the dual space of V, Ω is an open bounded subset of R^n, and the injections are continuous and dense. We denote in this Chapter by $(\cdot, \cdot)$ the $[L^2(\Omega)]^n$ inner product and the duality pairing, by $\|\cdot\|$ the norm of V and by $|\cdot|_2$ the $[L^2(\Omega)]^n$-norm. We recall [Aub79a] that the form $(\cdot, \cdot)$ extends uniquely from $V \times L^2[(\Omega)]^n$ to $V \times V^*$. Moreover let $L : V \to L^2(\Omega)$, $Lu = \hat{u}$, $\hat{u}(x) \in R$ be a linear continuous mapping. Further, assume that $l \in V^*$, that

$$L : V \to L^2(\Omega) \qquad \text{is compact} \tag{3.1.3}$$

and that

$$\tilde{V} = \{v \in V : \hat{v} \in L^\infty(\Omega)\} \qquad \text{is dense in } V \text{ for the } V-\text{norm}, \tag{3.1.4}$$

and has a Galerkin base in V. It is also assumed that $a(\cdot, \cdot): V \times V \to R$ is a bilinear symmetric continuous form which is coercive, i.e. there exists $c > 0$ constant such that

$$a(v, v) \geq c\|v\|^2 \qquad \forall v \in V. \tag{3.1.5}$$

We denote by $j : R \to R$, a locally Lipschitz function defined as in (1.2.49): let $\beta \in L^\infty_{\text{loc}}(R)$ and consider the functions $\bar{\beta}_\mu$ and $\bar{\bar{\beta}}_\mu$ defined by (1.2.45) (where ρ is replaced by μ) i.e.

$$\bar{\beta}_\mu(\xi) = \underset{|\xi_1 - \xi| \leq \mu}{\text{ess inf}} \beta(\xi_1) \text{ and } \bar{\bar{\beta}}_\mu(\xi) = \underset{|\xi_1 - \xi| \leq \mu}{\text{ess sup}} \beta(\xi_1). \tag{3.1.6}$$

They are decreasing and increasing functions of μ, respectively; therefore the limits for $\mu \to 0_+$ exist. We denote them by $\bar{\beta}(\xi)$ and $\bar{\bar{\beta}}(\xi)$ respectively; the multivalued function $\tilde{\beta}$ is defined by

$$\tilde{\beta}(\xi) = [\bar{\beta}(\xi), \bar{\bar{\beta}}(\xi)]. \tag{3.1.7}$$

If $\beta(\xi_{\pm 0})$ exists for every $\xi \in R$ then a locally Lipschitz function $j : R \to R$ can be determined up to an additive constant such that [Ch] (cf. (1.2.48))

$$\tilde{\beta}(\xi) = \partial j(\xi). \tag{3.1.8}$$

Now we formulate the following coercive hemivariational inequality (problem P^C): Find $u \in V$ such that

$$a(u, v - u) + \int_\Omega j^0(\hat{u}, \hat{v} - \hat{u})d\Omega \geq (l, v - u) \qquad \forall v \in V. \tag{3.1.9}$$

Let us define now the solution of the hemivariational inequality. An element $u \in V$ is said to be a solution of P^C if there exists $\chi \in L^1(\Omega)$ with $L^*\chi \in V^*$ (L^* denotes the transpose operator of L) such that

$$a(u,v) + (L^*\chi, v) = (l,v) \quad \forall v \in V \tag{3.1.9a}$$

and

$$\chi(x) \in \partial j(u(x)) \quad \text{a.e. on } \Omega, \tag{3.1.9b}$$

and where

$$(L^*\chi, v) = \int_\Omega \chi L v d\Omega = \int_\Omega \chi \hat{v} d\Omega \quad \text{if } v \in \tilde{V}. \tag{3.1.9c}$$

Therefore one could give the following definition. An element $u \in V$ is said to be a solution of P^C if there exists $\chi \in L^1(\Omega)$ such that

$$a(u,v) + \int_\Omega \chi \hat{v} d\Omega = (l,v) \quad \forall v \in \tilde{V} \tag{3.1.9d}$$

and (3.1.9b) hold. Obviously due to the assumption (3.1.4) the two above definitions are equivalent.

In order to define the regularized problem P_ε^C we consider the mollifier

$$p \in C_0^\infty(-1,+1), \; p \geq 0, \quad \text{with } \int_{-\infty}^{+\infty} p(\xi)d\xi = 1 \tag{3.1.10}$$

and let

$$\beta_\varepsilon = p_\varepsilon \star \beta \text{ with } p_\varepsilon(\xi) = \frac{1}{\varepsilon}p\left(\frac{\xi}{\varepsilon}\right) \quad 0 < \varepsilon < 1, \tag{3.1.11}$$

where $(\star)$ denotes the convolution product. The regularized problem P_ε^C reads: Find $u_\varepsilon \in V$ with $\beta_\varepsilon(\hat{u}_\varepsilon) \in L^1(\Omega)$, such as to satisfy the variational equality

$$a(u_\varepsilon, v) + \int_\Omega \beta_\varepsilon(\hat{u}_\varepsilon)\hat{v}d\Omega = (l,v) \quad \forall v \in \tilde{V}. \tag{3.1.12}$$

In order to define the corresponding finite dimensional problem $P_{\varepsilon n}^C$ we consider a Galerkin basis of $\tilde{V}$ in V and let V_n be the resulting n-dimensional subspace. This problem reads:

Problem $P_{\varepsilon n}^C$: Find $\hat{u}_{\varepsilon n} \in V_n$ such as to satisfy the variational equality

$$a(\hat{u}_{\varepsilon n}, v) + \int_\Omega \beta_\varepsilon(\hat{u}_{\varepsilon n})\hat{v}d\Omega = (l,v) \quad \forall v \in V_n. \tag{3.1.13}$$

Now we assume that there exists $\xi \in R^+$ such that

$$\operatorname*{ess\,sup}_{(-\infty,-\xi)} \beta(\xi_1) \leq 0 \leq \operatorname*{ess\,inf}_{(\xi,\infty)} \beta(\xi_1). \tag{3.1.14}$$

Roughly speaking we may say that the graph $\{\xi, \beta(\xi)\}$ ultimately increases. The existence proof is based on the following Lemmata.

Lemma 3.1 Suppose that (3.1.14) holds. Then we can determine $\rho_1 > 0$, $\rho_2 > 0$ such that for every $\hat{u}_{\varepsilon n} \in V_n$,

$$\int_\Omega \beta_\varepsilon(\hat{u}_{\varepsilon n})\hat{u}_{\varepsilon n}\,d\Omega \geq -\rho_1\rho_2\text{mes }\Omega. \tag{3.1.15}$$

Proof. From (3.1.11) we obtain that

$$\beta_\varepsilon(\xi) = (p_\varepsilon \star \beta)(\xi) = \int_{-\varepsilon}^{+\varepsilon} \beta(\xi - t)p_\varepsilon(t)dt \leq \operatorname*{ess\,sup}_{|t|\leq\varepsilon} \beta(\xi - t), \tag{3.1.16}$$

and analogously

$$\operatorname*{ess\,inf}_{|t|\leq\varepsilon} \beta(\xi - t) \leq \beta_\varepsilon(\xi). \tag{3.1.17}$$

In the above two inequalities we set $x = \xi - t$, $|x - \xi| \leq \varepsilon$ and enlarge the bounds for $-\infty < x \leq \varepsilon + \xi$ and $\xi - \varepsilon \leq x < \infty$, respectively. Then the supremum and the infimum for $\xi \in (-\infty, -\xi_1)$ and $\xi \in (\xi_1, \infty)$, respectively are formed and the bounds are enlarged by replacing $\varepsilon + \xi$ by $1 - \xi_1$ and $\xi - \varepsilon$ by $\xi_1 - 1$ ($\varepsilon < 1$); we obtain from (3.1.14) that there exists $\xi > 0$ such that

$$\sup_{(-\infty, -\xi)} \beta_\varepsilon(\xi_1) \leq 0 \leq \inf_{(\xi, \infty)} \beta_\varepsilon(\xi_1). \tag{3.1.18}$$

Thus we can determine $\rho_1 > 0$ and $\rho_2 > 0$ such that $\beta_\varepsilon(\xi) \geq 0$ if $\xi > \rho_1$, $\beta_\varepsilon(\xi) \leq 0$ if $\xi < -\rho_1$, and $|\beta_\varepsilon(\xi)| \leq \rho_2$ if $|\xi| \leq \rho_1$ and may write

$$\int_\Omega \beta_\varepsilon(\hat{u}_{\varepsilon n})\hat{u}_{\varepsilon n}\,d\Omega = \int_{|\hat{u}_{\varepsilon n}(x)|>\rho_1} \ldots d\Omega \tag{3.1.19}$$

$$+ \int_{|\hat{u}_{\varepsilon n}(x)|\leq\rho_1} \ldots d\Omega \geq 0 - \rho_1\rho_2\text{mes}\Omega. \qquad \text{q.e.d.}$$

Lemma 3.2 The problem $P_{\varepsilon n}^C$ has at least one solution $\hat{u}_{\varepsilon n} \in V_n$.

Proof. Equation (3.1.13) is written in the form

$$(\Lambda(u_{\varepsilon n}), v) = 0, \qquad \forall v \in V_n \tag{3.1.20}$$

and because of (3.1.5) and (3.1.15) we have the estimate

$$(\Lambda(u_{\varepsilon n}), u_{\varepsilon n}) \geq c\|u_{\varepsilon n}\|^2 - \rho_1\rho_2\text{mes}\Omega - c_1\|u_{\varepsilon n}\| \qquad c, c_1 > 0. \tag{3.1.21}$$

By applying Brouwer's fixed point theorem (cf. [Lio69] p.53) we obtain that (3.1.20) has a solution $u_{\varepsilon n}$ with $\|u_{\varepsilon n}\| \leq c$, where c is independent of ε and n. q.e.d.

Lemma 3.3 The sequence $\{\beta_\varepsilon(\hat{u}_{\varepsilon n})\}$ is weakly precompact in $L^1(\Omega)$.

Proof. Due to the Dunford-Pettis theorem (cf. [Eke], p.239) it suffices to show that for each $\mu > 0$ a $\delta(\mu) > 0$ can be determined such that for $\omega \subset \Omega$ with mes $\omega < \delta$

$$\int_\omega |\beta_\varepsilon(\hat{u}_{en})| d\Omega < \mu \, . \tag{3.1.22}$$

The inequality

$$\xi_0 |\beta_\varepsilon(\xi)| \leq |\beta_\varepsilon(\xi)\xi| + \xi_0 \sup_{|\xi| \leq \xi_0} |\beta_\varepsilon(\xi)|, \quad (\xi_0 > 0) \tag{3.1.23}$$

implies that

$$\int_\omega |\beta_\varepsilon(\hat{u}_{en})| d\Omega \leq \frac{1}{\xi_0} \int_\Omega |\beta_\varepsilon(\hat{u}_{en})\hat{u}_{en}| d\Omega + \int_\omega \sup_{|\hat{u}_{en}(x)| \leq \xi_0} |\beta_\varepsilon(\hat{u}_{en})| d\Omega \, . \tag{3.1.24}$$

But

$$\begin{aligned}
\int_\Omega |\beta_\varepsilon(\hat{u}_{en})\hat{u}_{en}| d\Omega &= \int_{|\hat{u}_{en}(x)| > \rho_1} |\beta_\varepsilon(\hat{u}_{en})\hat{u}_{en}| d\Omega + \int_{|\hat{u}_{en}(x)| \leq \rho_1} |\beta_\varepsilon(\hat{u}_{en})\hat{u}_{en}| d\Omega \\
&= \int_{|\hat{u}_{en}(x)| > \rho_1} |\beta_\varepsilon(\hat{u}_{en})\hat{u}_{en}| d\Omega - \int_{|\hat{u}_{en}(x)| \leq \rho_1} |\beta_\varepsilon(\hat{u}_{en})\hat{u}_{en}| d\Omega \\
&\quad + 2 \int_{|\hat{u}_{en}(x)| \leq \rho_1} |\beta_\varepsilon(\hat{u}_{en})\hat{u}_{en}| d\Omega \tag{3.1.25} \\
&\leq \int_{|\hat{u}_{en}(x)| > \rho_1} |\beta_\varepsilon(\hat{u}_{en})\hat{u}_{en}| d\Omega + \int_{|\hat{u}_{en}(x)| \leq \rho_1} \beta_\varepsilon(\hat{u}_{en})\hat{u}_{en} d\Omega \\
&\quad + 2 \int_{|\hat{u}_{en}(x)| \leq \rho_1} |\beta_\varepsilon(\hat{u}_{en})\hat{u}_{en}| d\Omega \\
&= \int_\Omega \beta_\varepsilon(\hat{u}_{en})\hat{u}_{en} d\Omega + 2 \int_{|\hat{u}_{en}(x)| \leq \rho_1} |\beta_\varepsilon(\hat{u}_{en})\hat{u}_{en}| d\Omega \\
&= (l, u_{en}) - a(u_{en}, u_{en}) + 2 \int_{|\hat{u}_{en}(x)| \leq \rho_1} |\beta_\varepsilon(\hat{u}_{en})\hat{u}_{en}| d\Omega \\
&\leq c + 2\rho_1 \rho_2 \text{mes} \, \Omega, \quad c \text{ constant.}
\end{aligned}$$

In the last two inequalities we have used the boundedness of $\|u_{en}\|$ and assumptions (3.1.5) and (3.1.14). Further, the relation

$$\sup_{|\xi| \leq \xi_0} |\beta_\varepsilon(\xi)| \leq \text{ess sup}_{|\xi| \leq \xi_0 + 1} |\beta(\xi)|, \tag{3.1.26}$$

is applied which can be easily verified by means of (3.1.11). Now choose ξ_0 such that for all ε and n

$$\frac{1}{\xi_0} \int_\omega |\beta_\varepsilon(\hat{u}_{en})\hat{u}_{en}| d\Omega \leq \frac{1}{\xi_0}(c + 2\rho_1 \rho_2 \text{mes} \Omega) \leq \frac{\mu}{2} \tag{3.1.27}$$

and δ such that

$$\operatorname*{ess\,sup}_{|\xi|\leq \xi_0+1} |\beta(\xi)| \leq \frac{\mu}{2\delta}. \tag{3.1.28}$$

Relation (3.1.26) implies with (3.1.27) that for mes $\omega < \delta$

$$\int_\omega \sup_{|\hat{u}_{\varepsilon n}(x)|\leq \xi_0} |\beta_\varepsilon(\hat{u}_{\varepsilon n})| d\Omega \leq \operatorname*{ess\,sup}_{|\hat{u}_{\varepsilon n}(x)|\leq \xi_0+1} |\beta(\hat{u}_{\varepsilon n})| \mathrm{mes}\,\omega \leq \frac{\mu}{2\delta}\cdot\delta = \frac{\mu}{2}. \tag{3.1.29}$$

From (3.1.24), (3.1.27) and (3.1.29), the relation (3.1.22) results. Thus $\{\beta_\varepsilon(\hat{u}_{\varepsilon n})\}$ is weakly precompact in $L^1(\Omega)$.

q.e.d.

Now the proof of the following theorem can be given.

Theorem 3.4 Problem P^C has at least one solution.

Proof. From Lemma 3.2 we have that $\|\hat{u}_{\varepsilon n}\| < c$, where c is independent of ε and n. Thus as $\varepsilon \to 0$, $n \to \infty$ (by considering subsequences if necessary) we may write that

$$u_{\varepsilon n} \to u \qquad \text{weakly in } V \tag{3.1.30}$$

and because of (3.1.3)

$$\hat{u}_{\varepsilon n} \to \hat{u} \qquad \text{strongly in } L^2(\Omega) \tag{3.1.31}$$

and thus

$$\hat{u}_{\varepsilon n} \to \hat{u} \qquad \text{a.e. on } \Omega. \tag{3.1.32}$$

Moreover due to Lemma 3.3 we can write that

$$\beta_\varepsilon(\hat{u}_{\varepsilon n}) \to \chi \qquad \text{weakly in } L^1(\Omega). \tag{3.1.33}$$

Using assumption (3.1.4) and the properties of the Galerkin basis and (3.1.9c) we can pass to the limit $\varepsilon \to 0$, $n \to \infty$ in (3.1.13) and obtain

$$a(u,v) + \int_\Omega \chi\hat{v}d\Omega = (l,v) \qquad \forall v \in \tilde{V}, \tag{3.1.34a}$$

from which it follows that a linear functional

$$(L^*\chi, v) = \int_\Omega \chi\hat{v}d\Omega \quad \forall v \in \tilde{V} \tag{3.1.34b}$$

can be uniquely extend to the whole space as $L^*\chi \in V^*$. Thus (3.1.34) can be written in the form

$$\alpha(u,v) + (L^*\chi, v) = (l,v) \quad \forall v \in V. \tag{3.1.34c}$$

In order to complete the proof it will be shown that

$$\chi \in \tilde{\beta}(\hat{u}) = \partial j(\hat{u}) \qquad \text{a.e. on } \Omega. \tag{3.1.35}$$

From (3.1.32) by applying Egoroff's theorem we can find that for any $\alpha > 0$ we can determine $\omega \subset \Omega$ with mes $\omega < \alpha$ such that

$$\hat{u}_{\varepsilon n} \to \hat{u} \qquad \text{uniformly on } \Omega - \omega \qquad (3.1.36)$$

with $\hat{u} \in L^\infty(\Omega - \omega)$. Thus for any $\alpha > 0$ we can find $\omega \subset \Omega$ with mes $\omega < \alpha$ such that for any $\mu > 0$ and for $\varepsilon < \varepsilon_0 < \mu/2$ and $n > n_0 > 2/\mu$ we have

$$|\hat{u}_{\varepsilon n} - \hat{u}| < \frac{\mu}{2}, \qquad \forall x \in \Omega - \omega. \qquad (3.1.37)$$

From (3.1.13) (3.1.14) we obtain that

$$\beta_\varepsilon(\hat{u}_{\varepsilon n}) \leq \operatorname*{ess\,sup}_{|\hat{u}_{\varepsilon n} - \xi| \leq \varepsilon} \beta(\xi) \leq \operatorname*{ess\,sup}_{|\hat{u}_{\varepsilon n} - \xi| < \frac{\mu}{2}} \beta(\xi) \leq \operatorname*{ess\,sup}_{|\hat{u} - \xi| \leq \mu} \beta(\xi) = \bar{\bar{\beta}}_\mu(\hat{u}), \quad (3.1.38)$$

where $\bar{\bar{\beta}}_\mu$ is defined in (3.1.6). Analogously we prove the inequality

$$\bar{\beta}_\mu(\hat{u}) = \operatorname*{ess\,inf}_{|\hat{u} - \xi| \leq \mu} \beta(\xi) \leq \beta_\varepsilon(\hat{u}_{\varepsilon n}). \qquad (3.1.39)$$

We take now $e \geq 0$ a.e. on $\Omega - \omega$ with $e \in L^\infty(\Omega - \omega)$, and we obtain from (3.1.38) and (3.1.39) the inequality

$$\int_{\Omega - \omega} \bar{\beta}_\mu(\hat{u}) e \, d\Omega \leq \int_{\Omega - \omega} \beta_\varepsilon(\hat{u}_{\varepsilon n}) e \, d\Omega \leq \int_{\Omega - \omega} \bar{\bar{\beta}}_\mu(\hat{u}) e \, d\Omega. \qquad (3.1.40)$$

Taking the limits as $\varepsilon \to 0$ and $n \to \infty$ we obtain that

$$\int_{\Omega - \omega} \bar{\beta}_\mu(\hat{u}) e \, d\Omega \leq \int_{\Omega - \omega} \chi e \, d\Omega \leq \int_{\Omega - \omega} \bar{\bar{\beta}}_\mu(\hat{u}) e \, d\Omega \qquad (3.1.41)$$

and as $\mu \to 0$ that

$$\int_{\Omega - \omega} \bar{\beta}(\hat{u}) e \, d\Omega \leq \int_{\Omega - \omega} \chi e \, d\Omega \leq \int_{\Omega - \omega} \bar{\bar{\beta}}(\hat{u}) e \, d\Omega. \qquad (3.1.42)$$

Since e is arbitrary we have that

$$\chi \in [\bar{\beta}(\hat{u}), \bar{\bar{\beta}}(\hat{u})] = \hat{\beta}(\hat{u}), \qquad \text{a.e. on } \Omega - \omega, \qquad (3.1.43)$$

where mes $\omega < \alpha$. For α as small as possible, we obtain (3.1.35).   q.e.d.

Several of the arguments applied in the proof of this theorem are borrowed from the method developed in [Rauch] for the existence proof for semilinear differential equations.

Further a proposition concerning the strong convergence of $u_{\varepsilon n}$ to u is proved.

Proposition 3.5 Suppose that there is c const > 0 such that

$$|\beta(\xi)| \leq c(1 + |\xi|) \qquad \forall \xi \in R. \tag{3.1.44}$$

Then $\hat{u}_{\varepsilon n} \to u$ strongly in V as $\varepsilon \to 0$, $n \to \infty$.

Proof. Let us put $v = \hat{u}_{\varepsilon n}$ in (3.1.9). It results that

$$a(u, u) \leq a(u, u_{\varepsilon n}) + \int_{\Omega} j^0(u, \hat{u}_{\varepsilon n} - u)d\Omega - (l, u_{\varepsilon n} - u). \tag{3.1.45}$$

Similarly we get from (3.1.13) that

$$a(u_{\varepsilon n}, u_{\varepsilon n}) = a(u_{\varepsilon n}, v_n) + \int_{\Omega} \beta_{\varepsilon}(\hat{u}_{\varepsilon n})(\hat{v}_n - \hat{u}_{\varepsilon n})d\Omega - (l, v_n - u_{\varepsilon n}), \quad \forall v_n \in V_n. \tag{3.1.46}$$

The coercivity of $a(\cdot, \cdot)$, and (3.1.45) and (3.1.46) imply that

$$
\begin{aligned}
c\|u - u_{\varepsilon n}\|^2 \;\leq\;& a(u - u_{\varepsilon n}, u - u_{\varepsilon n}) \tag{3.1.47} \\
\leq\;& a(u_{\varepsilon n}, v_n - u) + \int_{\Omega} j^0(\hat{u}, \hat{u}_{\varepsilon n} - \hat{u})d\Omega \\
& + \int_{\Omega} \beta_{\varepsilon}(\hat{u}_{\varepsilon n})(\hat{v}_n - \hat{u}_{\varepsilon n})d\Omega - (l, v_n - u).
\end{aligned}
$$

The definition of j^0 and (3.1.44) imply that

$$
\begin{aligned}
\int_{\Omega} j^0(\hat{u}, \hat{u}_{\varepsilon n} - \hat{u})d\Omega \;=\;& \int_{\Omega} \left(\limsup_{\substack{\lambda \to 0_+ \\ h \to 0}} \frac{1}{\lambda} \int_{\hat{u}+h}^{\hat{u}+h+\lambda(\hat{u}_{\varepsilon n}-\hat{u})} \beta(\xi)d\xi \right) d\Omega \tag{3.1.48} \\
\leq\;& c\int_{\Omega} \left[\limsup_{\substack{\lambda \to 0_+ \\ h \to 0}} \frac{1}{\lambda} \int_{\hat{u}+h}^{\hat{u}+h+\lambda(\hat{u}_{\varepsilon n}-\hat{u})} (1 + |\xi|)d\xi \right] d\Omega \\
=\;& c\int_{\Omega} (1 + |\hat{u}|)(\hat{u}_{\varepsilon n} - \hat{u})d\Omega.
\end{aligned}
$$

Due to (3.1.31) the right-hand side of (3.1.48) tends to zero for $\varepsilon \to 0$, $n \to \infty$. Moreover

$$
\begin{aligned}
\left| \int_{\Omega} \beta_{\varepsilon}(\hat{u}_{\varepsilon n})(\hat{v}_n - \hat{u}_{\varepsilon n})d\Omega \right| \;\leq\;& \|\beta_{\varepsilon}(\hat{u}_{\varepsilon n})\|_{L^2}\|\hat{v}_n - \hat{u}_{\varepsilon n}\|_{L^2} \tag{3.1.49} \\
\leq\;& \|\beta_{\varepsilon}(\hat{u}_{\varepsilon n})\|_{L^2}(\|\hat{v}_n - \hat{u}\|_{L^2} + \|\hat{u} - \hat{u}_{\varepsilon n}\|_{L^2})
\end{aligned}
$$

and $\|\beta_{\varepsilon}(\hat{u}_{\varepsilon n})\|_{L^2} < c$ (independently of ε, n) as it results from (3.1.11) (3.1.44) and (3.1.31). Now set into (3.1.47) a v_n such that

$$v_n \to u \qquad \text{strongly in } V. \tag{3.1.50}$$

Thus as $\varepsilon \to 0$ $n \to \infty$, the right-hand side of (3.1.49) tends to zero. From the above limits we obtain that the right-hand side of (3.1.47) tends to zero as $\varepsilon \to 0$, $n \to \infty$. Thus the strong convergence of $u_{\varepsilon n}$ to u in V is proved. q.e.d.

Note that a more abstract existence proof for problem P^C can be given as it will become obvious from the proofs in the next Chapters. Here we prefer the present proof, i.e. the use of problems P_ε^C and $P_{\varepsilon n}^C$, because it is directly applicable to most engineering and mechanical problems and because it permits the treatment of the semicoercive case under slightly more general assumptions. In the forthcoming Chapters more abstract existence results and more general types of hemivariational inequalities are given. Due to the lack of convexity a uniqueness result cannot be proved.

3.2 Semicoercive Hemivariational Inequalities

In this section the hemivariational inequality (3.1.9) will be studied on the assumption that $a(\cdot, \cdot)$ is no longer coercive but semicoercive, i.e. $a(\cdot, \cdot)$ is continuous and symmetric but it has a nonzero kernel, i.e.

$$\ker a(\cdot, \cdot) = \{q : a(q, q) = 0\} \neq \{0\}. \tag{3.2.1}$$

Moreover let

$$\ker a \text{ be finite dimensional.} \tag{3.2.2}$$

We assume that the norm $||v||$ on V is equivalent to $|||v||| = p(\tilde{v}) + |q|_2$, where $v = \tilde{v} + q$, $q \in \ker a$, $\tilde{v} \in \ker a^\perp$ (i.e. $(\tilde{v}, q) = 0$ $\forall q \in \ker a$) and $p(\tilde{v})$ is a seminorm on V such that $p(v) = p(v + q)$ $\forall v \in V$, $q \in \ker a$ and let

$$a(v, v) \geq c(p(v))^2, \qquad \forall v \in V, \ c = \text{const} > 0. \tag{3.2.3}$$

This semicoercivity inequality replaces (3.1.5). Further, we keep the assumptions (3.1.2), (3.1.3), (3.1.4), $l \in V^*$ and the assumption $\beta \in L^\infty_{\text{loc}}(R)$, which leads to (3.1.8). The semicoercive problem P^S, reads:
Find $u \in V$ such that

$$a(u, v - u) + \int_\Omega j^0(\hat{u}, \hat{v} - \hat{u})d\Omega \geq (l, v - u) \qquad \forall v \in V, \tag{3.2.4}$$

in the sense of (3.1.9a)−(3.1.9c).
We denote by q_+ and q_- the positive and the negative parts of $\hat{q}$, where $\hat{q} = Lq$, i.e. $q_+ = \max\{0, \hat{q}\}$, $q_- = \max\{0, -\hat{q}\}$ and the notation below is introduced

$$\beta(-\infty) = \limsup_{\xi \to -\infty} \beta(\xi) \qquad \text{and} \qquad \beta(\infty) = \liminf_{\xi \to \infty} \beta(\xi). \tag{3.2.5}$$

The following proposition gives a necessary condition for the existence of the solution.

Proposition 3.6 Let

$$\beta(-\infty) \leq \beta(\xi) \leq \beta(\infty), \qquad \forall \xi \in R. \tag{3.2.6}$$

Then a necessary condition for the existence of a solution $u \in V$ of problem P^S is the inequality

$$\int_{\Omega} [\beta(-\infty)q_+ - \beta(\infty)q_-]d\Omega \leq (l, q) \tag{3.2.7}$$

$$\leq \int_{\Omega} [\beta(\infty)q_+ - \beta(-\infty)q_-]d\Omega \quad \forall q \in \ker a.$$

If (3.2.6) holds strictly (with $<$ instead of $\leq$) then (3.2.7) also holds strictly.

Proof. We set in (3.2.4) $v - u = \pm q \in \ker a$, $q \neq 0$. We obtain

$$\int_{\Omega} j^0(\hat{u}, \pm \hat{q})d\Omega \geq \pm(l, q) \qquad \forall q \in \ker a, q \neq 0. \tag{3.2.8}$$

Then (3.2.8) is written as

$$\int_{\Omega} j^0(\hat{u}, \hat{q})d\Omega \geq (l, q) \geq -\int_{\Omega} j^0(\hat{u}, -\hat{q})d\Omega, \qquad \forall q \in \ker a, q \neq 0 \tag{3.2.9}$$

because $\hat{q} \rightarrow j^0(\hat{u}, \hat{q})$ is positively homogeneous. From the definition of j and j^0 we obtain due to (3.2.6) that

$$\int_{\Omega} j^0(\hat{u}, \hat{q})d\Omega \leq \int_{\Omega} [\beta(\infty)q_+ - \beta(-\infty)q_-]d\Omega, \qquad \forall q \in \ker a, q \neq 0 \tag{3.2.10}$$

and analogously we obtain the corresponding inequality for $\int_{\Omega} j^0(\hat{u}, -\hat{q})d\Omega$. Thus (3.2.7) is proved. Analogously we show the rest of the proposition.   q.e.d.

Theorem 3.7 Suppose that

$$\beta(-\infty) < \beta(\infty). \tag{3.2.11}$$

Then if

$$\int_{\Omega} [\beta(-\infty)q_+ - \beta(\infty)q_-]d\Omega < (l, q) \tag{3.2.12}$$

$$< \int_{\Omega} [\beta(\infty)q_+ - \beta(-\infty)q_-]d\Omega \quad \forall q \in \ker a, q \neq 0,$$

problem P^S has a solution.

Proof. The proof follows the same steps as the proof of Theorem 3.4. Estimate (3.1.15) is applied; it holds due to (3.2.11), but a more sharp estimate is necessary. As in the coercive case the regularized problem P_ϵ^S is defined and the finite dimensional problem P_{en}^S may be put in the form (3.1.20). From (3.1.20), (3.2.3) and (3.1.15) we obtain the estimate

$$(\Lambda(u_{en}), u_{en}) \geq c[p(\tilde{u}_{en})]^2 - c|||u_{en}||| - \rho_1\rho_2 \text{mes}\Omega, \quad c = \text{const} > 0. \quad (3.2.13)$$

We want now to apply Brouwer's theorem to prove that (3.1.20) has at least one solution u_{en} with $||u_{en}|| \leq c$ where c is a constant independent of ε and n. According to this theorem one has to show that $r > 0$ exists such that

$$||u_{en}|| \geq r \Longrightarrow (\Lambda(u_{en}), u_{en}) \geq 0. \quad (3.2.14)$$

Here we will prove that a number $M > 0$ can be determined such that

$$||\hat{u}_{en}|| > M \Longrightarrow (\Lambda(\hat{u}_{en}), \hat{u}_{en}) > 0 \quad (3.2.15)$$

and thus one may take $||u_{en}|| = r > M$. Instead of (3.2.15) we shall show equivalently that

$$(\Lambda(u_{en}), u_{en}) \leq 0 \Longrightarrow ||u_{en}|| \leq c. \quad (3.2.16)$$

From (3.2.13) if $(\Lambda(u_{en}), u_{en}) \leq 0$, then a constant $c > 0$ exists such that

$$p(\tilde{u}_{en}) \leq c(\sqrt{(|q_{en}|_2)} + 1) \quad (3.2.17)$$

where $u_{en} = \tilde{u}_{en} + q_{en}$. Thus it is enough to show that $(\Lambda(u_{en}), u_{en}) \leq 0$ implies $|q_{en}|_2 \leq c$, or equivalently, that a number $B > 0$ can be determined such that

$$|q_{en}|_2 > B \text{ and } (3.2.17) \Longrightarrow (\Lambda(u_{en}), u_{en}) > 0. \quad (3.2.18)$$

This last relation will be proved.

The definition of β_ε implies that

$$\beta_\varepsilon(\infty) = \lim_{\xi \to \infty} \beta_\varepsilon(\xi) = \lim_{\xi \to \infty} \int_{-\varepsilon}^{+\varepsilon} \beta(\xi - t) p_\varepsilon(t) dt \geq \lim_{\xi \to \infty} \operatorname*{ess\,inf}_{|x - \xi| \leq \varepsilon} \beta(x) \quad (3.2.19)$$

$$\geq \operatorname*{ess\,inf}_{\xi \to \infty, \, \xi - \varepsilon \leq x < \infty} \beta(x) = \liminf_{\xi \to \infty} \beta(x) = \beta(\infty).$$

Similarly $\beta_\varepsilon(-\infty) \leq \beta(-\infty)$. Thus (3.2.12) implies that

$$\int_\Omega [\beta_\varepsilon(-\infty)q_+ - \beta_\varepsilon(\infty)q_-] d\Omega < (l, q) \quad (3.2.20)$$

$$< \int_\Omega [\beta_\varepsilon(\infty)q_+ - \beta_\varepsilon(-\infty)q_-] d\Omega, \quad \forall q \in \ker a, q \neq 0.$$

We note now that (3.2.11) implies the exterior inequality of (3.1.14) as a strict inequality (i.e. esssup < essinf); a number $\tilde{M} > \rho_1$ (cf. (3.1.15)) can be chosen such that for any function $u \in V$ with $|\hat{u}(x)| > \tilde{M}$ and sign $\hat{u}(x) = $ sign $\hat{q}(x)$ for almost every $x \in \Omega$, we have from (3.2.20) that for $\hat{q} > 0$, $\hat{q} = q_+$, $\hat{u}(x) > \tilde{M}$ and thus

$$\int\limits_{\{x:\hat{q}(x)>0\}} \beta_\varepsilon(\hat{u}(x))\hat{q}(x)d\Omega - (l,q) > 0 \qquad (3.2.21)$$

$$- \int\limits_{\{x:\hat{q}(x)>0\}} \beta_\varepsilon(-\hat{u}(x))\hat{q}(x)d\Omega + (l,q) > 0. \qquad (3.2.22)$$

For $\hat{q} < 0$ we have $\hat{q} = -q_-$, $\hat{u}(x) < -M$ and thus (3.2.20) implies that

$$\int\limits_{\{x:\hat{q}(x)<0\}} \beta_\varepsilon(\hat{u}(x))\hat{q}(x)d\Omega - (l,q) > 0 \qquad (3.2.23)$$

$$- \int\limits_{\{x:\hat{q}(x)<0\}} \beta_\varepsilon(-\hat{u}(x))\hat{q}(x)d\Omega + (l,q) > 0. \qquad (3.2.24)$$

By an appropriate choice of the numbers $\delta \in (0,1]$, $N > 1$, $\eta > 0$, and $\alpha > 0$, and by taking into consideration that $\beta_\varepsilon(\hat{u}(x))\hat{u}(x) \geq 0$ and that sign $\hat{u}(x) =$ sign $\hat{q}(x)$ these inequalities imply for every u as above, the relations

$$\left(1 - \frac{1}{N}\right) \int\limits_{\{x:|\hat{q}(x)|>\delta\alpha\}} \beta_\varepsilon(\hat{u}(x))\hat{q}(x)d\Omega - (l,q) > \eta|q|_2 \qquad (3.2.25)$$

$$-\left(1 - \frac{1}{N}\right) \int\limits_{\{x:|\hat{q}(x)|>\delta\alpha\}} \beta_\varepsilon(-\hat{u}(x))\hat{q}(x)d\Omega + (l,q) > \eta|q|_2 \qquad (3.2.26)$$

as is obvious for $\delta \to 0_+$. Now we write $\hat{u}_{\varepsilon n} = \tilde{u}_{\varepsilon n} + q_{\varepsilon n}$, and let us take N as in (3.2.25), (3.2.26). Then for $\alpha > \alpha_0 = \tilde{M}\delta^{-1}(1 - 1/N)^{-1}$ it results

$$\int\limits_\Omega \beta_\varepsilon(\hat{u}_{\varepsilon n})\hat{u}_{\varepsilon n}d\Omega = \int\limits_{\substack{|\hat{u}_{\varepsilon n}(x)|<\delta\alpha/N \\ |\hat{q}_{\varepsilon n}(x)|>\delta\alpha}} \cdots + \int\limits_{\substack{|\hat{u}_{\varepsilon n}(x)|\geq\delta\alpha/N \\ |\hat{q}_{\varepsilon n}(x)|>\delta\alpha}} \cdots + \int\limits_{|\hat{q}_{\varepsilon n}(x)|\leq\delta\alpha} \cdots$$

$$\geq \int\limits_{\substack{|\hat{u}_{\varepsilon n}(x)|<\delta\alpha/N \\ |\hat{q}_{\varepsilon n}(x)|>\delta\alpha}} \beta_\varepsilon(\hat{u}_{\varepsilon n})\hat{u}_{\varepsilon n}d\Omega - \rho_1\rho_2 \operatorname{mes}\Omega \qquad (3.2.27)$$

$$\geq \left(1 - \frac{1}{N}\right) \int\limits_{\substack{|\hat{u}_{\varepsilon n}(x)|<\delta\alpha/N \\ |\hat{q}_{\varepsilon n}(x)|>\delta\alpha}} \hat{q}_{\varepsilon n}\beta_\varepsilon(\hat{u}_{\varepsilon n})d\Omega - \rho_1\rho_2 \operatorname{mes}\Omega.$$

Indeed for $|\hat{\tilde{u}}_{\varepsilon n}(x)| < \delta\alpha/N$ and $|\hat{q}_{\varepsilon n}(x)| > \delta\alpha$ one has that for $\alpha > \alpha_0$

$$|\hat{u}_{\varepsilon n}| = |\hat{\tilde{u}}_{\varepsilon n} + \hat{q}_{\varepsilon n}| > \left(1 - \frac{1}{N}\right)\delta\alpha > \tilde{M} \qquad (3.2.28)$$

and thus $\beta_\varepsilon(\hat{u}_{\varepsilon n})\hat{u}_{\varepsilon n} \geq 0$, and $\beta_\varepsilon(\hat{u}_{\varepsilon n})\hat{q}_{\varepsilon n} \geq 0$. Further it can be shown that for $\hat{q}_{\varepsilon n} > \delta\alpha$

$$\hat{u}_{\varepsilon n}(x) = \hat{\tilde{u}}_{\varepsilon n}(x) + \hat{q}_{\varepsilon n}(x) > -\frac{\delta\alpha}{N} + \hat{q}_{\varepsilon n}(x) > \hat{q}_{\varepsilon n}(x)\left(1 - \frac{1}{N}\right) \qquad (3.2.29)$$

and thus

$$\beta_\varepsilon(\hat{u}_{en})\hat{u}_{en} \geq \left(1 - \frac{1}{N}\right)\hat{q}_{en}\beta_\varepsilon(\hat{u}_{en}). \qquad (3.2.30)$$

Similarly for $\hat{q}_{en} < -\delta\alpha$. Thus

$$(\Lambda(u_{en}), u_{en}) \geq \left(1 - \frac{1}{N}\right)\int_{\substack{|\hat{u}_{en}(x)|<\delta\alpha/N \\ |\hat{q}_{en}(x)|>\delta\alpha}} \hat{q}_{en}\beta_\varepsilon(\hat{u}_{en})d\Omega - \rho_1\rho_2 \operatorname{mes}\Omega - (l, q_{en}) - (l, \tilde{u}_{en})$$

$$(3.2.31)$$

is obtained. For $\alpha > \alpha_0$ sufficiently large (3.2.31) and (3.2.25) imply that

$$(\Lambda(u_{en}), u_{en}) \quad > \quad \eta|q_{en}|_2 - \rho_1\rho_2 \operatorname{mes}\Omega - (l, \tilde{u}_{en}) \qquad (3.2.32)$$
$$\geq \quad \eta|q_{en}|_2 - c_1 - c_2\|\tilde{u}_{en}\| \geq \eta|q_{en}|_2 - c_1 - c_2'\|\|\tilde{u}_{en}\|\|$$
$$= \quad \eta|q_{en}|_2 - c_1 - c_2'p(\tilde{u}_{en}), \quad c_1, c_2, c_2' \text{ const} > 0.$$

From (3.2.32), assuming that (3.2.17) holds and that $\alpha > \alpha_0$, we get the estimate

$$(\Lambda(u_{en}), u_{en}) > \eta|q_{en}|_2 - c|q_{en}|_2^{1/2} - c', \quad c, c' = \text{const} > 0. \qquad (3.2.33)$$

The right-hand side of (3.2.33) is positive if $B > 0$ is such that

$$|q_{en}|_2 > B > \delta\alpha_0(\operatorname{mes}\Omega)^{1/2}. \qquad (3.2.34)$$

Thus we have proved (3.2.18) and therefore Brouwer's fixed point theorem implies that problem P_{en}^S has a solution u_{en} with $\|u_{en}\| < c$. The rest of the proof is the same as the proof of Theorem 3.4. q.e.d.

The reader is advised to compare firstly the sufficient conditions with the ones derived in the Landesman-Lazer theory [Land] and secondly this proof with the proof given in [Ken78] for semicoercive semilinear differential equations with dimension of ker a equal to one.

3.3 The Substationarity of the Energy

In this section the relation between the hemivariational inequality and the corresponding substationarity problem is investigated. Let us consider the following problem P_1: Find $u \in V$ such that the "energy" functional

$$\Pi(v) = \frac{1}{2}a(v, v) + \int_\Omega j(\hat{v})d\Omega - (l, v) \qquad (3.3.1)$$

is substationary at $v = u$.
Here the integral $\int_\Omega j(\hat{v})d\Omega$ is set equal to ∞ if it is not defined. This integral is finite on V if (3.1.44) holds.

This problem is by definition equivalent to the following problem: Find $u \in V$ which is a solution of the inclusion

$$0 \in \partial \Pi(u). \tag{3.3.2}$$

We recall that (3.3.2) is the definition of substationarity at v. Now we prove the following result.

Proposition 3.8 Suppose that j is locally Lipschitz and ∂-regular and that (3.1.44) holds. Then every solution of (3.3.2) satisfies the hemivariational inequality P^C (or P^S) and conversely[1].

Proof. Equation (3.3.2) may be put in the equivalent form

$$l \in \partial \Pi_1(v) \quad \text{for } v \in V, \tag{3.3.3}$$

where

$$\Pi(v) = \Pi_1(v) - (l, v). \tag{3.3.4}$$

Now compute directly $\partial \Pi_1(u)$ by using the definition of $\Pi^0(u, v)$. Note that $\frac{1}{2}a(u, u)$ is ∂-regular and that

$$\Pi_1^0(u, v) \le a(u, v) + J^0(\hat{u}, \hat{v}), \tag{3.3.5}$$

where J is the (finite) integral

$$J(\hat{u}) = \int_\Omega j(\hat{u})d\Omega. \tag{3.3.6}$$

We will first prove that

$$J^0(\hat{u}, \hat{v}) = \int_\Omega j^0(\hat{u}, \hat{v})d\Omega. \tag{3.3.7}$$

Let us denote by $b_{\lambda,h}$ the difference quotient

$$b_{\lambda,h}(\hat{u}, \hat{v}) = \frac{j(\hat{u} + h + \lambda\hat{v}) - j(\hat{u} + h)}{\lambda}. \tag{3.3.8}$$

Function $\xi \to j(\xi)$ is locally Lipschitz and therefore

$$|b_{\lambda,h}(\hat{u}, \hat{v})| \le c|\hat{v}|, \tag{3.3.9}$$

where c depends on the neighbourhood of $(\hat{u} + h)(x)$ and $|\cdot|$ denotes as usual the absolute value. Note that $\hat{u}, \hat{v} \in L^2(\Omega)$ and that $\xi \to j(\xi)$ is continuous. Thus $x \to b_{\lambda,h}(\hat{u}(x), \hat{v}(x))$ is measurable. We apply now Fatou's lemma for not integrable functions ([Dunf]). We get that

$$\int_\Omega \limsup_{\substack{\lambda \to 0_+ \\ h \to 0}} (b_{\lambda,h} - c|\hat{v}|)d\Omega \ge \limsup_{\substack{\lambda \to 0_+ \\ h \to 0}} \int_\Omega (b_{\lambda,h} - c|\hat{v}|)d\Omega. \tag{3.3.10}$$

---

[1]We mean in this Section the solution of the hemivariational inequality itself and not the solution of (3.1.9a,b,c).

Due to the growth assumption (3.1.44) c in (3.3.10) is a function of $L^2(\Omega)$ as it can be easily verified. Accordingly $\int c|\hat{v}|d\Omega$ is finite and may "disappear" from both sides in (3.3.10). Thus (3.3.10) implies that

$$\int_\Omega j^0(\hat{u},\hat{v})d\Omega \geq J^0(\hat{u},\hat{v}), \tag{3.3.11}$$

where the integrals are finite. Using the definition of limsup, Fatou's lemma, the ∂-regularity of j and (3.3.11) imply that

$$
\begin{aligned}
J^0(\hat{u},\hat{v}) &\geq \liminf_{\lambda\to 0+}\frac{J(\hat{u}+\lambda\hat{v})-J(\hat{u})}{\lambda} \geq \int_\Omega \liminf_{\lambda\to 0+}\frac{j(\hat{u}+\lambda\hat{v})-j(\hat{u})}{\lambda}d\Omega \\
&= \int_\Omega \lim_{\lambda\to 0+}\frac{j(\hat{u}+\lambda\hat{v})-j(\hat{u})}{\lambda}d\Omega = \int_\Omega j'(\hat{u},\hat{v})d\Omega \\
&= \int_\Omega j^0(\hat{u},\hat{v})d\Omega \geq J^0(\hat{u},\hat{v}).
\end{aligned}
\tag{3.3.12}
$$

From (3.3.11) and (3.3.12) we get (3.3.7). Thus (3.3.5) and (3.3.7) imply the hemivariational inequality (3.1.9).

Now the converse will be shown, i.e. that any u satisfying the hemivariational inequalities P^C or P^S is a solution of the substationarity problem (3.3.2). First we will show that J is ∂-regular. As in (3.3.12) Fatou's lemma implies that

$$
\begin{aligned}
J^0(\hat{u},\hat{v}) &\geq \liminf_{\lambda\to 0+}\frac{J(\hat{u}+\lambda\hat{v})-J(\hat{u})}{\lambda} \geq \int_\Omega \liminf_{\lambda\to 0+}\frac{j(\hat{u}+\lambda\hat{v})-j(\hat{u})}{\lambda}d\Omega \\
&= \int_\Omega j'(\hat{u},\hat{v})d\Omega = \int_\Omega \limsup_{\lambda\to 0+}\frac{j(\hat{u}+\lambda\hat{v})-j(\hat{u})}{\lambda} \\
&\geq \limsup_{\lambda\to 0+}\frac{J(\hat{u}+\lambda\hat{v})-J(\hat{u})}{\lambda} \geq \liminf_{\lambda\to 0+}\frac{J(\hat{u}+\lambda\hat{v})-J(\hat{u})}{\lambda}.
\end{aligned}
\tag{3.3.12a}
$$

Thus in (3.3.12a) we have equality everywhere, and $J'(\cdot,\cdot)$ exists, and

$$J'(\hat{u},\hat{v}) = \int_\Omega j'(\hat{u},\hat{v})d\Omega = \int_\Omega j^0(\hat{u},\hat{v})d\Omega = J^0(\hat{u},\hat{v}). \tag{3.3.13}$$

Because of the ∂-regularity of $J(\cdot)$ and $\frac{1}{2}a(\cdot,\cdot)$ in $\Pi_1(\cdot)$, Π_1 is obviously ∂-regular and therefore (3.3.5) holds as equality and

$$\Pi_1^0(u,v) = \Pi_1'(u,v). \tag{3.3.14}$$

Thus one may write P^C or P^S in the form (cf. (3.3.5), (3.3.13) and (3.3.14))

$$u \in V, \Pi_1'(u,v-u) = \Pi_1^0(u,v-u) \geq (l,v-u), \quad \forall v \in V \tag{3.3.15}$$

which yields the substationarity problem (3.3.2).   q.e.d.

3.4 Variational-Hemivariational Inequalities

Let us consider now a functional $\Phi: V \to (-\infty, +\infty]$, $\Phi \not\equiv \infty$, which is convex, l.s.c and proper. The following problem $\tilde{P}^C$ is formulated: Find $u \in V$ such as to satisfy

$$a(u, v - u) + \int_\Omega j^0(\hat{u}, \hat{v} - \hat{u})d\Omega + \Phi(v) - \Phi(u) \geq (l, v - u), \qquad \forall v \in V. \quad (3.4.1)$$

Here V, a and j have the properties $((3.1.2), (3.1.4), (3.1.5), (3.1.8))$ as in the case of problem $(3.1.9)$ in Sect. 3.1, i.e. as for coercive hemivariational inequalities. Moreover $l \in V^*$. We shall first study the "differentiable" case in which $\text{grad}\,\Phi(\cdot)$ exists everywhere and then the "nondifferentiable" case.

(a) *The "differentiable" problem* $\tilde{P}^C$. The following proposition holds:

Proposition 3.9 The inequality (3.4.1) is equivalent to the inequality

$$u \in V, \quad a(u, v - u) + \int_\Omega j^0(\hat{u}, \hat{v} - \hat{u})d\Omega + (\text{grad}\Phi(u), v - u) \quad (3.4.2)$$

$$\geq (l, v - u), \quad \forall v \in V.$$

Proof. From (3.4.1), we obtain (3.4.2) by setting $v = u + \lambda(w - u)$, $\lambda \in (0, 1)$, letting $\lambda \to 0_+$, and using the fact that $\xi \to j^0(\hat{u}, \xi)$ is positively homogeneous. Conversely (3.4.2) implies (3.4.1) by means of the inequality

$$\Phi(v) - \Phi(u) \geq (\text{grad}\Phi(u), v - u), \qquad \forall v \in V \quad (3.4.3)$$

which holds due to the convexity of Φ.   q.e.d.

Now the corresponding regularized problem $\tilde{P}_\varepsilon^C$ and the corresponding finite dimensional problem $\tilde{P}_{\varepsilon n}^C$ are defined: Find $u_\varepsilon \in V$ such as to satisfy the variational equality (cf. also (3.4.7))

$$a(u_\varepsilon, v) + \int_\Omega \beta_\varepsilon(\hat{u}_\varepsilon)\hat{v}d\Omega + (\text{grad}\Phi(u_\varepsilon), v) = (l, v), \qquad \forall v \in V. \quad (3.4.4)$$

Find $u_{\varepsilon n} \in V_n$ such that

$$a(u_{\varepsilon n}, v) + \int_\Omega \beta_\varepsilon(\hat{u}_{\varepsilon n})\hat{v}d\Omega + (\text{grad}\Phi(u_{\varepsilon n}), v) = (l, v), \qquad \forall v \in V_n. \quad (3.4.5)$$

Further the following theorem will be proved.

Theorem 3.10 Suppose that (3.1.14) holds and that

$$\text{grad}\Phi(0) = 0. \quad (3.4.6)$$

We assume that the linear continuous operator L has the property

$$L : V \to L^\infty(\Omega) \text{ is compact.} \qquad (3.4.7)$$

Then problem $\tilde{P}^C$ has at least one solution[2].

Proof. The monotonicity of $\text{grad}\Phi$ and the assumption (3.4.6) imply that

$$(\text{grad}\Phi(u_{\epsilon n}), u_{\epsilon n}) \geq 0, \qquad \forall u_{\epsilon n} \in V_n. \qquad (3.4.8)$$

Thus using (3.1.5), (3.1.15) and (3.4.8) we may write that

$$\begin{aligned}
(\Lambda(u_{\epsilon n}), u_{\epsilon n}) &= a(u_{\epsilon n}, u_{\epsilon n}) + (\text{grad}\Phi(u_{\epsilon n}), u_{\epsilon n}) \qquad (3.4.9) \\
&+ \int_\Omega \beta_\epsilon(\hat{u}_{\epsilon n})\hat{u}_{\epsilon n}d\Omega - (l, u_{\epsilon n}) \\
&\geq c_1||u_{\epsilon n}||^2 - \rho_1\rho_2\text{mes }\Omega - c_2||u_{\epsilon n}||, \quad c_1, c_2, \text{const} > 0.
\end{aligned}$$

From (3.4.9) by Brouwer's fixed point theorem, problem $\tilde{P}^C_{\epsilon n}$ has a solution $u_{\epsilon n}$ with $||u_{\epsilon n}|| < c$. Thus as $\epsilon \to 0$, $n \to \infty$ we may extract a subsequence denoted also by $\{u_{\epsilon n}\}$ such that (3.1.30), (3.1.31) and (3.1.32) hold (due to (3.4.7)).

A slight modification of the proof of Lemma 3.3 implies (3.1.22), i.e. the weak convergence of (3.1.33). The modification consists in the fact that in the last inequality of (3.1.25) the term $(\text{grad }\Phi(u_{\epsilon n}), u_{\epsilon n})$ appears, which, because of (3.4.8) does not change the result of (3.1.25). The foregoing estimates and (3.4.5) imply that

$$||\text{grad }\Phi(u_{\epsilon n})||_{V^*} < c \qquad (3.4.10)$$

and thus as $\epsilon \to 0$, $n \to \infty$

$$\text{grad }\Phi(u_{\epsilon n}) \to \psi \qquad \text{weakly in } V^*. \qquad (3.4.11)$$

Because of (3.1.4) and the properties of the Galerkin basis we may pass to the limit. Thus from (3.4.5) due to (3.1.4) the equality

$$a(u, v) + \int_\Omega \chi\hat{v}d\Omega + (\psi, v) = (l, v), \qquad \forall v \in V \qquad (3.4.12)$$

results since $\hat{u} \in L^\infty(\Omega)$. In order to complete the proof we have to show that

$$\psi = \text{grad }\Phi(u) \text{ in } V^* \qquad (3.4.13)$$

and that

$$\chi \in \hat{\beta}(\hat{u}), \quad \text{a.e. on } \Omega. \qquad (3.4.14)$$

The monotonicity of $\text{grad }\Phi$ implies that

---

[2]Note here that due to (3.4.7) relation (3.1.9c) holds for every $v \in V$ in the theory, as formulated here, of variational-hemivariational inequalities. Therefore a definition of the solution as the function satisfying relations analogous to (3.1.9a,b) is superfluous.

$$X_n = (\text{grad } \Phi(u_{\varepsilon n}) - (\text{grad } \Phi(\theta), u_{\varepsilon n} - \theta) \geq 0, \qquad \forall \theta \in V, \qquad (3.4.15)$$

and by means of (3.4.5) that

$$
\begin{aligned}
X_n = \ & -a(u_{\varepsilon n}, u_{\varepsilon n}) - \int_{\Omega} \beta_{\varepsilon}(\hat{u}_{\varepsilon n}) \hat{u}_{\varepsilon n} d\Omega + (l, u_{\varepsilon n}) \qquad (3.4.16) \\
& -(\text{grad } \Phi(u_{\varepsilon n}), \theta) - (\text{grad } \Phi(\theta), u_{\varepsilon n} - \theta) \geq 0 \qquad \forall \theta \in V.
\end{aligned}
$$

Due to (3.4.7) we may show that

$$\lim_{\substack{\varepsilon \to 0 \\ n \to \infty}} \int_{\Omega} \beta_{\varepsilon}(\hat{u}_{\varepsilon n}) \hat{u}_{\varepsilon n} d\Omega = \int_{\Omega} \chi \hat{u} d\Omega. \qquad (3.4.17)$$

Indeed $\hat{u} \in L^{\infty}(\Omega)$ and we have that

$$\int_{\Omega} \{\beta_{\varepsilon}(\hat{u}_{\varepsilon n})\hat{u}_{\varepsilon n} - \chi\hat{u}\} d\Omega = \int_{\Omega} \beta_{\varepsilon}(\hat{u}_{\varepsilon n})(\hat{u}_{\varepsilon n} - \hat{u})d\Omega + \int_{\Omega} \hat{u}(\beta_{\varepsilon}(\hat{u}_{\varepsilon n}) - \chi)d\Omega = A + B.$$
$$(3.4.18)$$

Since V is compactly imbedded into $L^{\infty}(\Omega)$, $\lim B = 0$. Again (3.1.22) implies that

$$\|\beta_{\varepsilon}(\hat{u}_{\varepsilon n})\|_{L^1(\Omega)} < c. \qquad (3.4.19)$$

From (3.4.19) and (3.4.7) we get that $\lim A = 0$.

For $\varepsilon \to 0$ and $n \to \infty$ we obtain from (3.4.16) using (3.4.17) and the inequality

$$\liminf a(u_{\varepsilon n}, u_{\varepsilon n}) \geq a(u, u) \qquad (3.4.20)$$

that

$$0 \leq \limsup X_n \leq -a(u, u) - \int_{\Omega} \chi\hat{u}d\Omega + (l, u) - (\psi, \theta) - (\text{grad } \Phi(\theta), u - \theta), \forall \theta \in V.$$
$$(3.4.21)$$

From (3.4.21) and (3.4.12) the inequality

$$(\psi - \text{grad } \Phi(\theta), u - \theta) \geq 0, \qquad \forall \theta \in V \qquad (3.4.22)$$

results. Now Minty's monotonicity argument is applied: setting in (3.4.22) $u - \theta = \lambda w$, $\lambda > 0$, we get the expression

$$(\psi - \text{grad } \Phi(u - \lambda w), w) \geq 0, \qquad \forall w \in V \qquad (3.4.23)$$

from which by passing to the limit $\lambda \to 0_+$, because of the monotonicity of the function $\lambda \to (\text{grad } \Phi(u - \lambda w), w)$, we obtain that

$$(\psi - \text{grad } \Phi(u), w) \geq 0, \qquad \forall w \in V. \qquad (3.4.24)$$

Then (3.4.13) results from (3.4.24) by setting $\pm w$. The proof of (3.4.14) is the same as in Theorem 3.4.

q.e.d.

(b) *The "nondifferentiable" problem* $\tilde{P}^C$. In this case $\operatorname{grad} \Phi$ does not exist everywhere. A regularization of Φ transforms the problem into a sequence of differentiable problems. The following assumption is made: there exist convex Gâteaux differentiable functionals $\Phi_\rho, \rho > 0$, which have the following properties:

i) $\Phi_\rho(v) \to \Phi(v)$, $\forall v \in V$ as $\rho \to 0$ (3.4.25)

ii) $\operatorname{grad} \Phi_\rho(0) = 0$ for every ρ (3.4.26)

iii) if $v_\rho \to v$ weakly in V for $\rho \to 0$ and $\Phi_\rho(v_\rho) < M$, where M is a constant, then

$$\liminf_{\rho \to 0} \Phi_\rho(v_\rho) \geq \Phi(v).   (3.4.27)$$

The regularized problem $\tilde{P}^C_{\epsilon\rho}$ is given below:
Find $u \in V$ such as to satisfy the variational equality

$$a(u_{\epsilon\rho}, v) + \int_\Omega \beta_\epsilon(\hat{u}_{\epsilon\rho}) \hat{v} d\Omega + (\operatorname{grad} \Phi(u_{\epsilon\rho}), v) = (l, v),   \forall v \in V.   (3.4.28)$$

From (3.4.28) the finite dimensional problem $\tilde{P}^C_{\epsilon\rho n}$ is obtained analogously to (3.4.5). We denote by $u_{\epsilon\rho n} \in V_n$ the solution of this problem if any exists. The following theorem will be now proved.

Theorem 3.11 Suppose that (3.1.14) and (3.4.7) hold and that Φ satisfies (3.4.25), (3.4.26), (3.4.27). Then the nondifferentiable problem $\tilde{P}^C$ has at least one solution.

Proof. For problem $\tilde{P}^C_{\epsilon\rho n}$ it can be shown using Brouwer's fixed point theorem (as in Theorem 3.10) that a solution exists and that   .

$$\|u_{\epsilon\rho n}\| \leq c,   (3.4.29)$$

where c is a constant independent of ϵ, ρ and n. Thus for $\epsilon \to 0$, $n \to \infty$

$$u_{\epsilon\rho n} \to u_\rho   \text{weakly in } V.   (3.4.30)$$

As in Theorem 3.10 an estimate similar to (3.1.22) is proved and thus

$$\beta_\epsilon(\hat{u}_{\epsilon\rho n}) \to \chi_\rho   \text{weakly in } L^1(\Omega).   (3.4.31)$$

Moreover

$$\|\operatorname{grad} \Phi_\rho(u_{\epsilon\rho n})\|_{V^*} \leq c   (3.4.32)$$

where c is independent of ϵ, n and ρ. Thus for $\epsilon \to 0$, $n \to \infty$

$$\operatorname{grad} \Phi_\rho(u_{\epsilon\rho n}) \to \psi_\rho   \text{weakly in } V^*.   (3.4.33)$$

Since the above estimates are independent of ϵ, n and ρ, for $\rho \to 0$

$$u_\rho \to u \qquad \text{weakly in } V \tag{3.4.34}$$

$$\chi_\rho \to \chi \qquad \text{weakly in } L^1(\Omega) \tag{3.4.35}$$

and

$$\psi_\rho \to \psi \qquad \text{weakly in } V^*. \tag{3.4.36}$$

Analogously to (3.4.17) it is shown that

$$\lim_{\substack{\varepsilon \to 0 \\ n \to \infty}} \int_\Omega \beta_\varepsilon(\hat{u}_{\varepsilon\rho n})\hat{u}_{\varepsilon\rho n}d\Omega = \int_\Omega \chi_\rho \hat{u}_\rho d\Omega. \tag{3.4.37}$$

The monotonicity argument applied as in the previous theorem implies that

$$\psi_\rho = \text{grad } \Phi_\rho(u_\rho). \tag{3.4.38}$$

Finally, for $\varepsilon \to 0$ and $n \to \infty$ the variational equality

$$a(u_\rho, v) + \int_\Omega \chi_\rho \hat{v}d\Omega + (\text{grad } \Phi_\rho(u_\rho), v) = (l, v), \qquad \forall v \in V \tag{3.4.39}$$

is obtained from $P^C_{\varepsilon\rho n}$. The next step in the proof is to pass to the limit in (3.4.39) for $\rho \to 0$. From (3.4.38) and (3.4.39) we obtain the inequality

$$\Phi_\rho(v) - \Phi_\rho(u_\rho) + a(u_\rho, v - u_\rho) + \int_\Omega \chi_\rho(\hat{v} - \hat{u}_\rho)d\Omega \geq (l, v - u_\rho), \quad \forall v \in V. \tag{3.4.40}$$

by applying (3.4.3) to Φ_ρ. If in (3.4.40) a v is chosen such that $\Phi(v) < \infty$, then (3.4.25) implies that a constant M_1 exists such that $\Phi_\rho(v) < M_1$, and from (3.4.40) that $\Phi_\rho(u_\rho) < M'_1$, where M'_1 is another constant. From this last inequality and from (3.4.39), relation (3.4.27) results. From (3.4.40) we obtain

$$D = \Phi_\rho(v) + a(u_\rho, v) + \int_\Omega \chi_\rho \hat{v}d\Omega \tag{3.4.41}$$

$$\geq \Phi_\rho(u_\rho) + a(u_\rho, u_\rho) + \int_\Omega \chi_\rho \hat{u}_\rho d\Omega + (l, v - u_\rho) = F, \quad \forall v \in V,$$

which for $\rho \to 0$ implies that

$$\liminf_{\rho \to 0} D = \lim_{\rho \to 0} \left\{ \Phi_\rho(v) + a(u_\rho, v) + \int_\Omega \chi_\rho \hat{v}d\Omega \right\} \tag{3.4.42}$$

$$= \Phi(v) + a(u, v) + \int_\Omega \chi \hat{v}d\Omega \geq \liminf_{\rho \to 0} F$$

$$= \liminf_{\rho \to 0} \left\{ \Phi_\rho(u_\rho) + a(u_\rho, u_\rho) + \int_\Omega \chi_\rho \hat{u}_\rho d\Omega + (l, v - u_\rho) \right\}$$

$$\geq \Phi(u) + a(u, u) + \int_\Omega \chi \hat{u}d\Omega + (l, v - u), \quad \forall v \in V.$$

Here we have used (3.4.27) and the property (3.4.20) of the bilinear form $a(u_\rho, u_\rho)$. Moreover we are prove analogously to (3.4.17) that

$$\lim_{\rho \to 0} \int_\Omega \chi_\rho \hat{u}_\rho d\Omega = \int_\Omega \chi \hat{u} d\Omega. \qquad (3.4.43)$$

In the final part of the proof we have to show (3.4.14). Its proof is the same as that in the proof of Theorem 3.4   q.e.d.

The semicoercive case of (3.4.1) is still an open problem. Note also that due to (3.4.7) certain parts of the proof (e.g. the proof of (3.4.14) and the proof of Lemma 3.3) can be simplified.

3.5 Applications in Mechanics and Engineering

The results of the previous sections can be applied directly to the study of B.V.Ps in mechanics and engineering involving nonmonotone multivalued boundary conditions, material laws, or interface laws (cf. e.g. Sect. 1.4). For these problems only the theory of hemivariational inequalities has for a first time made possible the derivation of variational formulations and the existence proof for their solution.

3.5.1 Contact of a Linear Elastic Body with an Adhesive Support

We refer first the reader to (1.4.44) for the notation applied in the sequel. Then we assume that on a subset Γ_S of the boundary Γ of $\Omega \subset R^3$ the nonmonotone multivalued law of Fig. 1.4.3a holds between normal displacements u_N and reactions $-S_N$. Moreover on Γ_S the tangential forces S_T are prescribed. Obviously we can write that

$$-S_N \in \bar{\partial}j(u_N), \quad S_T = C_T(x) \quad \text{on } \Gamma_S, \qquad (3.5.1)$$

where j is a locally Lipschitz functional. Let on $\Gamma_U = \Gamma - \Gamma_S$, $\text{mes}\,\Gamma_U > 0$, the displacements be zero. Accordingly we get a hemivariational inequality of the form (3.1.9) where $a(\cdot, \cdot)$ is the elastic energy (cf. (1.4.43)) of the body and l represents both the volume forces $f \in V^*$ in Ω and the given tangential forces $S_T = C_T$, $C_T \in L^2(\Gamma)$ (i.e. $(l, v) = (f, v) + \int_{\Gamma_S} C_{Ti} v_{Ti} d\Gamma$ where $v_T = \{v_{Ti}\}$ are the tangential displacements on Γ_S) Instead of $\int_\Omega j^0(\hat{u}, \hat{v} - \hat{u}) d\Omega$ we have now $\int_{\Gamma_S} j^0(u_N, v_N - u_N) d\Gamma$. For this problem $V = \{v : v \in [H^1(\Omega)]^3, v = 0 \text{ on } \Gamma_U\}$ and $a(\cdot, \cdot)$ is coercive on V. We have that $v \to v_N : V \to L^2(\Gamma)$ is compact. Indeed the trace theorem implies the continuous injection $v \to v_N : [H^1(\Omega)]^3 \to H^{\frac{1}{2}}(\Gamma)$ and because $H^{1/2}(\Omega) \subset L^2(\Gamma)$ is compact. Thus Theorem 3.4 can be proved for the problem under consideration. Moreover assumption (3.1.4) is replaced by the fact that $\{v \in V : v_N \in L^\infty(\Gamma_S)\}$ is dense in V for the H^1-norm. Note that if $\Gamma_U = \emptyset$ then $a(\cdot, \cdot)$ is semicoercive. In this case Theorem 3.7 holds, where $\hat{q} = q_N$, (q is a rigid body displacement), and in (3.2.7) and (3.2.12) the integrals are extended over Γ_S.

Analogously to the above cases Theorems 3.4 and 3.7 hold after minor modifications for the case of a plane linear elastic body, which on $\Gamma_S \subset \Gamma$ has an adhesive support obeying in the normal direction to the law of Fig. 1.4.3a and in the tangential direction to the stick-slip law of Fig. 1.4.3d. If mes $\Gamma_U > 0$ the problem is described by the hemivariational inequality (3.1.9) where V and $a(\cdot, \cdot)$ are as before, $l \in V^*$ are the volume forces in Ω, and $\int_\Omega j^0(\hat{u}, \hat{v} - \hat{u})d\Omega$ is replaced by

$$\int_{\Gamma_S} j_N^0(u_N, v_N - u_N)d\Gamma + \int_{\Gamma_S} j_T^0(u_T, v_T - u_T)d\Gamma,$$

where the j_N and j_T are the nonconvex superpotentials corresponding to the laws of the aforementioned figures. Then Theorems 3.4 and 3.7 may be proved for the present problem as well, with the difference that (3.2.12) takes the form

$$\int_{\Gamma_S} [(\beta_N(-\infty)q_{N+} + \beta_T(-\infty)q_{T+}) - (\beta_N(\infty)q_{N-} + \beta_T(\infty)q_{T-})]d\Gamma \quad (3.5.2)$$

$$< (l, q) < \int_{\Gamma_S} [(\beta_N(\infty)q_{N+} + \beta_T(\infty)q_{T+}) - (\beta_N(-\infty)q_{N-} + \beta_T(-\infty)q_{T-})]$$

$$\forall q \in \ker a, q \neq 0,$$

where q_{N+} and q_{N-} (respectively q_{T+} and q_{T-}) are the positive and negative parts of q_N (respectively of q_T), i.e. of the normal (respectively tangential) rigid displacement at the boundary.

3.5.2 Adhesively Connected Sandwich Plates

i) Hemivariational Inequality Formulations
Let us consider a layered plate consisting of m-layers. Each layer is an elastic plate and is referred to a right-handed orthogonal Cartesian coordinate system $Ox_1x_2x_3$ (Fig. 3.5.1). The plates have constant thicknesses $h_1, h_2, \ldots, h_m$, and the middle surface of each plate coincides with the respective Ox_1x_2-plane. Let $\Omega_j, j = 1, 2, \ldots, m$ be open, bounded and connected subsets of R^2 and suppose that their boundaries Γ_j are Lipschitzian ($C^{0,1}$-boundary). The domains Ω_j are occupied by the plates in their undeformed state. On $\Omega_j' \subset \Omega_j \cap \Omega_{j+1}$ (Ω_j is such that $\bar{\Omega}_j' \cap \Gamma_j = \emptyset$ and $\bar{\Omega}_j' \cap \Gamma_{j+1} = \emptyset$) the plates j and $j + 1$ are bonded together through an adhesive material. We denote by $\zeta_j(x)$ the deflection of the point $x = (x_1, x_2, x_3)$ and by $f_j = (0, 0, f_{3j})$, $f_{3j} = f_{3j}(x)$ (hereafter called f_j for simplicity) the distributed load of the considered plate per unit area of the middle surface. In order to describe the bonding action in the Ox_3-direction we split f_j into $\bar{\bar{f}}_j \in L^2(\Omega_j)$, which is the given external loading acting on

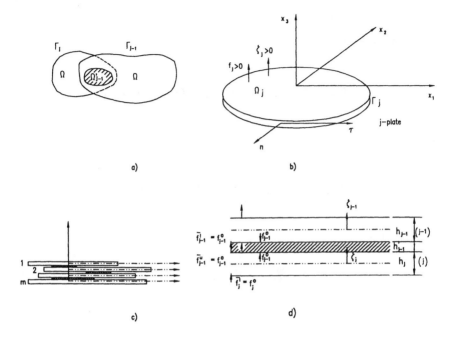

Fig. 3.5.1. Notations in the theory of layered plates

the j-th plate, and $\bar{f}_j$ which denotes the interaction between the plate under consideration (plate j) and the plates $j - 1$ and $j + 1$, caused by the bonding material, i.e.

$$f_j = \bar{f}_j + \bar{\bar{f}}_j \quad \text{on } \Omega_j. \tag{3.5.3}$$

$\bar{f}_j$ consists of two parts: the part $\bar{f}_j^u$ describing the influence due to the bonding with the plate $j - 1$ (upper plate) and the part $\bar{f}_j^l$ describing the influence of the bonding with the plate $j + 1$ (lower plate). Obviously $\bar{f}_1^u = 0$ and $\bar{f}_m^l = 0$, i.e. the upper (resp. the lower) surface of the first (resp. the last) lamina are not subjected to bonding forces, because they are free.

Then we make the general assumption that the force f_{j-1}^0 on the adhesive material between the $(j - 1)$- and the j-plate is generally a multivalued nonmonotone function $\tilde{\beta}_{j-1}$ of the relative displacement

$$[\zeta]_{j-1} = \zeta_{j-1} - \zeta_j \tag{3.5.4}$$

of the plates $j - 1, j$.

We write that

$$-f_{j-1}^0 \in \bar{\beta}_{j-1}(\zeta_{j-1} - \zeta_j) = \bar{\beta}_{j-1}([\zeta]_{j-1}) \quad \text{on } \Omega_{j-1}', \tag{3.5.5}$$

where

$$\Omega_{j-1}' \subset \Omega_{j-1} \cap \Omega_j, \quad \bar{\Omega}_{j-1}' \cap \Gamma_{j-1} = \emptyset, \quad \bar{\Omega}_{j-1}' \cap \Gamma_j = \emptyset. \tag{3.5.6}$$

We note that

$$\bar{f}_j^u = -f_{j-1}^0 \quad \text{and} \quad \bar{f}_j^l = f_j^0 \quad \text{on } \Omega_{j-1}' \tag{3.5.7}$$

and

$$\bar{f}_j^u = 0 \quad \text{on} \quad \Omega_j - \Omega_{j-1}'$$
$$\bar{f}_j^l = 0 \quad \text{on} \quad \Omega_j - \Omega_j'. \tag{3.5.8}$$

The simplest law describing the interlaminar forces and the impenetrability of the laminae is depicted in Fig. 3.5.2a. The binding material may sustain a small positive traction; then rupture (called also delamination) occurs, which is ideally brittle (AB) or semibrittle (AC). More realistic is the diagram in Fig. 3.5.2b which describes the behaviour of an interlaminar bonding sheet with initial thickness h_j', which can be compressed up to h_j^0. The condition of impenetrability holding for every two successive laminae is described by vertical branches OD in Fig. 3.5.2a. Here we can relax the plate theory assumption of the incompressibility of plate in the Ox_3-direction, by incorporating such a deformation into the $\bar{\beta}$-diagrams. Thus we allow the line OD of the interlaminar law to have a small slope (OD'). It is worth noting that the interlaminar laws can be more complicated (Fig. 3.5.2c) and may include local cracking and crushing effects of ideally brittle or semibrittle behaviour. Note the similarity of the sawtooth diagrams of Figs. 3.5.2c with Scanlon's diagram of reinforced concrete in tension [Flo]. Here we make the general assumption that $\bar{\beta}j$ is a nonmonotone multivalued function which may include "filled in" gaps of finite length, i.e. it results from a locally bounded measurable function $\beta_j : R \to R$, as indicated by means of the relations (1.2.45), (1.2.46), (1.2.47).

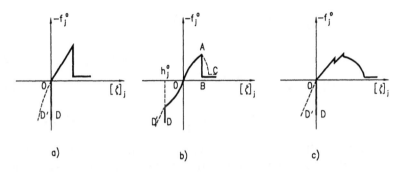

Fig. 3.5.2. Interlayer reaction-displacement diagrams

Then a locally Lipschitz function $\varphi_j : R \to R$ can be determined up to an additive constant such that

$$\tilde{\beta}_j(\xi) = \partial \varphi_j(\xi). \tag{3.5.9}$$

where ∂ is the generalized gradient and (1.2.48) holds. Thus (3.5.5) may be written as

$$-f_{j-1}^0 \in \tilde{\beta}_{j-1}(\xi) = \partial \varphi_{j-1}(\xi), \tag{3.5.10}$$

which leads by definition to a hemivariational inequality. We write for the j-plate, considered as completely free, the analogous relation to (1.4.38). It reads (for appropriately regular functions $\bar{Q}_j$, M_j and f_j)

$$\alpha_j(\zeta_j, z_j) = \int_{\Omega_j} f_j z_j d\Omega_j + \int_{\Gamma_j} \bar{Q}_j(\zeta_j) z_j d\Gamma_j \tag{3.5.11}$$

$$- \int_{\Gamma_j} M_j(\zeta_j) \frac{\partial z_j}{\partial n_j} d\Gamma_j, \quad \forall z_j \in Z_j, \quad j = 1, 2, \ldots, m,$$

assuming that $\alpha_j(\zeta_j, z_j)$ is the bilinear form of the elastic energy of the plate, $\bar{Q}_j$ and M_j are, respectively, the total shearing force [Gir] and the bending moment at the boundary Γ_j, $n = \{n_j\}$ is the outword unit normal vector to Γ_j and Z_j is the set of the kinematically admissible deflections ζ_j, which, since the plate is free, coincides with the classical Sobolev space $H^2(\Omega_j)$. Equation (3.5.11) holds in the framework of the Kirchhoff plate theory, but also for orthotropic or anisotropic, homogeneous or inhomogeneous plates; i.e. no plate stretching is considered here. In the case of an isotropic, homogeneous plate we have that

$$\alpha(\zeta, z) = K \int_\Omega [(1 - \nu)\zeta_{,\alpha\beta} z_{,\alpha\beta} + \nu \Delta\zeta \Delta z] d\Omega, \quad \alpha, \beta = 1, 2, \qquad (3.5.12a)$$

$$M(\zeta) = -K[\nu\Delta\zeta + (1 - \nu)(2n_1 n_2 \zeta_{,12} + n_1^2 \zeta_{,11} + n_2^2 \zeta_{,22})], \qquad (3.5.12b)$$

$$\bar{Q}(\zeta) = Q(\zeta) - \frac{\partial M(\zeta)}{\partial \tau} \qquad (3.5.12c)$$

$$= K \left[\frac{\partial \Delta\zeta}{\partial n} + (1 - \nu)\frac{\partial}{\partial \tau}[n_1 n_2 (\zeta_{,22} - \zeta_{,11}) + (n_1^2 - n_2^2)\zeta_{,12}] \right].$$

Here $K = Eh^3/12(1 - \nu^2)$ is the bending rigidity of the plate with E and ν the modulus of elasticity and the Poisson ratio respectively, h its thickness, and τ the tangential unit vector to the boundary.

From eqs. (3.5.11), we obtain through addition the expression

$$\sum_{j=1}^{m} \alpha_j(\zeta_j, z_j - \zeta_j) = \sum_{j=1}^{m} \int_{\Gamma_j} \bar{Q}_j(z_j - \zeta_j)d\Gamma_j \qquad (3.5.13)$$

$$- \sum_{j=1}^{m} \int_{\Gamma_j} M_j \frac{\partial z_j - \partial \zeta_j}{\partial n_j} d\Gamma_j + \sum_{j=1}^{m} \int_{\Omega_j} \bar{\bar{f}}_j (z_j - \zeta_j)d\Omega_j$$

$$+ \sum_{j=1}^{m-1} \int_{\Omega_j'} f_j^0([z]_j - [\zeta]_j)d\Omega_j', \quad \forall z_j \in Z_j.$$

Then (3.5.13) together with (3.5.10) implies the variational expression

$$\sum_{j=1}^{m} \alpha_j(\zeta_j, z_j - \zeta_j) + \sum_{j=1}^{m-1} \int_{\Omega_j'} \varphi_j^0([\zeta]_j, [z]_j - [\zeta]_j)d\Omega_j' \qquad (3.5.14)$$

$$\geq \sum_{j=1}^{m} \int_{\Gamma_j} \bar{Q}_j(z_j - \zeta_j)d\Gamma_j - \sum_{j=1}^{m} \int_{\Gamma_j} M_j \frac{\partial z_j - \partial \zeta_j}{\partial n_j} d\Gamma_j$$

$$+ \sum_{j=1}^{m} \int_{\Omega_j} \bar{\bar{f}}_j (z_j - \zeta_j)d\Omega_j, \quad \forall z_j \in Z_j,$$

which is a hemivariational inequality.

Until now we have not yet specified the boundary conditions of the problem. We shall assume that the boundary conditions are the classical ones of plate theory. Note that different boundary conditions can be assigned to each plate. Thus if the j-plate is clamped on Γ_j, then we assume that

$$Z_j = \left\{ z_j : z_j \in H^2(\Omega_j), z_j = 0, \frac{\partial z_j}{\partial n} = 0 \text{ on } \Gamma_j \right\} = \overset{0}{H}{}^2(\Omega_j). \qquad (3.5.15)$$

If the j-plate is simply supported on $\Gamma_j' \subset \Gamma_j$ ($z_j = 0$ and $M_j = 0$ on Γ_j) and free on $\Gamma_j - \Gamma_j'$ ($M_j = 0, \bar{Q}_j = 0$ - or more generally $M_j = M_0$ and $Q_j = Q_0$, M_0 and Q_0 given - on $\Gamma_j - \Gamma_j'$), then

$$Z_j = \{z_j : z_j \in H^2(\Omega_j), z_j = 0, \text{ on } \Gamma_j'\}, \tag{3.5.16}$$

and if $\Gamma_j \equiv \Gamma_j'$ then

$$Z_j = H^2(\Omega_j) \cap \overset{0}{H}{}^1(\Omega_j). \tag{3.5.17}$$

We will study the hemivariational inequality (3.5.14) on the assumption that for each plate the boundary conditions guarantee the coerciveness of the corresponding bilinear form. We make generally the following assumption for the elastic energy function $\{\zeta_j, z_j\} \to \alpha_j(\zeta_j, z_j)$ which is a continuous bilinear form on $H^2(\Omega_j) \times H^2(\Omega_j)$: the boundary conditions of each lamina guarantee that $\alpha_j(\zeta_j, z_j)$ is coercive, i.e. there is a constant $c > 0$ such that

$$\alpha_j(z_j, z_j) \geq c||z_j||^2, \quad \forall z_j \in H^2(\Omega_j). \tag{3.5.18}$$

Here $|| \cdot ||$ denotes the classical H^2-norm.

This assumption is satisfied for isotropic or orthotropic homogeneous plates, if the boundary conditions do not permit a "rigid-plate" deflection, which is a polynomial of degree one in x_1 and x_2 ($q = q_0 + q_1 x_1 + q_2 x_2$). This is guaranteed, for instance, in the case of a partially clamped plate, or in the case of a simply supported plate on Γ_j' on the assumption that Γ_j' is nonrectilinear. In this context we refer the reader to [Duv72]. Moreover in the case of nonhomogeneous plates we assume that the elasticity coefficients are functions from $L^\infty(\Omega_j)$.

The nonhomogeneous boundary conditions are incorporateed into the kinematically admissible sets Z_j. Until now we have assumed that $f_j, \bar{Q}_j, M_j$ in (3.5.11) are "appropriately regular". More precisely we assume that $\bar{\bar{f}}_j \in L^2(\Omega_j)$ and in the considered functional framework the foregoing integrals in (3.5.11) must be written as $\langle M_j(\zeta_j), \frac{\partial \zeta_j}{\partial k_j} \rangle_{1/2}$, and $\langle \bar{Q}_j(\zeta_j), \zeta_j \rangle_{3/2}$, where $\langle \cdot, \cdot \rangle_{1/2}$ denotes the duality pairing on $H^{1/2}(\Gamma_j) \times H^{-1/2}(\Gamma_j)$, and $\langle \cdot, \cdot \rangle_{3/2}$ the duality pairing on $H^{3/2}(\Gamma_j) \times H^{-3/2}(\Gamma_j)$. Note that $M_j(\zeta) \in H^{-1/2}(\Gamma_j)$, $\frac{\partial \zeta}{\partial n} \in H^{1/2}(\Gamma_j)$, $Q_j(\zeta) \in H^{-3/2}(\Gamma_j)$ and $\zeta_j \in H^{3/2}(\Gamma_j)$ by the trace theorem. For example, if plate Ω_j is subjected to the boundary conditions

$$z_j = 0 \quad \text{on} \quad \Gamma_j' \subset \Gamma_j, \quad \bar{Q} = Q_0 \quad \text{on} \quad \Gamma_j - \Gamma_j', \tag{3.5.19a}$$

$$\frac{\partial z_j}{\partial n} = g_j \quad \text{on} \quad \Gamma_j'' \subset \Gamma_j, \quad M = M_0 \quad \text{on} \quad \Gamma_j - \Gamma_j'', \tag{3.5.19b}$$

where Γ_j' is nonrectilinear, then $\alpha_j(\zeta_j, \zeta_j)$ is coercive. We have then that

$$Z_j = \left\{ z_j \in H^2(\Omega_j), \quad z_j = 0 \text{ on } \Gamma_j', \frac{\partial z_j}{\partial n} = g_j \text{ on } \Gamma_j'' \subset \Gamma_j \right\} \tag{3.5.20}$$

and let l_j be a linear continuous functional on $H^2(\Omega_j)$, i.e. $l_j \in [H^2(\Omega_j)]^*$, defined by

$$\langle l_j, z_j \rangle = \int_{\Omega_j} \bar{\bar{f}}_j \, z_j d\Omega + \langle Q_0, z \rangle_{3/2, \Gamma_j - \Gamma_j'} - \langle M_0, \frac{\partial z_j}{\partial n} \rangle_{1/2, \Gamma_j - \Gamma_j''}. \tag{3.5.21}$$

Here $\langle \cdot, \cdot \rangle$ denotes the duality pairing on $H^2(\Omega_j) \times [H^2(\Omega_j)]^*$ and $\langle \cdot, \cdot \rangle_{1/2, \Gamma_j - \Gamma_j'}$ denotes the restriction of the corresponding functional to $\Gamma_j - \Gamma_j''$ (i.e. if $M_0 \in L^2(\Gamma)$ then the last term in (3.5.21) becomes $\int\limits_{\Gamma_j - \Gamma_j''} M_0 \frac{\partial z_j}{\partial n} d\Gamma$).

We note finally that in the case of nonhomogeneous boundary conditions an appropriate translation is performed transforming the problem into a homogeneous one; thus we shall assume that Z_j is always a closed linear subspace of $H^2(\Omega_j)$. Now we can pose the general problem:

Find $\zeta_j \in Z_j, j = 1, 2, \ldots, m$ such as to satisfy the hemivariational inequality

$$\sum_{j=1}^{m} \alpha_j(\zeta_j, z_j - \zeta_j) + \sum_{j=1}^{m-1} \int_{\Omega_j'} \varphi_j^0([\zeta]_j, [z]_j - [\zeta]_j) d\Omega_j' \geq \sum_{j=1}^{m} \langle l_j, z_j - \zeta_j \rangle, \quad \forall z_j \in Z_j.$$
(3.5.22)

Due to the assumption (3.5.18) the Theorem 3.4 holds for the hemivariational inequality (3.5.22) as well as all the other results of Sect. 3.1. If the boundary conditions of the plates do not guarantee the coerciveness of the bilinear forms $\alpha_j(\cdot, \cdot)$, then the sufficient condition of Theorem 3.7 becomes

$$\sum_{j=1}^{m-1} \int_{\Omega_j'} \{\beta_j(-\infty)[q]_{j+} - \beta_j(\infty)[q]_{j-}\} d\Omega < \sum_{j=1}^{m} \langle l_j, q_j \rangle \qquad (3.5.23)$$

$$< \sum_{j=1}^{m-1} \int_{\Omega_j'} \{\beta_j(\infty)[q]_{j+} - \beta_j(-\infty)[q]_{j-}\} d\Omega \quad \forall q_j \in \ker a_j, \ q_j \neq 0.$$

Here $[q]_{j+}$ (resp. $[q]_{j-}$) denotes the positive part (resp. the negative part) of the relative rigid plate deflection $[q]_j = q_j - q_{j+1}$.

ii) Variational-Hemivariational Inequalities

Let us consider now the case in which the branch OD in Figs. 3.5.1a,b,c is vertical. Then analogously to the case described through the relation (1.4.27)–(1.4.30) the solution of the hemivariational inequality (3.5.22) will be sought in the convex closed set

$$K = \{z_j : z_j \in Z_j, \ [z]_j \geq 0 \quad j = 1, 2, \ldots, m - 1\}. \qquad (3.5.24)$$

Accordingly, by introducing the indicator I_K of K we get a variational-hemivariational inequality. Due to the compact imbeddings $H^2(\Omega_j) \subset C^0(\bar{\Omega}_j)$ and the imbeddings $C^0(\bar{\Omega}_j) \subset C^0(\Omega_j') \subset L^\infty(\Omega_j')$ we have the compact imbedding

$$H^2(\Omega_j) \subset L^\infty(\Omega_j'). \qquad (3.5.25)$$

Further we can verify easily all the other assumptions of Theorem 3.1 (Φ_ρ in (3.4.25)–(3.4.27) corresponds to the inclined line OD') and thus the above variational-hemivariational inequality has a solution. Analogous variational-hemivariational inequalities result if we consider monotone multivalued boundary conditions, at a subset of the boundary Γ (cf. e.g. [Pan88a]).

Note that all the problems expressed via hemivariational or variational-hemivariational inequalities are free boundary problems; for instance, in the case of the plate problem considered, the free boundary between the interface points which remain adhesively bonded and which not, is not *a priori* known.

4. Hemivariational Inequalities for Locally Lipschitz Functionals

In this Chapter we introduce two classes of locally Lipschitz functions on a reflexive Banach space V. The first class, denoted by $QPM(V)$, includes functions whose generalized gradient is quasi-pseudo-monotone, while the second one, $PM(V)$, contains functions with pseudo-monotone generalized gradients. It is shown that under certain conditions the classes $QPM(V)$ and $PM(V)$ are invariant under addition, and min or max operations. It is also proved that they include compositions of locally Lipschitz functions with linear compact operators.

The theory of pseudo-monotone operators permits the formulation of existence results for hemivariational inequalities involving functions from the classes $QPM(V)$ and $PM(V)$. The case in which both convex, proper, lower, semicontinuous and QPM or PM functions are involved will be also considered. The corresponding variational forms are then called variational-hemivariational inequalities, according to [Pan91]. Finally we examine the class of variational problems with nonlinearities which cannot be derived from functions from the classes $QPM(V)$ and $PM(V)$. In this context we deal with a new type of hemivariational inequalities which we call quasi-hemivariational inequalities. Applying the pseudo-monotonicity methods leads to the solution of such problems. The Chapter closes with applications of the theory to Mechanics. We give certain nonconvex superpotentials which belong to the classes $QPM(V)$ and $PM(V)$ and we discuss the corresponding hemivariational inequalities.

4.1 The Class of Locally Lipschitz Functions with Pseudo-Monotone and Quasi-Pseudo-Monotone Generalized Gradients

The aim of this Section is to study the main properties of locally Lipschitz functions whose generalized gradients are pseudo- or quasi-pseudo-monotone. Thus we will determine those nonconvex problems which can be studied by means of the theory of pseudo-monotone mappings. Consequently, we derive some existence results for hemivariational inequalities involving such functions.

Let V be a reflexive Banach space. By $PM(V)$ will be denoted the class of locally Lipschitz functions $f : V \to R$ with the property that $\partial f : V \to 2^{V^*}$ is pseudo-monotone. For the class of locally Lipschitz functions $f : V \to R$ whose generalized gradient $\partial f : V \to 2^{V^*}$ is quasi-pseudo-monotone we use the symbol $QPM(V)$. Throughout this Chapter the expression "f is locally Lipschitz on V" means that f is Lipschitz continuous on bounded subsets of V. It follows immediately by the definition that $QPM(V) \supset PM(V)$.

We start with the observation that a positive linear combination of functions from $QPM(V)$ belongs to $QPM(V)$ again.

Proposition 4.1 Let $f_1, f_2 \in QPM(V)$. Then for any $\alpha_1, \alpha_2 \in R^+$,

$$\alpha_1 f_1 + \alpha_2 f_2 \in QPM(V). \tag{4.1.1}$$

Proof. It is clear that for any $\alpha \in R^+$ and $f \in QPM(V)$ we have that $\alpha f \in QPM(V)$. Thus in order to prove the assertion it suffices to show that $\partial(f_1 + f_2)$ is quasi-pseudo-monotone. Let $\{u_i\}$ be a sequence in V converging weakly to u and let $u_i^* \in \partial(f_1 + f_2)(u_i)$ be such that $\limsup\langle u_i^*, u_i - u \rangle_V \leq 0$. According to the definition of $QPM(V)$ we have to deduce that $\lim\langle u_i^*, u_i - u \rangle_V = 0$ holds. Since for any locally Lipschitz functions f_1 and f_2, $\partial(f_1 + f_2) \subset \partial f_1 + \partial f_2$, we obtain that for each i the representation $u_i^* = \bar{u}_i^* + \bar{\bar{u}}_i^*$ holds, where $\bar{u}_i^* \in \partial f_1(u_i)$ and $\bar{\bar{u}}_i^* \in \partial f(u_i)$. Now, from

$$\limsup\langle u_i^*, u_i - u \rangle_V = \limsup\langle \bar{u}_i^* + \bar{\bar{u}}_i^*, u_i - u \rangle_V \leq 0 \tag{4.1.2}$$

we get

$$\limsup\langle \bar{u}_i^*, u_i - u \rangle_V \leq 0 \quad \text{and} \quad \limsup\langle \bar{\bar{u}}_i^*, u_i - u \rangle_V \leq 0. \tag{4.1.3}$$

Indeed, suppose that this is not true. Then we may assume that $\limsup\langle \bar{u}_i^*, u_i - u \rangle_V > 0$. Accordingly we can find $d > 0$ such that (by passing to a subsequence if necessary) $\lim\langle \bar{u}_i^*, u_i - u \rangle_V = d > 0$. Thus

$$\limsup\langle \bar{\bar{u}}_i^*, u_i - u \rangle_V \leq -d < 0. \tag{4.1.4}$$

On the other hand, the quasi-pseudo-monotonicity of ∂f_2 yields that

$$\lim\langle \bar{\bar{u}}_i^*, u_i - u \rangle_V = 0 \tag{4.1.5}$$

which is a contradiction to (4.1.4). Accordingly, (4.1.3) results. By virtue of the quasi-pseudo-monotonicity of ∂f_1 and ∂f_2, the following relations hold

$$\lim\langle \bar{u}_i^*, u_i - u \rangle_V = 0 \quad \text{and} \quad \lim\langle \bar{\bar{u}}_i^*, u_i - u \rangle_V = 0. \tag{4.1.6}$$

Consequently, $\lim\langle \bar{u}_i^* + \bar{\bar{u}}_i^*, u_i - u \rangle_V = \lim\langle u_i^*, u_i - u \rangle_V = 0$, as desired.   q.e.d.

For functions from $PM(V)$ we need an additional requirement to ensure that their positive linear combinations belong to $PM(V)$.

Proposition 4.2 Let $f_1, f_2 \in PM(V)$. Suppose that $\partial(\alpha_1 f_1 + \alpha_2 f_2) = \alpha_1 \partial f_1 + \alpha_2 \partial f_2$, for $\alpha_1, \alpha_2 \in R^+$. Then

$$\alpha_1 f_1 + \alpha_2 f_2 \in PM(V). \qquad (4.1.7)$$

Proof. It is obvious that in order to show the assertion it is enough to establish the pseudo-monotonicity of $\partial(f_1 + f_2)$. By Proposition 4.1 the generalized gradient of the sum $f_1 + f_2$ is quasi-pseudo-monotone. Thus for any sequence $\{u_i\}$ as in the proof of Proposition 4.1, we have to show that each weak limit of the corresponding sequence $\{u_i^\star\}$ belongs to $\partial(f_1 + f_2)(u)$. Since $u_i^\star = \bar{u}_i^\star + \bar{\bar{u}}_i^\star$ for some $\bar{u}_i^\star \in \partial f_1(u_i)$ and $\bar{\bar{u}}_i^\star \in \partial f_2(u_i)$ one may assume without loss of generality that there exist $\bar{u}^\star$ and $\bar{\bar{u}}^\star$ in $V^\star$ such that the sequences $\{\bar{u}_i^\star\}$ and $\{\bar{\bar{u}}_i^\star\}$ converge weakly to $\bar{u}^\star$ and $\bar{\bar{u}}^\star$ respectively (passing to a subsequence if necessary). From the pseudo-monotonicity of ∂f_1 and ∂f_2, and from (4.1.3) it results that $\bar{u}^\star$ and $\bar{\bar{u}}^\star$ belong to $\partial f_1(u)$ and $\partial f_2(u)$, respectively. Due to the hypothesis, $\partial f_1(u) + \partial f_2(u) = \partial(f_1 + f_2)(u)$. Thus $u^\star = \bar{u}^\star + \bar{\bar{u}}^\star \in \partial f_1(u) + \partial f_2(u) = \partial(f_1 + f_2)(u)$, and the proof is complete. q.e.d.

Arguing by induction we obtain the following generalizations of Propositions 4.1 and 4.2.

Corollary 4.3 Let $f_n : V \to R$, $n = 1, \ldots, N$, be locally Lipschitz functions from a reflexive Banach space V into R. Then the following conditions hold:

(i) If $f_1, \ldots, f_N \in QPM(V)$ then for any $\alpha_1, \ldots, \alpha_N \in R^+$,

$$\alpha_1 f_1 + \ldots + \alpha_N f_N \in QPM(V). \qquad (4.1.8)$$

(ii) If $f_1, \ldots, f_N \in PM(V)$ and $\partial(\alpha_1 f_1 + \ldots + \alpha_N f_N) = \alpha_1 \partial f_1 + \ldots + \alpha_N \partial f_N$ for $\alpha_1, \ldots, \alpha_N \in R^+$, then

$$\alpha_1 f_1 + \ldots + \alpha_N f_N \in PM(V). \qquad (4.1.9)$$

In particular, the equality in the assumption (ii) of Corollary 4.3 holds true if all the functions are ∂-regular, i.e., $f_n^0(\cdot, \cdot) = f_n'(\cdot, \cdot), n = 1, \ldots, N$.

Further we deal with the products of functions belonging to the classes $QPM(V)$ and $PM(V)$. The following proposition holds.

Proposition 4.4 Let $f_1, f_2 \in QPM(V)$. Suppose that $f_1 \geq 0$ and $f_2 \geq 0$. Then $f_1 f_2 \in QPM(V)$.

Proof. Suppose that $\{u_i\}$ is a sequence in V converging weakly to u, for which there exists a sequence $u_i^\star \in \partial(f_1 f_2)(u_i)$ such that $\limsup \langle u_i^\star, u_i - u \rangle_V \leq 0$. We have to show that the foregoing conditions imply $\lim \langle u_i^\star, u_i - u \rangle_V = 0$. As it is well known for any $v \in V$ we have that $\partial(f_1 f_2)(v) \subset f_1(v) \partial f_2(v) + f_2(v) \partial f_1(v)$.

This implies the representation $u_i^* = f_1(u_i)\,\bar{\bar{u}}_i^* + f_2(u_i)\bar{u}_i^*$ for each i, where $\bar{u}_i^* \in \partial f_1(u_i)$ and $\bar{\bar{u}}_i^* \in \partial f_2(u_i)$. We claim that the condition

$$\limsup\langle f_1(u_i)\,\bar{\bar{u}}_i^* + f_2(u_i)\bar{u}_i^*, u_i - u\rangle_V \le 0 \qquad (4.1.10)$$

implies

$$\limsup\langle f_1(u_i)\,\bar{\bar{u}}_i^*, u_i - u\rangle_V \le 0 \qquad (4.1.11)$$

and

$$\limsup\langle f_2(u_i)\bar{u}_i^*, u_i - u\rangle_V \le 0. \qquad (4.1.12)$$

Indeed, suppose that this is not true. Then we may assume that

$$\limsup\langle f_1(u_i)\,\bar{\bar{u}}_i^*, u_i - u\rangle_V > 0. \qquad (4.1.13)$$

Thus we can determine a positive number $d > 0$ and subsequence again denoted by $\bar{\bar{u}}_i^*$, such that

$$\lim\langle f_1(u_i)\,\bar{\bar{u}}_i^*, u_i - u\rangle_V = d > 0. \qquad (4.1.14)$$

But then, for the corresponding subsequence of the other summand, we will have

$$\limsup\langle f_2(u_i)\bar{u}_i^*, u_i - u\rangle_V \le -d < 0. \qquad (4.1.15)$$

Hence, taking into account the boundedness of the sequence $\{f_2(u_i)\}$ and $f_2(u_i) \ge 0$, we easily deduce the existence of $\bar{d} > 0$ such that

$$\limsup\langle \bar{u}_i^*, u_i - u\rangle_V \le -\bar{d} < 0. \qquad (4.1.16)$$

By virtue of the quasi-pseudo-monotonicity of ∂f_1 we arrive at the relation $\lim\langle \bar{u}_i^*, u_i - u\rangle_V = 0$ which is the contradiction. Thus (4.1.11), (4.1.12) have been established. Now we show that (4.1.11), (4.1.12) imply that

$$\lim\langle f_1(u_i)\,\bar{\bar{u}}_i^*, u_i - u\rangle_V = 0 \quad \text{and} \quad \lim\langle f_2(u_i)\bar{u}_i^*, u_i - u\rangle_V = 0. \qquad (4.1.17)$$

It suffices to show that (4.1.11) implies the equality $\lim\langle f_1(u_i)\,\bar{\bar{u}}_i^*, u_i - u\rangle_V = 0$. Analogously we proceed with (4.1.12). Let us assume that

$$\liminf\langle f_1(u_i)\,\bar{\bar{u}}_i^*, u_i - u\rangle_V = -d < 0 \qquad (4.1.18)$$

for some positive $d > 0$. Then by taking into account the boundedness and the nonnegativity of $\{f_1(u_i)\}$ and by choosing an appropriate subsequence (denoted again by $\bar{\bar{u}}_i^*$) we are led to the conclusion that $\lim\langle \bar{\bar{u}}_i^*, u_i - u\rangle_V = -\bar{d} < 0$ for some $\bar{d} > 0$, which contradicts the quasi-pseudo-monotonicity of ∂f_2. Accordingly, we obtain $\liminf\langle f_1(u_i)\,\bar{\bar{u}}_i^*, u_i - u\rangle_V \ge 0$, which together with (4.1.11) is equivalent to (4.1.17). Hence we deduce immediately that

$$\lim\langle u_i^*, u_i - u\rangle_V = \lim\langle f_1(u_i)\,\bar{\bar{u}}_i^* + f_2(u_i)\bar{u}_i^* - u\rangle_V = 0. \qquad (4.1.19)$$

Thus we have proved the quasi-pseudo-monotonicity of $\partial(f_1 f_2)$ and the proof is complete. $\qquad$ q.e.d.

For functions from the PM-class some additional requirements are needed in order to ensure that their finite product belongs to $PM(V)$.

Proposition 4.5 Let $f_1, f_2 \in PM(V)$ be nonnegative functions satisfying the (C) condition and $\partial(f_1 f_2)(\cdot) = f_1(\cdot)\partial f_2(\cdot) + f_2(\cdot)\partial f_1(\cdot)$. Moreover, suppose that one of the following two conditions holds

(i) $f_N(v) > 0$ for all $v \in V, N = 1, 2$;
or
(ii) For any $v \in V$, if $f_N(v) = 0$, then $\partial f_N(v) = \{0\}, N = 1, 2$.

Then the product $f_1 f_2$ satisfies the (C) condition and belongs to $PM(V)$.

Proof. The first part of the proof coincides with that of Proposition 4.4. Namely, with respect to a sequence $\{u_i\}$ in V converging weakly to u, for which there exists a sequence $u_i^* \in \partial(f_1 f_2)(u_i)$ such that $\limsup\langle u_i^*, u_i - u\rangle_V \leq 0$, we prove that $\lim\langle u_i^*, u_i - u\rangle_V = 0$. It remains to show that each weak limit of $\{u_i^*\}$ in V^* lies in $\partial(f_1 f_2)(u)$. By passing to a subsequence (again denoted by the same symbol) we may assume that $\{u_i^*\}$ converges weakly to some u^*. Recalling that $u_i^* = f_1(u_i)\,\bar{\bar{u}}_i^* + f_2(u_i)\bar{u}_i^*$, where $\bar{u}_i^* \in \partial f_1(u_i)$ and $\bar{\bar{u}}_i^* \in \partial f_2(u_i)$, we may also assume that $\{\bar{u}_i^*\}$ and $\{\bar{\bar{u}}_i^*\}$ converge weakly to some $\bar{u}^*$ and $\bar{\bar{u}}^*$, respectively. Moreover, following the lines of the previous proof we arrive at the equalities

$$\lim\langle f_1(u_1)\,\bar{\bar{u}}_i^*, u_i - u\rangle_V = 0 \quad \text{and} \quad \lim\langle f_2(u_i)\bar{u}_i^*, u_i - u\rangle_V = 0 \qquad (4.1.20)$$

which constitute a convenient starting point for the further steps of the proof.

Suppose that (i) holds. In view of the weak lower semicontinuity of f_N, and the fact that $f_N(u) > 0$, one can determine a $\delta > 0$ such that $f_N(u_i) \geq \delta > 0$ for sufficiently large i, $N = 1, 2$. Hence, due to (4.1.20) we are led to the equalities $\lim\langle\bar{\bar{u}}_i^*, u_i - u\rangle_V = 0$ and $\lim\langle\bar{u}_i^*, u_i - u\rangle_V = 0$. Now by virtue of the pseudo-monotonicity of ∂f_N and the fact that f_N satisfies the (C) condition we are allowed to conclude that $\bar{u}^* \in \partial f_1(u), \bar{\bar{u}}^* \in \partial f_2(u)$ and $\lim f_1(u_i) = f_1(u)$, $\lim f_2(u_i) = f_2(u)$. Accordingly, the representation $u^* = f_1(u)\,\bar{\bar{u}}^* + f_2(u)\bar{u}^*$ is obtained, from which, owing to the hypothesis

$$\partial(f_1 f_2)(\cdot) = f_1(\cdot)\partial f_2(\cdot) + f_2(\cdot)\partial f_1(\cdot) \qquad (4.1.21)$$

we arrive at $u^* \in \partial(f_1 f_2)(u)$. Note that from our considerations the equality $\lim f_1(u_i)f_2(u) = f_1(u)f_2(u)$ has been established. Thus $f_1 f_2$ satisfies the (C) condition and the assertion results.

Now, let (ii) holds. In view of the foregoing investigations to complete the proof it remains to consider the case when one of the limits $\lim f_1(u_i)$ or $\lim f_2(u_i)$ is equal to 0 (the existence of these limits is obvious by passing to a suitable subsequence). By symmetry we may assume first that $\lim f_1(u_i) = 0$. Denote $\lim f_2(u_i) = \beta, \beta \geq 0$. From the weak lower semicontinuity of $f_1, f_1(u) = 0$. If $\beta = 0$ then we obtain again $f_2(u) = 0$ and consequently $u^* = 0$, which is of course an element of

$$\partial(f_1 f_2)(u) = f_1(u)\partial f_2(u) + f_2(u)\partial f_1(u) = \{0\}. \qquad (4.1.22)$$

If $\beta > 0$, then from the weak lower semicontinuity of f_2 it follows that $f_2(u_i) \geq \frac{1}{2}\beta > 0$ for sufficiently large i. Hence, due to (4.1.20), one gets the relation $\lim\langle \bar{u}_i^*, u_i - u \rangle_V = 0$. Further, due to the pseudo-monotonicity of $\partial f_1, \bar{u}^* \in \partial f_1(u)$. Finally, we are led to the representation $u^* = \beta \bar{u}^*$. Since, $f_1(u) = 0$, the hypothesis (ii) implies that $\bar{u}^* = 0$, and consequently $u^* = 0$. But then $\partial(f_1 f_2)(u) = f_1(u)\partial f_2(u) + f_2(u)\partial f_1(u) = \{0\}$ which implies the desired inclusion $u^* \in \partial(f_1 f_2)(u)$. Thus $f_1 f_2 \in PM(V)$. The fact that $f_1 f_2$ satisfies the (C) condition follows immediately from the equality $\lim f_1(u_i) = 0$. The proof of Proposition 4.5 is complete.   q.e.d.

Corollary 4.6 Let $f \in PM(V)$ be a nonnegative function satisfying the (C) condition. Then $f^2 \in PM(V)$.

Proof. It suffices to check that for functions $f_1 = f_2 = f$ the hypotheses of Proposition 4.5 hold. Note that $(f^2)^0(u,v) = 2f(u)f^0(u,v)$ for any $u, v \in V$. Hence $\partial(f^2)(\cdot) = 2f(\cdot)\partial f(\cdot)$ and the condition (ii) of Proposition 4.5 is satisfied. q.e.d.

It is well known that if $f_1, \ldots, f_N$ are ∂-regular then

$$\partial f(\cdot) = \sum_j [f_1(\cdot) \cdot \ldots \cdot f_{j-1}(\cdot) \cdot \partial f_j(\cdot) f_{j+1}(\cdot) \cdot \ldots \cdot f_N(\cdot)] \qquad (4.1.23)$$

for arbitrary integer N. Arguing analogously we arrive at the following generalizations of Propositions 4.4, 4.5, and Corollary 4.6.

Corollary 4.7 Let $f_n : V \to R, n = 1, \ldots, N$, be locally Lipschitz, nonnegative functions from a reflexive Banach space V into R. Then the following hold:

(i) If $f_1, \ldots, f_N \in QPM(V)$, then $f_1 \cdot \ldots \cdot f_N \in QPM(V)$.

(ii) If $f_1, \ldots, f_N \in PM(V)$ and satisfy the (C) condition, (4.1.23) holds and if one of the two conditions

   (i′) $f_n(v) > 0$ for all $v \in V, n = 1, \ldots, N$;

or

   (ii′) For any $v \in V$, if $f_n(v) = 0$, then $\partial f_n(v) = \{0\}, n = 1, \ldots, N$,

is satisfied, then the product $f_1 \cdot \ldots \cdot f_N \in PM(V)$ and satisfies the (C) condition. Moreover, if $f \geq 0$, $f \in PM(V)$ and f has the (C) property, then for an arbitrary natural number k, $f^k \in PM(V)$.

Since all convex locally Lipschitz functions belong to $PM(V)$, have the (C) property and are ∂-regular, the following Corollary can be deduced.

Corollary 4.8 Let $f_n : V \to R, n = 1, \ldots, N$, be convex, nonnegative, locally Lipschitz functions from a reflexive Banach space V into R. The following hold:

(i) $f_1 \cdot \ldots \cdot f_N \in QPM(V)$;

(ii) If we suppose in addition that one of the two conditions holds

    (i') $f_n(v) > 0$ for all $v \in V, n = 1, \ldots, N$;

or

    (ii') for each $v \in V$ $f_n(v) = 0$ implies $\partial f_n(v) = \{0\}, n = 1, \ldots, N$;

then the product $f_1 \cdot \ldots \cdot f_n \in PM(V)$ and satisfies the (C) condition.

(iii) If $f \geq 0$ is a convex, locally Lipschitz function on V, then for any natural number k, $f^k \in PM(V)$.

Now we show that the condition (ii') of Corollary 4.8 is essential for the corresponding product to remain within the PM-class.

Example 4.9 Let $V = l^2$, where l^2 is the Hilbert space of all real-valued sequences such that

$$l^2 = \{v = (x_j), \sum_j |x_j|^2 < +\infty\}, \tag{4.1.24}$$

with the norm

$$\|v\| = (\sum_j |x_j|^2)^{1/2}, v = (x_j) \in l^2. \tag{4.1.25}$$

Define

$$f_1(v) = \|v\|^2 + 1, \tag{4.1.26}$$

$$f_2(v) = \begin{cases} 0 & \text{if } x_1 \geq 1 \\ -x_1 + 1 & \text{if } x_1 < 1, \end{cases} \quad v = (x_j) \in l^2 \tag{4.1.27}$$

and

$$f(v) = f_1(v) f_2(v) \quad v \in l^2. \tag{4.1.28}$$

It can be easily verified that for

$$e = (1, 0, \ldots), \tag{4.1.29}$$

$$e_i = (1, 0, \ldots, 0, \underset{i}{1}, 0, \ldots) \tag{4.1.30}$$

we have

$$\left.\begin{array}{c} f_1(e_i) = 3, \ i = 2, 3, \ldots, \\[4pt] f_1(e) = 2, \\[4pt] f_2(e_i) = 0, \ i = 2, 3, \ldots, \\[4pt] f_2(e) = 0. \end{array}\right\} \tag{4.1.31}$$

After simple calculations we also get that

$$\partial f_1(v) = 2v, \tag{4.1.32}$$

$$\partial f_2(e) = \partial f_2(e_i) = \{(\beta, 0, \ldots) : -1 \leq \beta \leq 0\}. \tag{4.1.33}$$

Notice that $f_2(e) = 0$ and $\partial f_2(e) \neq \{0\}$. Thus the condition (ii') of Corollary 4.8 does not hold. Further, we obtain

$$
\begin{aligned}
\partial f(e_i) &= f_1(e_i)\partial f_2(e_i) + f_2(e_i)\partial f_1(e_i) \\
&= \{(3\beta, 0, \ldots) : -1 \leq \beta \leq 0\},
\end{aligned}
\tag{4.1.34}
$$

and

$$
\begin{aligned}
\partial f(e) &= f_1(e)\partial f_2(e) + f_2(e)\partial f_1(e) \\
&= \{(2\beta, 0, \ldots) : -1 \leq \beta \leq 0\},
\end{aligned}
\tag{4.1.35}
$$

Moreover, setting

$$e_i^* = (-3, 0, \ldots) \in \partial f(e_i), i = 2, 3, \ldots, \tag{4.1.36}$$

we may easily check that for the sequences $\{e_i\}$ and $\{e_i^*\}$ all the requirements in the definition of the pseudo-monotonicity are satisfied. Namely,

$$e_i \to e \quad \text{weakly in } l^2,$$

$$e_i^* \to e^* = (-3, 0, \ldots) \quad \text{weakly in } l^2, \tag{4.1.37}$$

$$\limsup\langle e_i^*, e_i - e \rangle = 0 \leq 0,$$

where $\langle \cdot, \cdot \rangle$ is the scalar product in l^2. But the element $e^* = (-3, 0, \ldots)$ does not belong to $\partial f(e) = \{(2\beta, 0, \ldots) : -1 \leq \beta \leq 0\}$ and this means that ∂f is not pseudo-monotone.

4.2 Pointwise Minima and Maxima of Functions from the Classes $QPM(V)$ and $PM(V)$ and Compositions with Linear Compact Operators

This part is devoted first to the study of the quasi-pseudo-monotone and pseudo-monotone properties of the generalized gradient of nonconvex functions which are the pointwise maxima or minima of finite collections of functions from the QPM and PM classes. Then we will examine whether a composition of a locally Lipschitz function with a linear compact operator belongs to the PM or the QPM class of functions.

Let $f_i : V \to R$, $i = 1, \ldots, N$, be a finite collection of functions from a reflexive Banach space V into R. We define $\max(f_1, \ldots, f_N) : V \to R$ and $\min(f_1, \ldots, f_N) : V \to R$ by the expressions

$$\max(f_1, \ldots, f_N)(v) = \max\{f_i(v) : i = 1, \ldots, N\}, v \in V, \tag{4.2.1}$$

$$\min(f_1, \ldots, f_N)(v) = \min\{f_i(v) : i = 1, \ldots, N\}, v \in V. \tag{4.2.2}$$

Proposition 4.10 Let $f_1, f_2 \in QPM(V)$. Then both $\max(f_1, f_2) \in QPM(V)$ and $\min(f_1, f_2) \in QPM(V)$.

Proof. We start with the function $\max(f_1, f_2)$. Suppose that $\{u_i\}$ is a sequence in V converging weakly to u, and that there exists a sequence $u_i^* \in \partial[\max(f_1 f_2)](u_i)$ with $\limsup\langle u_i^*, u_i - u\rangle_V \leq 0$. We have to show that condition $\lim\langle u_i^*, u_i - u\rangle_V = 0$ results. It is well known (cf. (1.2.40)) that $\partial[\max(f_1, f_2)](v) \subset \mathrm{co}\{\partial f_n(v) : n \in I(v)\}$ for each $v \in V$, where $I(v)$ denotes the set of all indices n for which $f_n(v) = \max(f_1, f_2)(v)$. Thus we obtain the representation $u_i^* = \lambda_i \bar{u}_i^* + (1 - \lambda_i) \bar{\bar{u}}_i^*$, for some $\bar{u}_i^* \in \partial f_1(u_i)$, $\bar{\bar{u}}_i^* \in \partial f_2(u_i)$ and $\lambda_i \in [0, 1]$. We claim that the condition

$$\limsup\langle \lambda_i \bar{u}_i^* + (1 - \lambda_i) \bar{\bar{u}}_i^*, u_i - u\rangle_V \leq 0 \qquad (4.2.3)$$

implies

$$\limsup\langle \lambda_i \bar{u}_i^*, u_i - u\rangle_V \leq 0$$

and

$$\qquad (4.2.4)$$

$$\limsup\langle (1 - \lambda_i) \bar{\bar{u}}_i^*, u_i - u\rangle_V \leq 0.$$

Indeed, suppose that this is not true. Then we may assume that

$$\limsup\langle \lambda_i \bar{u}_i^*, u_i - u\rangle_V > 0. \qquad (4.2.5)$$

Thus we can find a positive $d > 0$ and a subsequence of $\{u_i\}$ (again denoted by $\{u_i\}$) such that

$$\lim\langle \lambda_i \bar{u}_i^*, u_i - u\rangle_V = d > 0. \qquad (4.2.6)$$

For the corresponding subsequence of the other summand, we have

$$\limsup\langle (1 - \lambda_i) \bar{\bar{u}}_i^*, u_i - u\rangle_V \leq -d < 0. \qquad (4.2.7)$$

Hence, taking into account the boundedness of $\{\lambda_i\} \subset [0, 1]$ we easily deduce the existence of $\bar{d} > 0$ with

$$\limsup\langle \bar{u}_i^*, u_i - u\rangle_V \leq -\bar{d} < 0 \qquad (4.2.8)$$

from which, by virtue of the quasi-pseudo-monotonicity of ∂f_1, the equality $\lim\langle \bar{u}_i^*, u_i - u\rangle_V = 0$ follows which is a contradiction. Thus (4.2.4) has been proved. Now we show that this implies that

$$\lim\langle \lambda_i \bar{u}_i^*, u_i - u\rangle_V = 0 \qquad (4.2.9a)$$

$$\lim\langle (1 - \lambda_i) \bar{\bar{u}}_i^*, u_i - u\rangle_V = 0. \qquad (4.2.9b)$$

By symmetry it suffices to show that $\limsup\langle \lambda_i \bar{u}_i^*, u_i - u\rangle_V \leq 0$ implies $\lim\langle \lambda_i \bar{u}_i^*, u_i - u\rangle_V = 0$. If we suppose that $\liminf\langle \lambda_i \bar{u}_i^*, u_i - u\rangle_V = -d < 0$ for some positive $d > 0$, then by taking into account that $\lambda_i \in [0, 1]$ and by choosing an appropriate subsequence (denoted again by the same symbol) we arrive at the conclusion that $\lim\langle \bar{u}_i^*, u_i - u\rangle_V = -\bar{d} < 0$ for some $\bar{d} > 0$, which contradicts the quasi-pseudo-monotonicity of ∂f_1. Accordingly, we are led to the conclusion that $\liminf\langle \lambda_i \bar{u}_i^*, u_i - u\rangle_V \geq 0$ which implies (4.2.9a). Hence we deduce immediately that

$$\lim\langle u_i^*, u_i - u\rangle_V = \lim\langle \lambda_i \bar{u}_i^* + (1 - \lambda_i)\,\bar{\bar{u}}_i^*, u_i - u\rangle_V = 0. \qquad (4.2.10)$$

Thus the quasi-pseudo-monotonicity of $\partial \max(f_1, f_2)$ has been shown. In order to prove that $\min(f_1, f_2) \in QPM(V)$ we argue similarly by using the same method. This is justified by the fact that, like in the previous case, $\partial[\min(f_1, f_2)](v) \subset \mathrm{co}\{\partial f_n(v) : n \in I(v)\}$ for each $v \in V$. The proof of Proposition 4.10 is now complete. \hfill q.e.d.

Proposition 4.11 Suppose that $f_1, f_2 \in PM(V)$ have the (C) property. Then the following conditions hold:

(i) If $\partial[\max(f_1, f_2)](\cdot) = \mathrm{co}\{\partial f_n(v) : n \in I(\cdot)\}$, then $\max(f_1, f_2) \in PM(V)$ and $\max(f_1, f_2)$ satisfies the (C) condition.

(ii) If $\partial[\min(f_1, f_2)](\cdot) = \mathrm{co}\{\partial f_n(v) : n \in I(\cdot)\}$ and $\min(f_1, f_2)$ satisfies the (C) condition, then $\min(f_1, f_2) \in PM(V)$.

Proof. We start with the proof of (i). For this purpose let us suppose that $\{u_i\}$ is a sequence in V converging weakly to u, for which there exists $u_i^* \in \partial[\max(f_1, f_2)](u_i)$ such that $\limsup\langle u_i^*, u_i - u\rangle_V \leq 0$. Following the lines of the proof of Proposition 4.10. we deduce that $\lim\langle u_i^*, u_i - u\rangle_V = 0$. Thus in order to complete the proof it remains to show that each weak limit of $\{u_i^*\}$ lies in $\partial[\max(f_1, f_2)](u)$. By passing to a subsequence (again denoted by the same symbol) we ensure the weak convergence of $\{u_i^*\}$ to some u^*. Moreover, we may suppose that $\lim f_1(u_i) = \alpha_1, \lim f_2(u_i) = \alpha_2$ where $\alpha_1, \alpha_2 \in R$. Recalling that $u_i^* = \lambda_i \bar{u}_i^* + (1 - \lambda_i)\,\bar{\bar{u}}_i^*$, for some $\bar{u}_i^* \in \partial f_1(u_i)$, $\bar{\bar{u}}_i^* \in \partial f_2(u_i)$ and $\lambda_i \in [0, 1]$, we may also assume that: $\bar{u}_i^* \xrightarrow{w} \bar{u}^*, \bar{\bar{u}}_i^* \xrightarrow{w} \bar{\bar{u}}^*$ and $\lim \lambda_i = \lambda$ for some $\bar{u}^*, \bar{\bar{u}}^* \in V^*$ and a certain $\lambda \in [0, 1]$. As in the previous case we arrive easily at the equalities

$$\lim\langle \lambda_i \bar{u}_i^*, u_i - u\rangle_V = 0 \quad \text{and} \quad \lim\langle (1 - \lambda_i)\,\bar{\bar{u}}_i^*, u_i - u\rangle_V = 0 \qquad (4.2.11)$$

that constitute a convenient starting point for the further part of the proof. Without loss of generality we may assume that $I(u_i) = \mathrm{const.}$ and $1 \in I(u_i)$. Thus

$$f_1(u_i) \geq f_2(u_i). \qquad (4.2.12)$$

In the case $\lambda \in (0, 1)$ we argue as follows. There exists $\delta > 0$ such that $\lambda_i \geq \delta$ and $1 - \lambda_i \geq \delta$ for sufficiently large i. Thus from (4.2.11) it follows that $\lim\langle \bar{u}_i^*, u_i - u\rangle_V = 0$ and $\lim\langle \bar{\bar{u}}_i^*, u_i - u\rangle_V = 0$ which, due to the (C) property of f_1 and f_2, implies $\alpha_1 = f_1(u)$ and $\alpha_2 = f_2(u)$. Since in the case considered (4.2.12) must hold as an equality, $\max(f_1, f_2)(u) = f_1(u) = f_2(u)$. Moreover, by the pseudo-monotonicity of ∂f_1 and ∂f_2 it results that $\bar{u}^* \in \partial f_1(u)$ and

$\bar{\bar{u}}^* \in \partial f_2(u)$. Consequently, $u^* = \lambda \bar{u}^* + (1-\lambda)\,\bar{\bar{u}}^* \in \text{co}\{\partial f_n(u) : n = 1, 2\}$ which by the hypothesis coincides with $\partial[\max(f_1, f_2)](u)$.

If we suppose that $\lambda = 0$, then, due to (4.2.12), only the case that $f_1(u_i) = f_2(u_i)$ is possible (if $f_1(u_i) > f_2(u_i)$ then we would have $\lambda_i = 1$, which excludes $\lambda = 0$). Thus $\alpha_1 = \alpha_2$ and from (4.2.11) we obtain $\lim\langle \bar{u}_i^*, u_i - u \rangle_V = 0$. Further, by the (C) property and the pseudo-monotonicity of ∂f_2, $\alpha_2 = f_2(u)$ and $\bar{\bar{u}}^* \in \partial f_2(u)$. Hence $u^* = \bar{\bar{u}}^*$. To complete this case let us notice that the weak lower semicontinuity of f_1 implies $f_2(u) = \alpha_2 = \alpha_1 \geq f_1(u)$. Therefore $2 \in I(u)$, and

$$u^* = \bar{\bar{u}}^* \in \partial f_2(u) \subset \text{co}\{\partial f_n(u) : n \in I(u)\} = \partial[\max(f_1, f_2)](u). \qquad (4.2.13)$$

For the last case $\lambda = 1$ we argue similarly. From (4.2.11) one gets $\lim\langle \bar{u}_i^*, u_i - u \rangle_V = 0$. Hence, by virtue of the pseudo-monotonicity of ∂f_1 and the (C) property of f_1, $\bar{u}^* \in \partial f_1(u)$ and $\alpha_1 = f_1(u)$. Taking into account (4.2.12) we are led to $f_1(u) = \alpha_1 \geq \alpha_2$ which due to the weak lower semicontinuity of f_2 yields $f_1(u) \geq f_2(u)$. Hence $1 \in I(u)$ and consequently

$$u^* = \bar{u}^* \in \partial f_1(u) \subset \text{co}\{\partial f_n(u) : n \in I(u)\} = \partial[\max(f_1, f_2)](u). \qquad (4.2.14)$$

Notice that in all the cases considered we have obtained the convergence $\max(f_1, f_2)(u_i) \to \max(f_1, f_2)(u)$. Thus $\max(f_1, f_2)$ satisfies the (C) property. The proof of (i) is complete.

Now we proceed to the proof of (ii). The same representation of the generalized gradient for $\min(f_1, f_2)$ permits to argue as in the previous case with slight modifications. The only differences arise when considering the cases $\lambda = 0$ and $\lambda = 1$. When dealing with $\min(f_1, f_2)$, we replace (4.2.12) by the inequality

$$f_1(u_i) \leq f_2(u_i). \qquad (4.2.15)$$

in order to keep $1 \in I(u)$. If $\lambda = 0$ then only the case that $f_1(u_i) = f_2(u_i)$ is possible (for $f_1(u_i) < f_2(u_i)$ we would have $\lambda_i = 1$ which excludes $\lambda = 0$). This implies $\alpha_1 = \alpha_2$. Moreover, from (4.2.11) we arrive at $\lim\langle \bar{\bar{u}}_i^*, u_i - u \rangle_V = 0$. Hence, by the (C) property of f_2 and the pseudo-monotonicity of ∂f_2 it results that $\bar{\bar{u}}^* \in \partial f_2(u)$ and $\alpha_2 = f_2(u)$. Hence, $f_2(u) = \alpha_2 = \alpha_1 \geq f_1(u)$ due to the weak lower semicontinuity of f_1. Thus $\min(f_1, f_2)(u) = f_1(u)$. What we need now is to show that the equality $f_2(u) = f_1(u)$ holds. But this is a consequence of the hypothesis that $\min(f_1, f_2)$ has the (C) property. Indeed, the (C) condition requires the sequence $\{u_i\}$ to satisfy the relations

$$\min(f_1, f_2)(u_i) \to \min(f_1, f_2)(u) = \alpha_2 = \alpha_1 = f_2(u) \leq f_1(u). \qquad (4.2.16)$$

Hence, $f_2(u) = f_1(u)$ and we may write that

$$u^* = \bar{\bar{u}}^* \in \partial f_2(u) \subset \text{co}\{\partial f_n(u) : n = 1, 2\} = \partial[\min(f_1, f_2)](u). \qquad (4.2.17)$$

The assertion follows.

For $\lambda = 1$ we argue as follows. From (4.2.11) we obtain $\lim\langle \bar{u}_i^*, u_i - u \rangle_V = 0$. Hence, by the pseudo-monotonicity of ∂f_1 and the (C) property of f_1 it results that $\bar{u}^* \in \partial f_1(u)$ and $\alpha_1 = f_1(u)$. Taking into account the (C) property of $\min(f_1, f_2)$ we arrive at the relations

$$\min(f_1, f_2)(u_i) = f_1(u_i) \to \alpha_1 = f_1(u) = \min(f_1, f_2)(u) \leq f_2(u). \qquad (4.2.18)$$

Accordingly, $1 \in I(u)$ and therefore

$$u^* = \bar{u}^* \in \operatorname{co}\{\partial f_n(u) : n \in I(u)\} = \partial[\min(f_1, f_2)](u). \qquad (4.2.19)$$

The proof of Proposition 4.11 is complete.   q.e.d.

Pointwise maxima and minima of a finite collection of functions from the QPM- and PM-classes still remain within these classes provided the conditions below hold. Arguing similarly we are led to the following generalization of Propositions 4.10 and 4.11.

Corollary 4.12 Let $f_n : V \to R, n = 1, \ldots, N$, be locally Lipschitz functions from a reflexive Banach space V into R. The following conditions hold:

(i) If $f_1, \ldots, f_N \in QPM(V)$ then $\max(f_1, \ldots, f_N) \in QPM(V)$ and $\min(f_1, \ldots, f_N) \in QPM(V)$;

(ii) If $f_1, \ldots, f_N \in PM(V)$ satisfy the (C) condition, and $\partial[\max(f_1, \ldots, f_N)](\cdot) = \operatorname{co}\{\partial f_n(\cdot) : n \in I(\cdot)\}$, then $\max(f_1, \ldots, f_N) \in PM(V)$ and satisfies the (C) condition;

(iii) If $f_1, \ldots, f_N \in PM(V)$ satisfy the (C) condition, $\min(f_1, \ldots, f_N)$, satisfies the (C) condition and $\partial[\min(f_1, \ldots, f_N)](\cdot) = \operatorname{co}\{\partial f_n(\cdot) : n \in I(\cdot)\}$, then $\min(f_1, \ldots, f_N) \in PM(V)$.

It is well known that a sufficient condition for the equality

$$\partial[\max(f_1, \ldots, f_N)](\cdot) = \operatorname{co}\{\partial f_n(\cdot) : n \in I(\cdot)\} \qquad (4.2.20)$$

to hold is that $f_1, \ldots, f_N$ are ∂-regular. Hence we can easily check that in order to get

$$\partial[\min(f_1, \ldots, f_N)](\cdot) = \operatorname{co}\{\partial f_n(v) : n \in I(\cdot)\} \qquad (4.2.21)$$

it is enough to suppose that $(-f_1), \ldots, (-f_N)$ are ∂-regular. But for convex $f_1, \ldots, f_N$ this kind of regularity implies the Gateaux differentiability of the functions involved. Therefore we are led to the following conclusion.

Corollary 4.13 Let $f_n : V \to R, n = 1, \ldots, N$, be locally Lipschitz convex functions from a reflexive Banach space V into R. Then the following conditions hold:

(i) $\max(f_1, \ldots, f_N) \in PM(V)$;

(ii) $\min(f_1, \ldots, f_N) \in QPM(V)$;

(iii) If $\min(f_1, \ldots, f_N)$ satisfies the (C) condition and if $f_i, i = 1, \ldots, N$, are Gâteaux differentiable, then $\min(f_1, \ldots, f_N) \in PM(V)$.

All the hypotheses mentioned in (iii) of Corollary 4.13 are essential in order to establish the pseudo-monotonicity of $\partial \min(f_1, \ldots, f_N)$. To show this we consider two examples.

Example 4.14 Let $V = H$ be an infinite dimensional Hilbert space. Define two differentiable convex functions f_1 and f_2 by

$$f_1(v) = 0 \qquad\qquad v \in H,$$
$$f_2(v) = \tfrac{1}{2}\|v + v_0\|^2 - 1 \quad v \in H, \tag{4.2.22}$$

with some $v_0 \in H$ such that $\|v_0\| = \tfrac{1}{2}$. By the differentiability of f_1 and f_2, the required regularity conditions hold. Because of the convexity, f_1 and f_2 possess the (C) property. Set

$$f(v) = \min\{f_n(v) : n = 1, 2\}. \tag{4.2.23}$$

Let $\{e_i\} \subset H$ be a sequence in H fulfilling the conditions: $\|e_i\| = 2$ and $e_i \xrightarrow{w} e = 0$. It can be easily verified that for the sequences $\{e_i\}$ and $\{e_i^*\}$ with $e_i^* = 0$, all the requirements of the definition of pseudo-monotonicity hold, namely $e_i^* = 0 \in \partial f(e_i)$, $e_i^* \xrightarrow{w} e^* = 0$ and $\limsup \langle e_i^*, e_i - e \rangle_H = 0 \leq 0$. Now, let us check that the function $f = \min(f_1, f_2)$ does not fulfill the (C) condition. Indeed, $f(e_i) = 0 = f_1(e_i) < \tfrac{1}{8} \leq f_2(e_i)$ and $f(e) = f_2(e) = \tfrac{1}{2}\|v_0\|^2 - 1 = -\tfrac{7}{8} < f_1(e) = 0$. Hence $\lim f(e_i) = 0 > -\tfrac{7}{8} = f(e)$. Finally, notice that $e^* = 0 \notin \partial f(0) = \{v_0\}$ and therefore ∂f is not pseudo-monotone.

Example 4.15 Let $V = l^2$, where l^2 is the Hilbert space of all real-valued sequences such that

$$l^2 = \{v = (x_i), i = 1, 2, \ldots, \sum_i |x_i|^2 < +\infty\}, \tag{4.2.24}$$

with the norm

$$\|v\| = \left(\sum_i |x_i|^2\right)^{1/2}, v = (x_i) \in l^2. \tag{4.2.25}$$

Define the convex functions

$$f_1(v) = \|v\|^2 - 1, \tag{4.2.26}$$

$$f_2(v) = \begin{cases} 0 & \text{if } x_1 \geq 1 \\ -x_1 + 1 & \text{if } x_1 < 1, \end{cases} \quad v = (x_i) \in l^2$$

and the corresponding pointwise minimum function

$$f(v) = \min\{f_i(v) : i = 1, 2\}. \tag{4.2.27}$$

It is obvious that f_2 is not Gateaux-differentiable at the point $e = (1, 0, \ldots)$ in which $f_1(e) = f_2(e)$ and therefore, the regularity condition of (ii) in Corollary 4.13 does not hold. Further, we can verify that for

$$e_i = (1, 0, \ldots, 0, \underset{i}{2}, 0, \ldots), \quad i = 2, 3, \ldots, \tag{4.2.28}$$

we have

$$\left. \begin{array}{l} f(e_i) = f_2(e_i) = 0 < f_1(e_i) = 4, \quad i = 2, 3, \ldots, \\[2mm] f(e) = f_1(e) = f_2(e) = 0, \end{array} \right\} \tag{4.2.29}$$

and

$$\left. \begin{array}{c} e_i \xrightarrow{w} e, \\[2mm] e_i^\star = -e \in \partial f(e_i) = \partial f_2(e_i), i = 2, 3, \ldots, \\[2mm] e_i^\star \xrightarrow{w} e^\star = -e, \end{array} \right\} \tag{4.2.30}$$

Moreover,

$$\limsup(e_i^\star, e_i - e) = 0 \le 0. \tag{4.2.31}$$

Here $(\cdot, \cdot)$ denotes the scalar product in l^2. Thus the sequences $\{e_i\}$ and $\{e_i^\star\}$ fulfill the requirements in the definition of pseudo-monotonicity. Moreover the convergence $f(e_i) \to f(e)$ related to the (C) property holds as well. On the other hand, it can be easily deduced that

$$f^0(e, k) = \left\{ \begin{array}{ll} 2x_1 & \text{if } x_1 \ge 0 \\[4mm] & \\[4mm] 0 & \text{if } x_1 < 0 \end{array} \right. \quad k = (x_1, x_2, \ldots), \tag{4.2.32}$$

from which it results that

$$\partial f(e) = \{(\nu, 0, \ldots) : 0 \le \nu \le 2\}. \tag{4.2.33}$$

Thus $e^\star = -e \notin \partial f(e)$ and therefore ∂f cannot be pseudo-monotone.

Untill now we have dealt with functions from the PM-class satisfying the (C) condition. Now we construct a locally Lipschitz function which does not have the (C) property, but whose generalized gradient is pseudo-monotone. By making use of this function we shall show that the (C) property is essential for $f_1, f_2 \in PM(V)$ in order that $\max(f_1, f_2) \in PM(V)$ and $\min(f_1, f_2) \in PM(V)$ as shown in Corollary 4.12. It will provide also the example of a function which shows that the pseudo-monotonicity of ∂f does not imply the generalized pseudo-monotonicity of $N_{\text{epi}f}$ (cf. Proposition 2.18 and Corollary 2.17).

Example 4.16 Let $V = l^2$, where l^2 is the Hilbert space of all real-valued sequences such that

$$l^2 = \{v = (x_j), j = 1, 2, \ldots, \sum_j x_j^2 < +\infty\}, \qquad (4.2.34)$$

with the norm

$$\|v\| = (\sum_j x_j^2)^{1/2}, \quad v = (x_j) \in l^2, \qquad (4.2.35)$$

and the inner product given by

$$(u, v) = \sum_j x_j y_j \quad u = (x_j), v = (y_j) \in l^2. \qquad (4.2.36)$$

By means of the differentiable convex functions

$$\left.\begin{array}{l} f_0(v) = \sum_j x_j^2 \\[2mm] f_n(v) = 2\sum_{j\neq n} x_j^2 + 100(x_n - 1)^2 + 1, \quad n = 1, 2, \ldots, \end{array}\right\} \qquad (4.2.37)$$

we define the corresponding pointwise minimum function

$$\Psi(v) = \inf\{f_n(v) : n = 0, 1, \ldots\}, \quad v = (x_j) \in l^2. \qquad (4.2.38)$$

Now we show that $\partial\Psi$ is pseudo-monotone. First we observe that for each $n = 1, 2, \ldots$, the set

$$A_n = \{v \in l^2 : f_n(v) \leq f_0(v)\}, \qquad (4.2.39)$$

is contained in the ball

$$B_n = \left\{v \in l^2 : \|v - e_n\| < \frac{1}{3}\right\}, \qquad (4.2.40)$$

where

$$e_n = (0, \ldots, 0, \underset{n}{1}, 0, \ldots) \quad n = 1, 2, \ldots. \qquad (4.2.41)$$

Indeed, from the condition $f_n(v) \leq f_0(v), v = (x_j) \in l^2$, we get

$$\sum_j x_j^2 \geq 2\sum_{j\neq n} x_j^2 + 100(x_n - 1)^2 + 1, \qquad (4.2.42)$$

which leads to

$$99x_n^2 - 200x_n + 101 \leq -\sum_{j\neq n} x_j^2 \leq 0. \qquad (4.2.43)$$

Hence, we easily deduce that

$$x_n \in \left[1, \frac{202}{198}\right] \quad \text{and} \quad \sum_{j\neq n} x_j^2 \leq \frac{1}{99}, \qquad (4.2.44)$$

and consequently

$$\|v - e_n\|^2 = |x_n - 1|^2 + \sum_{j\neq n} x_j^2 \leq (\frac{2}{99})^2 + \frac{1}{99} < \frac{1}{9}. \qquad (4.2.45)$$

But this means that $||v - e_n|| < \frac{1}{3}, n = 1, 2, \ldots$, as desired. Because $||e_n - e_k|| = \sqrt{2}$ for $n \neq k$, the obtained result permits us to conclude that $\{A_n\}$ is a disjoint family of sets. By the definition of Ψ this fact implies that for any $v \in l^2$ the corresponding set $I(v)$ of all indices which are activ at v contains no more than two indices. Moreover, if it contains two indices then one of them must be 0.

It is clear that in order to prove the pseudo-monotonicity of $\partial\Psi$ it suffices to show that $\partial\Psi$ is generalized pseudo-monotone (see Proposition 2.2). Suppose that $\{u_i\}$ is a sequence in l^2 converging weakly to u and that there exists a sequence $u_i^* \in \partial\Psi(u_i)$ converging weakly to some u^*, such that $\limsup(u_i^*, u_i - u) \leq 0$. We have to show that $\lim(u_i^*, u_i - u) = 0$ and that $u^* \in \partial\Psi(u)$. Let us introduce the notations: $u_i = (x_j^{(i)}), u_i^* = (y_j^{(i)})$ and $u = (x_j), u^* = (y_j)$. To show the assertion it is enough to consider the following three cases (by passing to a subsequence of $\{u_i\}$ and by making the suitable renumbering, if necessary):

(i) The sequence $\{u_i\}$ does not have any common points with the set $\cup_j A_j$;

(ii) There exists a natural number N such that $\{u_i\} \subset A_N$;

(iii) For all $i = 1, 2, \ldots$, one has $u_i \in A_i$.

(i) In this case we have $I(u_i) = \{0\}$. Thus $\partial\Psi(u_i) = \partial f_0(u_i) = \{2u_i\}$ and consequently, $u_i^* = 2u_i$. From the condition $\limsup(2u_i, u_i - u) = \limsup(u_i^*, u_i - u) \leq 0$ we deduce the strong convergence of $\{u_i\}$ to u. Then, the upper semicontinuity of $\partial\Psi$ from l^2 to l^2 endowed with the weak topology, yields $u^* = 2u \in \partial\Psi(u)$, and the assertion follows.

(ii) If $\{u_i\} \subset A_N$ then we can find a sequence $\{\lambda_i\} \subset [0,1]$ such that $u_i^* \in \lambda_i \partial f_0(u_i) + (1 - \lambda_i)\partial f_N(u_i)$. After simple calculations one obtains

$$y_j^{(i)} = 2\lambda_i x_j^{(i)} + 4(1 - \lambda_i)x_j^{(i)} = 2x_j^{(i)}(2 - \lambda_i), \quad \text{for } j \neq N, \left.\begin{array}{l}\\ \\\end{array}\right\}$$
$$y_N^{(i)} = 2\lambda_i x_N^{(i)} + 200(1 - \lambda_i)(x_N^{(i)} - 1). \tag{4.2.46}$$

From the weak convergence of $\{u_i\}$ to u it follows that for any fixed $j = 1, 2, \ldots$, $\lim_i x_j^{(i)} = x_j$. We show now that the condition $\limsup(u_i^*, u_i - u) \leq 0$ implies the strong convergence of $\{u_i\}$ to u, i.e.,

$$\lim_i \sum_j |x_j^{(i)} - x_j|^2 = 0. \tag{4.2.47}$$

By making use of (4.2.46) we obtain

$$\limsup_i \{2(2 - \lambda_i) \sum_{j\neq N} x_j^{(i)}(x_j^{(i)} - x_j) \tag{4.2.48}$$
$$+ [2\lambda_i x_N^{(i)} + 200(1 - \lambda_i)(x_N^{(i)} - 1)](x_N^{(i)} - x_N)\} \leq 0.$$

Since $\lim_i x_N^{(i)} = x_N$ and $2(2 - \lambda_i) \geq 2$, we get that

$$\limsup_i \sum_{j \neq N} x_j^{(i)}(x_j^{(i)} - x_j) \leq 0. \tag{4.2.49}$$

Hence, taking into account the weak convergence of $\{u_i\}$ to u, we may write the relation

$$\lim_i \sum_{j \neq N} x_j(x_j^{(i)} - x_j) = 0 \tag{4.2.50}$$

from which it results

$$\lim_i \sum_{j \neq N} |x_j^{(i)} - x_j|^2 = 0. \tag{4.2.51}$$

This combined with $\lim_i x_N^{(i)} = x_N$ is equivalent to the strong convergence of $\{u_i\}$ to u in l^2, as desired. Hence we easily deduce that $\lim(u_i^*, u_i - u) = 0$. In order to show that $u^* \in \partial \Psi(u)$ we use again the upper semicontinuity of $\partial \Psi$ from l^2 to the weak topology of l^2. The proof of (ii) is complete.

(iii) If $u_i \in A_i$ then there exists a sequence $\{\lambda_i\} \subset [0,1]$ such that $u_i^* \in \lambda_i \partial f_0(u_i) + (1 - \lambda_i)\partial f_i(u_i)$. We may assume (by choosing a further subsequence) that $\lambda_i \to \lambda$ for some $\lambda \in [0,1]$. The following conditions hold:

$$\left.\begin{aligned} y_j^{(i)} &= 2\lambda_i x_j^{(i)} + 4(1 - \lambda_i)x_j^{(i)} = 2x_j^{(i)}(2 - \lambda_i), \quad \text{for } j \neq i, \\ y_i^{(i)} &= 2\lambda_i x_i^{(i)} + 200(1 - \lambda_i)(x_i^{(i)} - 1). \end{aligned}\right\} \tag{4.2.52}$$

From the weak convergence of $\{u_i\}$ to u it follows that for all $j = 1, 2, \ldots, \lim_i x_j^{(i)} = x_j$. Now we show that the condition $\limsup(u_i^*, u_i - u) \leq 0$ implies the weak convergence of $\{u_i\}$ to 0. Indeed, from (4.2.52) it results that

$$\limsup_i \left\{ 2(2 - \lambda_i) \sum_{j \neq i} x_j^{(i)}(x_j^{(i)} - x_j) \right. \tag{4.2.53}$$
$$\left. + [2\lambda_i x_i^{(i)} + 200(1 - \lambda_i)(x_i^{(i)} - 1)](x_i^{(i)} - x_i) \right\} \leq 0.$$

Since $u_i \in A_i$, $x_i^{(i)} \in [1, \frac{202}{198}]$ and we may assume that $x_i^{(i)} \to \beta$ for some $\beta \in [1, \frac{202}{198}]$. We claim that all the coordinates x_j of u belong to the interval $[-\frac{1}{9}, \frac{1}{9}]$. Indeed, from (4.2.44) we have

$$\sum_{k \neq i} |x_k^{(i)}|^2 \leq \frac{1}{99}. \tag{4.2.54}$$

Hence, for $i > j$ one obtains $|x_j^{(i)}|^2 \leq \frac{1}{99} \leq \frac{1}{81}$, and $|x_j^{(i)}| \leq \frac{1}{9}$. Thus the limit x_j of $x_j^{(i)}$ as $i \to \infty$ must remain in $[-\frac{1}{9}, \frac{1}{9}]$, as required. This fact allows us to write

$$x_i^{(i)} - x_i \geq \frac{8}{9}. \tag{4.2.55}$$

Taking into account the weak convergence of $\{u_i\}$ to u we have

$$\lim_i \sum_j x_j(x_j^{(i)} - x_j) = 0. \tag{4.2.56}$$

Hence, by combining this with (4.2.53) one obtains

$$\limsup_i \{2(2 - \lambda_i) \sum_{j \neq i} |x_j^{(i)} - x_j|^2 - 2(2 - \lambda_i)x_i(x_i^{(i)} - x_i) \quad (4.2.57)$$
$$+ \quad [2\lambda_i x_i^{(i)} + 200(1 - \lambda_i)(x_i^{(i)} - 1)](x_i^{(i)} - x_i)]\} \leq 0.$$

Since $x_i \to 0$ as $u = (x_i) \in l^2$, $x_i^{(i)} - x_i \geq \frac{8}{9}$ and

$$2(2 - \lambda_i) \sum_{j \neq i} |x_j^{(i)} - x_j|^2 \geq 0, \quad (4.2.58)$$

the following condition can be deduced from (4.2.57)

$$\limsup(2\lambda_i x_i^{(i)} + 200(1 - \lambda_i)(x_i^{(i)} - 1)) \leq 0. \quad (4.2.59)$$

This way we arrive at the equality

$$2\lambda\beta + 200(1 - \lambda)(\beta - 1) = 0. \quad (4.2.60)$$

In view of $\beta \in [1, \frac{102}{98}]$ and $\lambda \in [0,1]$ the above equality is valid only for $\beta = 1$ and $\lambda = 0$. Hence we conclude that $u = 0$. Indeed, from (4.2.43) the estimate

$$-99(x_i^{(i)})^2 + 200x_i^{(i)} - 101 \geq \sum_{j \neq i} |x_j^{(i)}|^2 \geq |x_k^{(i)}|^2 \quad (4.2.61)$$

holds for any fixed k and $i > k$. Let us note that due to $x_i^{(i)} \to \beta = 1$,

$$\lim\{-99(x_i^{(i)})^2 + 200x_i^{(i)} - 101\} = 0. \quad (4.2.62)$$

Thus, letting i tend to infinity in (4.2.61) one obtains

$$x_k = \lim_i x_k^{(i)} = 0, \quad k = 1, 2, \ldots, \quad (4.2.63)$$

and finally, $u = 0$. Now we show that $u^* = (y_j) = 0$, as well. To this end it suffices to prove that $\lim_i y_k^{(i)} = y_k = 0$ for all k. This fact follows directly from the equality $y_k^{(i)} = 2x_k^{(i)}(2 - \lambda_i)$ which is true for $k \neq i$, and from the fact that $\lim_i x_k^{(i)} = 0$.

To summarize, we have proved that $\{u_i\}$ converges weakly to $u = 0$ and the corresponding sequence $\{u_i^*\}$ converges weakly to $u^* = 0$. Further, it is easily seen that $\partial\Psi(u) = \partial f_0(u) = \{2u\} = \{0\}$. Thus $u^* \in \partial\Psi(u)$; to establish the pseudo-monotonicity of $\partial\Psi$ it remains to show that

$$\lim_i(u_i^*, u_i - u) = \lim_i \sum_j y_j^{(i)}(x_j^{(i)} - x_j) = 0. \quad (4.2.64)$$

From (4.2.52) and (4.2.63) we have

$$(u_i^*, u_i - u) = 2(2 - \lambda_i) \sum_{j \neq i} |x_j^{(i)}|^2 + [2\lambda_i x_i^{(i)} + 200(1 - \lambda_i)(x_i^{(i)} - 1)]x_i^{(i)}. \quad (4.2.65)$$

Taking into account (4.2.61) and (4.2.62) it results that

$$\lim_i [2(2 - \lambda_i) \sum_{j \neq i} |x_j^{(i)}|^2] = 0, \qquad (4.2.66)$$

while due to $x_i^{(i)} \to 1$ and $\lambda_i \to 0$ one can deduce

$$\lim_i [2\lambda_i x_i^{(i)} + 200(1 - \lambda_i)(x_i^{(i)} - 1)]x_i^{(i)} = 0. \qquad (4.2.67)$$

Hence we obtain (4.2.64) and finally the proof of the pseudo-monotonicity of $\partial \Psi$ is complete.

The pseudo-monotonicity of $\partial \Psi$ does not imply the (C) property of $\partial \Psi$. Indeed, if we take a sequence $\{e_i\}$ given by (4.2.41), then we have $\Psi(e_i) = f_0(e_i) = f_i(e_i) = 1$. Moreover, $\{e_i\}$ is a sequence converging weakly to $e = 0$. The convexity and the Gâteaux differentiability of the functions involved imply that $\partial \Psi(e_i) = \lambda \partial f_0(e_i) + (1 - \lambda) \partial f_i(e_i) = \{\lambda e_i : \lambda \in [0, 1]\}$. Thus setting $e_i^* = 0$ implies that $e_i^* \in \partial \Psi(e_i)$ and $\limsup(e_i^*, e_i - e) = 0 \leq 0$. Should Ψ had the (C) property we would obtain $\lim \Psi(e_i) = \Psi(e)$. But this is not true as $\lim \Psi(e_i) = 1$ and $\Psi(e) = 0$. Accordingly, Ψ does not satisfy the (C) condition.

Example 4.17 The function Ψ provides an example of a function whose generalized gradient is pseudo-monotone, whereas $N_{\text{epi}\Psi}$ is not generalized pseudo-monotone. To show this fact is suffices to consider the sequence $(e_i, \Psi(e_i)) = (e_i, 1) \in \text{epi}\,\Psi \subset l^2 \times R$ for which we have $(0, -1) \in N_{\text{epi}\Psi}(e_i, 1)$, $(e_i, 1) \to (0, 1)$ weakly in $l^2 \times R$ and $\limsup((0, -1), (e_i, 1) - (0, 1)) = 0$. Obviously, $(0, -1) \to (0, -1)$ weakly in $l^2 \times R$. Thus, if $N_{\text{epi}\Psi}$ were generalized pseudo-monotone, then $(0, -1)$ would belong to $N_{\text{epi}\Psi}(0, 1)$. But this is not true because $N_{\text{epi}\Psi}(0, 1) = \{(0, 0)\}$ due to $\Psi(0) = 0 < 1$. Accordingly, $N_{\text{epi}\Psi}$ cannot have the generalized pseudo-monotone property.

If the (C) condition is not satisfied, then, generally, $f_1, f_2 \in PM(V)$ does not imply that $\max(f_1, f_2) \in PM(V)$ and $\min(f_1, f_2) \in PM(V)$. To show this let us consider a function $\max(\Psi, f)$ in l^2, where

$$f(v) = \frac{1}{2}(x_1 - 0, 1)^2 + \frac{1}{2} \sum_{j \geq 2} x_j^2 + \frac{1}{4}, \quad v = (x_i) \in l^2. \qquad (4.2.68)$$

Since $f(0) > \Psi(0)$ and $f(e_i) < \Psi(e_i)$, $\partial \max(\Psi, f)(0) = \partial f(0) = \{(\frac{-1}{10}, 0, \ldots)\}$ and $\partial \max(\Psi, f)(e_i) = \partial \Psi(e_i) = \{\lambda e_i : \lambda \in [0, 1]\}$. The pseudo-monotone condition does not hold for the sequences $\{e_i\}$ and $\{e_i^*\}$, with $e_i^* = 0$, because $e^* = 0 \notin \partial \max(\Psi, f)(0)$.

Choosing the same sequences we can check that $\partial \min(\Psi, f)$ is not pseudo-monotone if

$$f(v) = 2(x_1 - 0, 1)^2 + 2 \sum_{j \geq 2} x_j^2 - \frac{1}{4}, \quad v = (x_i) \in l^2. \qquad (4.2.69)$$

Now we consider functions which can be expressed as a composition of a locally Lipschitz function with a linear compact operator. We suppose that there exists a linear compact operator $L : V \to Y$ from V into Y, and a locally Lipschitz function $F : Y \to R$ from Y into R, such that $f = F \circ L$. It turns out that $f \in QPM(V)$. To get $f \in PM(V)$ we need some regularity conditions.

Proposition 4.18   Let $F : Y \to R$ be a locally Lipschitz function from a Banach space Y into R and let $L : V \to Y$ be a linear compact operator from a reflexive Banach space V into Y. Then the following conditions hold:

(i) $F \circ L \in QPM(V)$;

(ii) If $F^0(Lu, Lv) = (F \circ L)(u, v)$ for any $u, v \in V$, then $F \circ L \in PM(V)$.

Proof. (i) Let us choose a sequence $\{u_i\}$ in V converging weakly to u. We show that if $u_i^* \in \partial(F \circ L)(u_i)$, then we have $\lim \langle u_i^*, u_i - u \rangle_V = 0$. Indeed, by the definition,

$$(F \circ L)^0(u_i, v) \geq \langle u_i^*, v \rangle_V \quad \forall v \in V. \qquad (4.2.70)$$

Thus, taking into account the relation

$$
\begin{aligned}
F^0(Lu, Lv) &= \limsup_{\substack{h \to_Y 0 \\ \lambda \to 0+}} \frac{F(Lu + h + \lambda Lv) - F(Lu + h)}{\lambda} \qquad (4.2.71) \\
&\geq \limsup_{\substack{h \to_Y 0 \\ \lambda \to 0+}} \frac{F(Lu + Lh + \lambda Lv) - F(Lu + Lh)}{\lambda} \\
&= (F \circ L)^0(u, v) \quad u, v \in V,
\end{aligned}
$$

we arrive at

$$F^0(Lu_i, Lv) \geq (F \circ L)^0(u_i, v) \geq \langle u_i^*, v \rangle_V \quad \forall v \in V, \qquad (4.2.72)$$

from which it follows that

$$|\langle u_i^*, v \rangle_V| \leq K \|Lv\|_Y \quad \forall v \in V, \qquad (4.2.73)$$

where K is a positive constant. In particular we have

$$|\langle u_i^*, u_i - u \rangle_V| \leq K \|Lu_i - Lu\|_Y. \qquad (4.2.74)$$

The compactness of L and the weak convergence of $\{u_i\}$ to u implies the strong convergence of $\{Lu_i\}$ to Lu in Y. Hence, the required equality $\lim \langle u_i^*, u_i - u \rangle_V = 0$ follows. Accordingly, $F \circ L \in QPM(V)$ and (i) has been established.

To show (ii) let us suppose in addition that for any $u, v \in V$, $F^0(Lu, Lv) = (F \circ L)(u, v)$ and that $u_i^* \to u^*$ weakly in V^*. We have to show that $u^* \in \partial(F \circ L)(u)$. It is well known that a mapping $Y \times Y \ni (y, z) \to F^0(y, z)$ is upper semicontinuous. Thus, in view of the compactness of L we have that

$$V \times V \ni (u, v) \to F^0(Lu, Lv) \qquad (4.2.75)$$

is weakly upper semicontinuous. Hence

$$\limsup F^0(Lu_i, Lv - Lu_i) \leq F^0(Lu, Lv - Lu) \quad \forall v \in V, \tag{4.2.76}$$

for any sequence $\{u_i\}$ weakly converging to u, and thus by the hypothesis we may write that

$$\limsup (F \circ L)^0(u_i, v - u_i) \leq (F \circ L)^0(u, v - u) \quad \forall v \in V. \tag{4.2.77}$$

If $\{u_i^*\}$ converges weakly to u^*, then letting $i \to \infty$ in the inequality

$$(F \circ L)^0(u_i, v - u_i) \geq \langle u_i^*, v - u_i \rangle_V \quad v \in V \tag{4.2.78}$$

yields

$$(F \circ L)^0(u, v - u) \geq \langle u^*, v - u \rangle_V \quad \forall v \in V. \tag{4.2.79}$$

But it means that $u^* \in \partial(F \circ L)(u)$. The proof is complete.   q.e.d.

Remark 4.19 For the equality $F^0(Lu, Lv) = (F \circ L)^0(u, v), u, v \in V$, it suffices to suppose that either F or $-F$ is ∂-regular. Indeed, if F is ∂-regular, then we obtain $F^0(Lu, Lv) = F'(Lu, Lv) = (F \circ L)'(u, v) \leq (F \circ L)^0(u, v)$ which due to the opposite inequality leads to the assertion. Similarly, if $-F$ is ∂-regular then one obtains $F^0(Lu, Lv) = (-F)^0(Lu, -Lv) = (-F)'(Lu, -Lv) = (-F \circ L)'(u, -v) \leq (-F \circ L)^0(u, -v) = (F \circ L)^0(u, v)$. Thus the assertion is proved.

4.3 Hemivariational Inequalities Involving Functions from QPM- and PM-Classes

When studying variational problems involving nonconvex, nondifferentiable functions one arrives at hemivariational inequalities which are the generalization of the well known in the literature, variational inequalities. As, for variational inequalities the theory of maximal monotone mappings has proved to be the main mathematical tool for establishing existence results, the theory of pseudo-monotone and generalized pseudo-monotone mappings seems to be the main mathematical tool in the theory of hemivariational inequalities. In particular, the results of the previous section characterize those nonconvex functions that can be treated by means of the theory of pseudo-monotone mappings and allow us to recognize whether the nonconvexity, we deal with, can be treated by means of the developed theory. The properties of functions belonging to the PM- and QPM-classes make it possible to formulate some existence results for hemivariational inequalities.

Before passing to the study of hemivariational inequalities let us survey the main classes of nonconvex functions belonging to $QPM(V)$ and $PM(V)$. Suppose that V is a reflexive Banach space. Then

(i) $\min(f_1, \ldots, f_N) \in QPM(V)$, if $f_n : V \to R, n = 1, \ldots, N$, are convex, locally Lipschitz functions;

(ii) $f_1 \cdot \ldots \cdot f_N \in QPM(V)$, if $f_n : V \to R$, $n = 1, \ldots, N$, are convex, locally Lipschitz, nonnegative functions;

(iii) $F \circ L \in QPM(V)$, if $F : Y \to R$ is a locally Lipschitz function and $L : V \to Y$ is a linear compact operator;

(iv) $\min(f_1, \ldots, f_N) \in PM(V)$, if $f_n : V \to R$, $n = 1, \ldots, N$, are convex, locally Lipschitz functions such that $\min(f_1, \ldots, f_N)$ satisfies the (C) property, and $\partial[\min(f_1, \ldots, f_N)](\cdot) = \text{co}\{\partial f_n(\cdot) : n \in I(\cdot)\}$;

(v) $f_1 \cdot \ldots \cdot f_N \in PM(V)$, if $f_1, \ldots, f_N : V \to R, n = 1, \ldots, N$, are convex, locally Lipschitz, nonnegative functions and one of the two conditions holds

(1) $f_n(v) > 0$ for all $v \in V, n = 1, \ldots, N$;

or

(2) For any $v \in V$, if $f_n(v) = 0$, then $\partial f_n(v) = \{0\}, n = 1, \ldots, N$.

(vi) $F \circ L \in PM(V)$, if $F : Y \to R$ is a locally Lipschitz function, $L : V \to Y$ a linear compact operator and if $F^0(Lu, Lv) = (F \circ L)^0(u, v)$ for any $u, v \in V$.

Now we can state an existence result for hemivariational inequalities involving nonconvex locally Lipschitz functions belonging to $PM(V)$ and $QPM(V)$.

Theorem 4.20 Let $A : V \to V^*$ be a pseudo-monotone operator from a reflexive Banach space V into V^*. Suppose that the following hypotheses hold:

(H1) either $f \in PM(V)$, or $f \in QPM(V)$ and A satisfies the $(S)_+$ condition;

(H2) There exists a function $c : R^+ \to R$ with $c(r) \to \infty$ as $r \to \infty$, such that for all $v \in V$ and $v^* \in \partial f(v)$, $\langle Av + v^*, v - u_0 \rangle \geq c(\|v\|_V)\|v\|_V$ for some $u_0 \in V$.

Then the hemivariational inequality: For a given $g \in V^*$ find $u \in V$ such that

$$\langle Au - g, v - u \rangle + f^0(u, v - u) \geq 0 \quad \forall v \in V \tag{4.3.1}$$

has at least one solution.

Proof. Hemivariational inequality (4.3.1) can be written equivalently as $g \in Au + \partial f(u)$. Thus the problem is reduced to the question whether g belongs to the range of $A + \partial f$. Propositions 2.4 and 2.20 ensure the pseudo-monotonicity of this mapping. Hence, due to its coercivity, the range $R(A + \partial f) = V^*$ (Theorem 2.6 and Remark 2.7) Thus, in particular, $g \in Au + \partial f(u)$ for some $u \in V$. q.e.d.

Remark 4.21 If A is coercive in the sense that for any $v \in V$ $\langle Av, v - u_0 \rangle_V \geq c(\|v\|_V)\|v\|_V$ holds for some $u_0 \in V$, then for the coercivity of the sum $A + \partial f$ it suffices to assume that

$$f^0(v, u_0 - v) \leq k(1 + ||v||_V) \quad \forall v \in V, \quad k = \text{const.} \tag{4.3.2}$$

Indeed, for each $v^* \in \partial f(v)$ the estimate

$$\langle v^*, v - u_0 \rangle_V \geq -k(1 + ||v||_V), \quad \forall v \in V, \tag{4.3.3}$$

holds, which ensures the coercivity of $A + \partial f$.

Remark 4.22 Let $f_i : V \to R, i = 1, \ldots, N$, be a finite collection of locally Lipschitz convex and nonnegative functions from a reflexive Banach space V into R. Define $f : V \to R$ by the product

$$f(v) = f_1(v) \cdot \ldots \cdot f_N(v) \quad v \in V. \tag{4.3.4}$$

The following formula can be derived:

$$\partial f(v) = \sum_i [f_1(v) \cdot \ldots \cdot f_{i-1}(v) \cdot \partial f_i(v) \cdot f_{i+1}(v) \cdot \ldots \cdot f_N(v)] \tag{4.3.5}$$

from which we obtain that

$$u^* \in \partial f(u) \iff \langle u^*, v - u \rangle_V \tag{4.3.6}$$
$$\leq \sum_i [f_1(u) \cdot \ldots \cdot f_{i-1}(u) \cdot (f_i(v) - f_i(u)) \cdot f_{i+1}(u) \cdot \ldots \cdot f_N(u)] \quad \forall v \in V.$$

In such a case the hemivariational inequality (4.3.1) can be written equivalently as

$$\langle Au - g, v - u \rangle_V \tag{4.3.7}$$
$$+ \sum_i [f_1(u) \cdot \ldots \cdot f_{i-1}(u) \cdot (f_i(v) - f_i(u)) \cdot f_{i+1}(u) \cdot \ldots \cdot f_N(u)] \geq 0 \quad \forall v \in V.$$

If A is coercive then for the coercivity of the sum $A + \partial f$ it suffices to assume that

$$\sum_i [f_1(v) \cdot \ldots \cdot f_{i-1}(v) \cdot (f_i(v) - f_i(u_0)) \cdot f_{i+1}(v) \cdot \ldots \cdot f_N(v)]$$

$$\geq -k(1 + ||v||_V) \forall v \in V, \tag{4.3.8}$$

where k is a constant.

Now we consider the case in which a locally Lipschitz function $f : V \to R$ from a reflexive Banach space V into R can be represented as the composition of a Lipschitz function with a linear compact operator. Namely, we suppose that there exist a linear compact operator $L : V \to Y$ from V into Y, where Y is a Banach space, and a locally Lipschitz function $F : Y \to R$ from Y into R, such that

$$f(v) = F(Lv) \quad v \in V. \tag{4.3.9}$$

On the assumption that $g \in V^*$ we shall deal with the following hemivariational inequality. Find $u \in V$ such that

$$\langle Au - g, v - u \rangle_V + F^0(Lu, Lv - Lu) \geq 0 \quad \forall v \in V. \tag{4.3.10}$$

The theorem below provides some conditions which guarantee the existence of solutions to (4.3.10).

Theorem 4.23 Let $A : V \to V^*$ be a pseudo-monotone operator from a reflexive Banach space V into V^*, $F : Y \to R$ a locally Lipschitz function from a Banach space Y into R and let $L : V \to Y$ be linear, compact operator from V into Y. Suppose that there exists a function $c : R^+ \to R$ with $c(r) \to \infty$ as $r \to \infty$, such that for all $v \in V$, $\langle Av, v - u_0 \rangle_V \geq c(||v||)||v||$ for some $u_0 \in V$, and that

$$F^0(Lv, Lu_0 - Lv) \leq k(1 + ||v||_V) \quad \forall v \in V \tag{4.3.11}$$

with a constant $k \geq 0$. Then the hemivariational inequality (4.3.10) has at least one solution.

Proof. Define a multivalued mapping $\partial^L F : V \to 2^{V^*}$ by the formula

$$\partial^L F(u) = \{u^* \in V : F^0(Lu, Lv) \geq \langle u^*, v \rangle_V \quad \forall v \in V\}. \tag{4.3.12}$$

Then the problem (4.3.10) can be written as $g \in Au + \partial^L Fu$. Thus to obtain the result it remains to examine the range property of the multivalued mapping $A + \partial^L F$. For this purpose we check whether $\partial^L F$ is pseudo-monotone. It is clear that in order to prove the pseudo-monotonicity of $\partial^L F$ it is enough to check that $\partial^L F$ is generalized pseudo-monotone. Let us choose a sequence $\{u_i\}$ in V converging weakly to u, for which there exists $u_i^* \in \partial^L Fu_i$ with $u_i^* \xrightarrow{w} u^*$ for some $u^* \in V$ and such that $\limsup \langle u_i^*, u_i - u \rangle_V \leq 0$. Taking into account the strong convergence of $\{Lu_i\}$ to Lu in Y and the Lipschitz property of F, we obtain immediately from the condition

$$F^0(Lu_i, Lv) \geq \langle u_i^*, v \rangle_V \tag{4.3.13}$$

that

$$|\langle u_i^*, u_i - u \rangle_V| \leq K||Lu_i - Lu||_Y. \tag{4.3.14}$$

Consequently,

$$\lim \langle u_i^*, u_i - u \rangle_V = 0. \tag{4.3.15}$$

Finally, it remains to verify that $u^* \in \partial^L F(u)$. The upper semicontinuity of the function

$$Y \times Y \ni (y, z) \to F^0(y, z) \tag{4.3.16}$$

implies that

$$\limsup F^0(Lu_i, Lv - Lu_i) \leq F^0(Lu, Lv - Lu) \quad \forall v \in V. \tag{4.3.17}$$

Combining the foregoing properties with

$$F^0(Lu_i, Lv - Lu_i) \geq \langle u_i^*, v - u_i \rangle_V \quad \forall v \in V, \tag{4.3.18}$$

and letting i tend to infinity in (4.3.18) we deduce easily that

$$F^0(Lu, Lv - Lu) \geq \langle u^*, v - u \rangle_V \quad \forall v \in V, \tag{4.3.19}$$

which is equivalent to $u^* \in \partial^L Fu$. Thus the generalized pseudo-monotonicity of $\partial^L F$ is established. Hence, by Proposition 2.2 the pseudo-monotonicity of $\partial^L F$ follows. Consequently, due to Proposition 2.4, $A + \partial^L F$ is pseudo-monotone. The coercivity of $A + \partial^L F$ is guaranteed by (4.3.11). Accordingly, Range $(A + \partial^L F) = V^*$ (Theorem 2.6). This completes the proof.   q.e.d.

Remark 4.24 Let us observe that $\partial^L F(v) \supset \partial(F \circ L)(v)$ for each $v \in V$, where $\partial(F \circ L)$ stands for the generalized gradient of the composition $F \circ L : V \to R$. Indeed, this fact follows directly from the inequality

$$F^0(Lu, Lv) \geq (F \circ L)^0(u, v) \quad u, v \in V. \tag{4.3.20}$$

Accordingly, each solution of (4.3.1) is also a solution of (4.3.10) with $f = F \circ L$. If F or $-F$ is ∂-regular, the foregoing inequality holds as an equality. In such a case $\partial^L F$ and $\partial(F \circ L)$ coincide and u is a solution of (4.3.10), if and only if it is a solution of (4.3.1).

The results obtained in this Section can be applied to hemivariational inequalities involving nonconvex functionals expressed by means of integrals.

Let V be a reflexive Banach space and let $\Omega \subset R^n$ be an open bounded subset of $R^n, n \geq 1$. It will be supposed that the injection $V \subset L^p(\Omega; R^N), N \geq 1$, is compact for some $1 < p < \infty$, and that $A : V \to V^*$ is a pseudo-monotone coercive operator from V into V^*. Let $j : \Omega \times R^N \to R$ be a function defined for almost all $x \in \Omega$ and all $\xi \in R^N$, satisfying the following conditions:

(i) For all $\xi \in R^N$ the function

$$\Omega \ni x \to j(x, \xi) \tag{4.3.21}$$

(as a function of the variable x) is measurable on Ω.

(ii) For almost all $x \in \Omega$ the function

$$R^N \ni \xi \to j(x, \xi) \tag{4.3.22}$$

(as a function of the variable ξ) is locally Lipschitz on R^N.

(iii) The function $j(\cdot, 0)$ is finitely integrable in Ω, i.e. $j(\cdot, 0) \in L^1(\Omega)$.

For a given $g \in V^*$ let us consider the following hemivariational inequality: Find $u \in V$ such that

$$\langle Au - g, v - u \rangle_V + \int_\Omega j^0(u, v - u) d\Omega \geq 0 \quad \forall v \in V. \tag{4.3.23}$$

In order to apply the results of this Section certain conditions guaranteeing that the functional

$$J(v) = \int_\Omega j(v)d\Omega \quad v \in L^p(\Omega; R^N) \tag{4.3.24}$$

will be locally Lipschitz on $L^p(\Omega; R^N)$ are needed. With respect to this the following growth condition is introduced:

(iv) For almost all $x \in \Omega$, and each $\xi \in R^N$,

$$\eta \in \partial j(x, \xi) \Longrightarrow |\eta| \le c(1 + |\xi|^{p-1}) \tag{4.3.25}$$

for some constant $c \ge 0$ not depending on $x \in \Omega$.

As it is well known [Aub79, Clar83], the conditions (i)–(iv) imply that J is locally Lipschitz on $L^p(\Omega; R^N)$, (if fact, Lipschitz continuous on each bounded subset of $L^p(\Omega; R^N)$) and that for each $u \in L^p(\Omega; R^N)$ and $\chi \in L^q(\Omega; R^N)$,

$$\chi \in \partial J(u) \Longrightarrow \chi(x) \in \partial j(x, u(x)) \text{ for a.e. } x \in \Omega, \tag{4.3.26}$$

where $1/p + 1/q = 1$. Now we are in a position to formulate the following existence results for hemivariational inequality (4.3.23).

Theorem 4.25 Let A be a pseudo-monotone operator from a reflexive Banach space V into $V^\star$. Suppose that there exists a function $c : R^+ \to R$ with $c(r) \to \infty$ as $r \to \infty$, such that for all $v \in V$, $\langle Av, v \rangle_V \ge c(\|v\|_V)\|v\|_V$, that the injection $V \subset L^p(\Omega; R^N), N \ge 1$, is compact for some $1 < p < \infty$, and that $j : \Omega \times R^N \to R$ fulfills the requirements (i)–(iv). Moreover, let us assume that

$$j^0(x; \xi, -\xi) \le \alpha(x)|\xi| \tag{4.3.27}$$

for some nonnegative function $\alpha \in L^q(\Omega)$ with $1/p + 1/q = 1$. Then the hemivariational inequality (4.3.23) has at least one solution.

Proof. Let us denote by $i : V \to L^p(\Omega; R^N)$ the compact injection of V into $L^p(\Omega; R^N)$ and define the mapping $\partial^i J : V \to 2^{V\star}$ by

$$\partial^i J(u) = \{u^\star \in V^\star : J^0(iu; iv) \ge \langle u^\star, v \rangle_V \quad \forall v \in V\}, \tag{4.3.28}$$

where

$$J^0(y, z) = \limsup_{\substack{h \to 0 \\ \lambda \to 0+}} \frac{J(y + h + \lambda z) - J(y + h)}{\lambda}, \quad y, z \in L^p(\Omega; R^N). \tag{4.3.29}$$

Here $h \to 0$ in the L^p-norm. From (4.3.27) and Fatou's lemma it results

$$J^0(iu; -iu) \le \int_\Omega j^0(u; -u)d\Omega \le \|\alpha\|_{L^q(\Omega)}\|u\|_{L^p(\Omega)} \le c\|\alpha\|_{L^q(\Omega)}\|u\|_V \quad \forall u \in V. \tag{4.3.30}$$

Thus (4.3.11) holds with $u_0 = 0$ and $k = c\|\alpha\|_{L^q(\Omega)}$, and according to Theorem 4.23 there exists a solution to hemivariational inequality

$$\langle Au - g, v - u \rangle_V + J^0(iu; iv - iu) \geq 0 \quad \forall v \in V. \qquad (4.3.31)$$

At the same time, Fatou's lemma yields that

$$J^0(iu; iv - iu) \leq \int_\Omega j^0(u; v - u) d\Omega, \quad \forall u, v \in V, \qquad (4.3.32)$$

Accordingly each solution of (4.3.31) is also solution of (4.3.23) and thus the proof of Theorem 4.25 is complete. q.e.d.

If operator A is pseudo-monotone and coercive with the coercivity function $c(r) \approx r$ as $r \to \infty$, and $p \geq 2$, then the condition (4.3.27) can be replaced by a weaker one, namely

$$j^0(x; \xi, -\xi) \leq \beta(x)(1 + |\xi|^s), \qquad (4.3.33)$$

where $0 \leq s < 2$ and $\beta(\cdot)$ is a nonnegative function from $L^{\bar{q}}(\Omega)$ with $\bar{q} = p/(p - s)$. In such a case the coercivity of $A + \partial^i J$ again easily follows and we arrive at the following result.

Theorem 4.26 Let A be a pseudo-monotone operator from a reflexive Banach space V into V^*. Let us also assume that A is coercive with the coercivity function $c(r) \approx r$ as $r \to \infty$. Moreover, we suppose that the injection $V \subset L^p(\Omega; R^N), N \geq 1$, is compact for some $2 \leq p < \infty$, and that $j : \Omega \times R^N \to R$ fulfills the requirements (i)–(iv) together with (4.3.33). Then the hemivariational inequality (4.3.23) has at least one solution.

In Chapter 5 it will be shown that the condition (iv) ensuring the local Lipschitz continuity of J can be made weaker under some additional requirements concerning A. This condition can be replaced by a certain directional growth condition which does not guarantee that J is locally Lipschitz and even finite on V; in spite of this the existence of a solution for the corresponding hemivariational inequalities is ensured.

4.4 Variational-Hemivariational Inequalities

The results obtained in this Chapter combined with Theorem 2.12 concerning the range of the sum of pseudo-monotone and maximal monotone mappings, permit the formulation of some existence theorems for variational-hemivariational inequalities. For the reader's convenience let us recall that a multivalued mapping $T : V \to 2^{V^*}$ is said to be quasi-bounded, if for each $M > 0$ there exists $K(M) > 0$ such that, whenever (u^*, u) lies on the graph $G(T)$ of T and

$$\langle v^*, v \rangle_V \leq M \|v\|_V, \quad \|v\|_V \leq M, \qquad (4.4.1)$$

then

$$\|v^*\|_{V^*} \leq K(M). \qquad (4.4.2)$$

T is said to be strongly quasi-bounded if for each $M > 0$ there exists $K(M) > 0$ such that for all $(v^*, v) \in G(T)$ with

$$\langle v^*, v \rangle_V \leq M, \quad ||v||_V \leq M, \tag{4.4.3}$$

we have

$$||v^*||_{V^*} \leq K(M). \tag{4.4.4}$$

The following result can be formulated.

Theorem 4.27 Let $A : V \to V^*$ be a pseudo-monotone operator from a reflexive Banach space V into V^*, $f : V \to R$ a locally Lipschitz function from V into R and let φ be a convex, lower semicontinuous, proper function from V into $\bar{R} = R \cup \{+\infty\}$. Suppose that the following hypotheses hold:

(H1) either $f \in PM(V)$, or $f \in QPM(V)$ and A satisfies the $(S)_+$ condition;

(H2) $u_0 \in D(\partial\varphi)$;

(H3) either A_{u_0} is quasi-bounded, or $\partial\varphi_{u_0}$ is strongly quasi-bounded;

(H4) There exists a function $c : R^+ \to R$ with $c(r) \to \infty$ as $r \to \infty$, such that for all $v \in V$ and $v^* \in \partial f(v)$, $\langle Av + v^*, v - u_0 \rangle_V \geq c(||v||)||v||_V$.

Then the variational-hemivariational inequality: For a given $g \in V^*$ find $u \in V$ such that

$$\langle Au - g, v - u \rangle_V + \varphi(v) - \varphi(u) + f^0(u, v - u) \geq 0 \quad \forall v \in V; \tag{4.4.5}$$

has at least one solution.

Proof. The assumptions made imply that according to the results obtained previously $A + \partial f$ is pseudo-monotone. Since $A + \partial f$ is coercive, so is $A + \partial f + \partial\varphi$ because of the existence of an affine minorant for φ. Theorem 2.12 implies that the range of $A + \partial f + \partial\varphi$ coincides with the whole V^*. This establishes the existence of a solution of (4.4.5).   q.e.d.

Another consequence of our results can be formulated in the form of the following theorem.

Theorem 4.28 Let $A : V \to V^*$ be a pseudo-monotone operator from a reflexive Banach space V into V^*, $F : Y \to R$ a locally Lipschitz function from a Banach space Y into R, and let $L : V \to Y$ be a linear compact operator from V into Y. Further, assume that $\varphi : V \to \bar{R} = R \cup \{+\infty\}$ is convex lower semicontinuous and proper. Suppose that the following hypotheses hold:

(H1) $u_0 \in D(\partial\varphi)$;

(H2) either A_{u_0} is quasi-bounded, or $\partial\varphi_{u_0}$ is strongly quasi-bounded;

(H3) There exists a function $c : R^+ \to R$ with $c(r) \to \infty$ as $r \to \infty$, such that for all $v \in V$ and $v^* \in \partial^L F(v)$, $\langle Av + v^*, v - u_0 \rangle_V \geq c(||v||)||v||_V$.

Then the variational-hemivariational inequality problem: for a given $g \in V^*$ find $u \in V$ such that

$$\langle Au - g, v - u \rangle_V + \varphi(v) - \varphi(u) + F^0(Lu, Lv - Lu) \geq 0 \quad \forall v \in V, \qquad (4.4.6)$$

has at least one solution.

Remark 4.29 It is clear that each solution of the variational-hemivariational inequality (4.4.5) with $f = F \circ L$ is a solution of (4.4.6). If, additionally, $\partial(F \circ L) = \partial^L F$, then u is a solution of (4.4.5), if and only if it is a solution of (4.4.6). Such a situation takes a place when the equality $F^0(Lu, Lv) = (F \circ L)^0(u, v)$ is valid for any $u, v \in V$, for instance.

4.5 Quasi-Hemivariational Inequalities

The theory of pseudo-monotone mappings leads to the formulation of existence results for nonlinear problems in which nonlinearities are determined by mappings of the form $h\partial f$. In such a case we cannot deal with hemivariational inequalities because there is not, in general, a function G with $\partial G = h\partial f$. The problem we are going to investigate consists in proving the existence of a solution to the following quasi-hemivariational inequality

$$\langle Au - g, v - u \rangle_V + h(u)f^0(u, v - u) \geq 0 \quad \forall v \in V. \qquad (4.5.1)$$

By making use of the results obtained we are able to give conditions which guarantee the existence of solutions to this nonlinear problem.

Theorem 4.30 Let $A : V \to V^*$ be a pseudo-monotone operator from a reflexive Banach space V into V^*, and let $h : V \to R$ be a continuous nonnegative function from V into R. Suppose that the following hypotheses hold:

(H1) either $f \in QPM(V)$, A satisfies the $(S)_+$ condition and h maps bounded sets into bounded sets, or $f \in PM(V)$ and h is weakly continuous;

(H2) There exist a function $c : R^+ \to R$ with $c(r) \to \infty$ as $r \to \infty$, and $u_0 \in V$, such that for all $v \in V$, $\langle Av, v - u_0 \rangle_V \geq c(\|v\|_V)\|v\|_V$;

(H3) The estimate

$$h(v)f^0(v, u_0 - v) \leq k(1 + \|v\|_V) \quad \forall v \in V, k = \text{const}, \qquad (4.5.2)$$

holds.

Then the quasi-hemivariational inequality: For a given $g \in V^*$ find $u \in V$ such that

$$\langle Au - g, v - u \rangle_V + h(u)f^0(u, v - u) \geq 0 \quad \forall v \in V; \qquad (4.5.3)$$

has at least one solution.

Proof. We shall prove that $A + h\partial f$ is pseudo-monotone. Let $\{u_i\}$ be a sequence converging weakly to u, for which there exists $\{Au_i + h(u_i)u_i^*\}$, with $u_i^* \in \partial f(u_i)$, such that

$$\limsup\langle Au_i + h(u_i)u_i^*, u_i - u\rangle_V \leq 0. \qquad (4.5.4)$$

Due to the boundedness of $\{u_i^*\}$ we may choose a subsequence (for the sake of simplicity we keep the same symbol for it) such that $\{u_i^*\}$ converges weakly to some u^* in V^*. We can also suppose that $h(u_i) \to \alpha$ for some real number $\alpha \geq 0$ ((H1) ensures the boundedness of $\{h(u_i)\}$). Along the lines of the previous proofs we deduce from (4.5.4) the inequalities:

$$\limsup\langle Au_i, u_i - u\rangle_V \leq 0 \text{ and } \limsup\langle h(u_i)u_i^*, u_i - u\rangle \leq 0. \qquad (4.5.5)$$

From the pseudo-monotonicity of A it results that $\liminf\langle Au_i, u_i - v\rangle_V \geq \langle Au, u - v\rangle_V$ for any $v \in V$. By making use of the second inequality, in both cases $f \in QPM(V)$ and $f \in PM(V)$, we obtain easily that

$$\lim\langle h(u)u_i^*, u_i - u\rangle_V = 0. \qquad (4.5.6)$$

Now let us consider the case in which A satisfies the $(S)_+$ condition. Then the strong convergence $u_i \to u$ in V follows. Since $f \in QPM(V)$ and h is continuous in V, we arrive at $h(u_i)u_i^* \to h(u)u^*$ weakly in V^*, with $u^* \in \partial f(u)$. Thus

$$\liminf\langle Au_i + h(u_i)u_i^*, u_i - v\rangle_V \geq \langle Au + h(u)u^*, u - v\rangle_V \qquad (4.5.7)$$

for any $v \in V$.

Now we suppose that $f \in PM(V)$ and h is weakly continuous. If $\alpha > 0$, then (4.5.6) yields $\lim\langle u_i^*, u_i - u\rangle_V = 0$ and from the pseudo-monotonicity of ∂f, $u^* \in \partial f(u)$. Further, the weak continuity of h leads to $\alpha = h(u)$ and we obtain (4.5.7). In the case $\alpha = 0$ we have that $h(u) = 0$ and (4.5.7) follows again. Finally, the pseudo-monotonicity of the sum $A + h\partial f$ has been established. The coercivity of $A + h\partial f$ follows from (4.5.2). Finally we can invoke Theorem 2.12 to complete the proof.   q.e.d.

Another result can be formulated for quasi-hemivariational inequalities involving compositions of locally Lipschitz functions with linear compact operators.

Theorem 4.31 Let $A : V \to V^*$ be a pseudo-monotone operator from a reflexive Banach space V into V^*, $F : Y \to R$ a locally Lipschitz function from a Banach space Y into R, and let $L : V \to Y$ be a linear compact operator from V into Y. Moreover, we assume that $h : V \to R$ is a continuous nonnegative function from V into R. Suppose that the following conditions hold:

(H1)   either A satisfies the $(S)_+$ condition and h maps bounded sets into bounded sets; or h is weakly continuous;

(H2)   There exists a function $c : R^+ \to R$ with $c(r) \to \infty$ as $r \to \infty$, and $u_0 \in V$, such that for all $v \in V, \langle Av, v - u_0 \rangle_V \geq c(||v||_V)||v||_V$;

(H3)   The estimate

$$h(v)F^0(Lv, Lu_0 - Lv) \leq k(1 + ||v||_V) \quad \forall v \in V, k = \text{const}, \quad (4.5.8)$$

is fulfilled.

Then the quasi-hemivariational inequality: for a given $g \in V^*$ find $u \in V$ such that

$$\langle Au - g, v - u \rangle_V + h(u)F^0(Lu, Lv - Lu) \geq 0 \quad \forall v \in V, \quad (4.5.9)$$

has at least one solution.

Because of the similarity with previous cases the proof of this theorem will be omitted here.

Remark 4.32 Let us consider the case in which the function $h : V \to R^+$ in Theorem 4.30 is assumed to take the form

$$h(u) = \gamma(f(u)) \quad u \in V, \quad (4.5.10)$$

where $\gamma : R \to R^+$ is a continuous nonnegative function on R. Then in the hypothesis (H1) of this theorem the condition "$f \in PM(V)$ and h is weakly continuous" can be replaced by "$f \in PM(V)$ and f satisfies the (C) property." Moreover, the condition "h maps bounded sets into bounded sets" can be ommited. Analogously, in Theorem 4.31, if

$$h(u) = \gamma(F(Lu)) \quad u \in V, \quad (4.5.11)$$

in the hypothesis (H1) the condition: "h is weakly continuous" can be omitted.

From Theorem 4.30 and Remark 4.32 we obtain immediately the following results.

Corollary 4.33 Let $A : V \to V^*$ be a pseudo-monotone operator from a reflexive Banach space V into V^*, $f : V \to R$ a locally Lipschitz function from V into R and let $\gamma : R \to R^+$ be a nonnegative continuous function on R. Suppose that the following hypotheses hold:

(H1)   either $f \in QPM(V)$ and A satisfies the (S_+) condition, or $f \in PM(V)$ and f has the (C) property;

(H2)   There exists a function $c : R^+ \to R$ with $c(r) \to \infty$ as $r \to \infty$, and $u_0 \in V$, such that $\langle Av, v - u_0 \rangle_V \geq c(||v||_V)||v||_V \quad \forall v \in V$;

(H3) $\gamma(f(v))f^0(v; u_0 - v) \leq k(1 + ||v||_V) \quad \forall v \in V, \ k = \text{const} \geq 0$.

Then the problem: find $u \in V$ such that

$$\langle Au - g, v - u \rangle_V + \gamma(f(u))f^0(u; v - u) \geq 0 \quad \forall v \in V, \qquad (4.5.12)$$

has at least one solution.

Corollary 4.34 Let $A : V \to V$ be a pseudo-monotone operator from a reflexive Banach space V into V^*, $F : Y \to R$ a locally Lipschitz function from a Banach space Y into R. Moreover, we assume that $L : V \to Y$ is a linear compact operator from V into Y and $\gamma : R \to R^+$ is a nonnegative continuous function on R. Suppose that the following conditions hold:

(H1)   There exist a function $c : R^+ \to R$ with $c(r) \to \infty$ as $r \to \infty$, and $u_0 \in V$, such that $\langle Av, v - u_0 \rangle_V \geq c(||v||_V)||v||_V$;

(H2) $\gamma(F(Lv))F^0(Lv, Lu_0 - Lv) \leq k(1 + ||v||_V) \, \forall v \in V$, $k = \text{const} \geq 0$.

Then the quasi hemivariational inequality: find $u \in V$ such that

$$\langle Au - g, v - u \rangle_V + \gamma(F(Lu))F^0(Lu, Lv - Lu) \geq 0 \quad \forall v \in V, \qquad (4.5.13)$$

has at least one solution.

4.6 Applications to Mechanics and Engineering

In this Section certain applications of the foregoing theory to problems arising in Mechanics and Engineering will be given. We should point out here that the theory of this Chapter and of the next ones permits the study of hemivariational inequalities related to multidimensional nonconvex superpotential laws (e.g. the law (1.4.24) or (1.4.25)), whereas in the previous Chapter we deal with one-dimensional nonconvex superpotential laws (e.g. the law (1.4.23)). One very important application of the theory of this Chapter is that it permits the formulation of the three-dimensional analoga of the one-dimensional nonmonotone possibly multivalued laws depicted in Sect. 1.4 and the study of the corresponding hemivariational inequalities which was not possible by the method of Chapter 3. Concerning the notations we refer to Sect. 1.4.

4.6.1   Two- and Three-dimensional Nonconvex Superpotential Laws

i) Let us begin with a simple case. We consider the nonmonotone one-dimensional friction law of Fig. 4.6.1 which holds at the boundary Γ of a body $\Omega \subset R^2$ (i.e. a plane body). If $\Omega \subset R^3$ then the friction law is two-dimensional in a local coordinate system on the tangential plane to each point of Γ, or three-dimensional in the global orthogonal cartesian coordinate system $Ox_1x_2x_3$ of R^3. In the first case it relates $\{S_{T_\alpha}\}$ with $\{u_{T_\alpha}\}$, where $\alpha = 1, 2$ denotes the local coordinates, and in the second case it relates $\{S_{T_i}\}$ with $\{u_{T_i}\}$, where $i = 1, 2, 3$ denotes the global coordinates. Therefore we must consider a law of the form

$$-S_T \in \partial j_T(u_T), \qquad (4.6.1)$$

where $S_T = \{S_{T_i}\}$ or $\{S_{T_\alpha}\}$, $u_T = \{u_{T_i}\}$ or $\{u_{T_\alpha}\}$ and j_T is a nonconvex super-potential. Obviously j_T must have the form (cf. Fig. (4.6.1b))

$$j_T(u_T) = \max\left(f_1(u_{T_1}, u_{T_2}, u_{T_3}), f_2(u_{T_1}, u_{T_2}, u_{T_3})\right), \qquad (4.6.2)$$

where f_1 and f_2 do not have a common tangent plane at zero. If moreover f_1, f_2 are quadratic functions then they correspond completely to the linear decreasing branches AB and A′B′. Note that $\alpha = \mu|S_N|$, where μ is the friction coefficient, if one wants to "imitate" a Coulomb friction law. Moreover in dynamic problems u_T will be replaced by the tangential velocity v_T. Obviously for f_1 and f_2 quadratic, j_T is locally Lipschitz and fulfills the growth assumption (4.3.25) and (4.3.33).

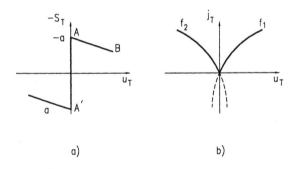

a)   b)

Fig. 4.6.1. On a nonmonotone friction law

ii) Analogously we proceed in the friction law of Fig. 4.6.2 with a local locking effect.

In the general case we can write it in the form (4.6.1) with

$$j_T(u_T) = \max\left(\alpha|u_T|, f_1(u_{T_1}, u_{T_2}, u_{T_3}), f_2(u_{T_1}, u_{T_2}, u_{T_3})\right) \qquad (4.6.3)$$

Again for f_1 and f_2 quadratic, j_T is locally Lipschitz and satisfies the growth assumptions (4.3.25) and (4.3.33). Let us now assume that b is small enough and let us appropriately choose f_1 and f_2 is order to reproduce in three dimensions the law of Fig. 4.6.3. In this case we may pose

$$j_T(u_T) = \min(a|u_T|, c|u_T| + d) \qquad (4.6.3a)$$

where a, c and d may be appropriately chosen. Again j_T is locally Lipschitz and fulfills (4.3.25) and (4.3.33).

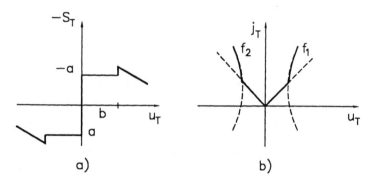

Fig. 4.6.2. Another nonmonotone interface friction law

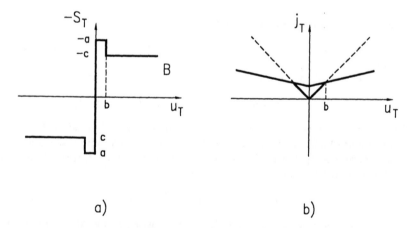

Fig. 4.6.3. Friction law with different adhesive and sliding friction thresholds

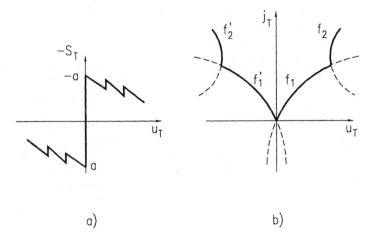

Fig. 4.6.4. A zig-zag friction law

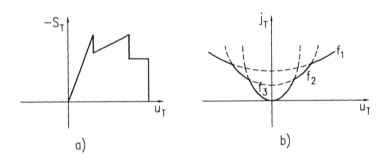

Fig. 4.6.5. Friction laws without adhesion area

iii) Analogously we may work with the three-dimensional generalization of the law of Fig. 4.6.4a. In this case we may pose that

$$j_T(u_T) = \max(f_1, f_2, f_1', f_2'), \tag{4.6.4}$$

where f_1, f_2, f_1', f_2' are quadratic and f_1 and f_1' are in Fig. 4.6.1 functions. Obviously j_T is locally Lipschitz and satisfies the growth assumptions (4.3.25) and (4.3.33). Analogously we may describe the support action where adhesion does not occur. For instance, the law of Fig.4.6.5a can be written in the three dimensions by setting

$$j_T(u_T) = \min(f_1, f_2, f_3), \tag{4.6.5}$$

where f_1, f_2, f_3 are quadratic functions of the three coordinates $u_{T_1}, u_{T_2}, u_{T_3}$. We recall (see Sect. 1.4) that we may add to (4.6.1) the expression $\partial I_K(u_T)$, where K is a convex closed subset of $u_T = \{u_{T_i}\}$, or the expression $\partial f_i(u_{T_i})$, $i = 1, 2, 3$, as in (1.4.35). In the first case we have infinite branches in the law, which lead to variational-hemivariational inequalities, and in the second fuzzy effects.

Obviously j_T in (4.6.5) is locally Lipschitz and satisfies the growth assumption (4.3.25) and (4.3.33). The same is true if fuzzy effects are considered.

iv) We shall give now some three-dimensional extensions of the adhesive contact law. In the one dimension, i.e. between S_N and u_N this law is depicted in Fig. 4.6.6

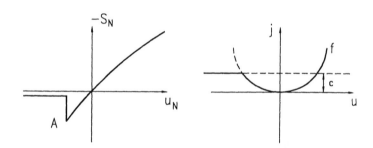

Fig. 4.6.6. The adhesive contact law

and in the three-dimensional case it is written as (1.4.25) with

$$j(u) = \min(f, c). \tag{4.6.6}$$

Here f is a convex function if AB is monotone increasing, and c is a constant. Obviously j in (4.6.6) satisfies the condition (4.3.25) and is locally Lipschitz, as well as the condition (4.3.33).

4.6.2 The General Adhesive Contact and Friction Problem

Let us consider a linear elastic body $\Omega \subset R^3$ with a Lipschitzian boundary Γ. We pose ourselves in the framework of (1.4.38) and we assume that on $\Gamma_S \subset \Gamma$ the general adhesive contact boundary condition

$$-S \in \partial j(x, u), \quad S(x) \in R^3, \quad u(x) \in R^3, \quad x \in \Gamma_S \qquad (4.6.7)$$

holds, whereas on $\Gamma_F \subset \Gamma$ the boundary forces are given, i.e.

$$S_i = F_i, \quad F_i = F_i(x), \quad i = 1, 2, 3 \text{ on } \Gamma_F. \qquad (4.6.8)$$

Finally we assume that on the rest part Γ_U of the boundary the displacements are zero i.e.,

$$u_i = 0 \text{ on } \Gamma_U. \qquad (4.6.9)$$

We introduce the admissible subset for the displacements

$$V = \{v : v = \{v_i\}, \ i = 1, 2, 3, \quad v_i \in H^1(\Omega), \quad v_i = 0 \text{ on } \Gamma_U\} \qquad (4.6.10)$$

and analogously to (1.4.49) we may formulate the following problem: find $u \in V$ such that

$$a(u, v - u) + \int_{\Gamma_S} j^0(x, u, v - u) d\Gamma \geq (l, v - u), \quad \forall v \in V, \qquad (4.6.11)$$

where

$$(l, v) = \int_{\Omega} f_i v_i d\Omega + \int_{\Gamma_F} F_i v_i d\Gamma, \qquad (4.6.12)$$

$a(\cdot, \cdot)$ is given by (1.4.43) and $C_{ijhk} \in L^\infty(\Omega)$. We recall that the trace operator $v \to v|_\Gamma$ is a linear continuous operator from $[H^1(\Omega)]^3$ to $[H^{1/2}(\Gamma)]^3$ that the imbedding $H^{1/2}(\Gamma) \subset L^2(\Gamma)$ is compact, and that $a(\cdot, \cdot)$ is coercive and strongly monotone on V. On the assumption that $f_i \in L^2(\Omega)$, $F_i \in L^2(\Gamma)$ and that $j : \Gamma_S \times R^3 \to R$ is a function satisfying i), ii), iii) and iv) in Remark 4.24 (Ω is replaced by Γ_S) we may apply Theorem 4.26 and we obtain that if j fulfills a condition of the type (4.3.33), then the hemivariational inequality (4.6.11) has at least one solution.

We may proceed analogously if (4.6.7) is replaced by the general friction condition with given normal force

$$-S_T \in \partial j_T(x, u_T), \quad S_T(x) \in R^3, \quad u_T(x) \in R^3, \quad x \in \Gamma_S \qquad (4.6.13)$$

$$S_N = C_N, \quad C_N(x) \in R \quad x \in \Gamma_S. \qquad (4.6.14)$$

Then we have a hemivariational inequality similar to (4.6.12) but with $\int_{\Gamma_S} j^0(x, u, v - u) d\Gamma$ replaced by $\int_{\Gamma_S} j^0(x, u_T, v_T - u_T) d\Gamma$ and

$$(l, v) = \int_{\Omega} f_i v_i d\Omega + \int_{\Gamma_F} F_i v_i d\Gamma + \int_{\Gamma_S} C_N v_N d\Gamma. \qquad (4.6.15)$$

On the same assumptions as previously we can conclude that this hemivariational inequality has at least one solution. The corresponding semicoercive cases are still open problems.

In most engineering problems the normal forces are not prescribed as in (4.6.14) but they obey the Signorini boundary condition (1.4.11). They read

$$\text{if } u_N < 0 \text{ then } S_N = 0 \text{ and } S_T = 0 \tag{4.6.16}$$

$$\text{if } u_N = 0 \text{ then } S_N \leq 0 \text{ and } (4.6.13) \text{ holds.} \tag{4.6.17}$$

In (4.6.13) j_T depends on S_N. Note that this also is a still open problem although methods have been already developed for its efficient numerical treatment [Pan93, Mis92,93]. These methods are based on a scheme developed for the classical Coulomb friction problem by the second author (cf. [Pan75] and [Ka88,90]). In the first step we assume that S_N has a given value and we solve numerically the arising hemivariational inequality for the tangential contact problem. Its solution supplies a value for S_T which is used in the second step for the solution of the variational inequality arising when the Signorini boundary conditions hold in the normal direction and the tangential forces S_T have the value obtained in the previous step. The normal contact problem supplies a new value for the normal forces which is used for the solution of the tangential contact problem and so on until the differences $|S_N^{j+1} - S_N^j|$ and $|S_{T_i}^{j+1} - S_{T_i}^j|$ become appropriately small. This algorithm (called PANA for the Coulomb problem see [Ka88,90]) is a fixed point type algorithm. Analogously one should probably proceed for the existence proof of the B.V.P. defined via (4.6.16) and (4.6.17). In this respect one should also compare the existence proof given for the classical Coulomb friction problem coupled with the Signorini boundary condition [Neč, Ja83,84].

4.6.3 On the Fuzzy Friction and Adhesive Contact Problem. The Case of Locking Support

In Sect. 1.4 we have introduced nonmonotone possibly multivalued laws with fuzzy regions by means of the superpotential f defined through (1.4.32)−(1.4.34). Let us assume that we deal with a friction law of the type (4.6.13), (4.6.14) which for $-\varepsilon \leq u_{T_i} \leq \varepsilon$, $i = 1, 2, 3$, ε is small and given, presents fuzzy behaviour. Then in (4.6.13) the multivalued relation should be replaced by the relation

$$-S_T \in \partial j_T(x, u_T) + \partial f(u_{T_1}) + \partial f(u_{T_2}) + \partial f(u_{T_3}). \tag{4.6.18}$$

Here f is the same as in (1.4.35) and ∂ denotes the generalized gradient. The corresponding hemivariational inequality reads: Find $u \in V$ such as to satisfy

$$a(u, v-u) + \int_{\Gamma_S} [j_T^0(x, u_T, v_T - u_T) + \sum_{i=1}^{3} f^0(u_{T_i}, v_{T_i} - u_{T_i})] d\Gamma \geq (l, v-u) \quad \forall v \in V. \tag{4.6.19}$$

Under the same assumptions for j_T as in the previous Subsection j_T and f fulfills all the assumptions of Theorem 4.26 (with Ω replaced by Γ_S) and therefore

(4.6.19) has at least one solution. Analogously we proceed for the adhesive contact superpotential.

An interesting problem results if the adhesive contact is with a locking support, e.g. a rubber support with limited compressibility [Prag57,58, Léné73,74]. In this case we assume that the displacement vector u must belong to a given convex closed set K at each point of the boundary Γ_S. Then (4.6.7) will be replaced by the law (cf. (1.4.31))

$$-S \in \partial j(x, u) + \partial I_K(u) \tag{4.6.20}$$

and we have to solve the following problem: find $u \in V$ such as to satisfy the variational-hemivariational inequality

$$a(u, v-u) + \int_{\Gamma_S} [j^0(x, u, v-u) d\Gamma + I_K(v) - I_K(u) \geq (l, v-u) \quad \forall v \in V. \tag{4.6.21}$$

Combining the proofs of Theorems 4.26 and 4.28 we directly obtain for the above variational-hemivariational inequality a result analogous to Theorem 4.26 for K convex and closed subset of V. Indeed the strongly monotone operator A corresponding to the bilinear form $a(\cdot, \cdot)$ $((Au, v) = a(u, v))$ is bounded and therefore (H2) of Theorem 4.28 is fulfilled. Then the assumptions (i)−(iv) and (4.3.33) of Theorem 4.26 concerning j guarantee the existence of at least one solution of (4.6.21).

4.6.4 Nonmonotone Multivalued Relations in Structural Analysis

Structural analysis deals with discretized models of structures obtained by means of a discretization procedure, e.g. the Finite Element Method (F.E.M.). Suppose that a structure is discretized by n-finite elements by using the method of the natural generalized stresses and strains, i.e. the strains of each element are not affected by rigid body displacements and the stresses within each element are selfequilibrated [Argy65,66,69]. We denote the stress (resp. the strain) vector of the i-element by s_i (resp. e_i) and the displacement (resp. the load) vector of the k-node by u_k (resp. p_k). Let m be the total number of nodes. We assume that the behaviour of each element contains a linear part and a strongly nonlinear part with nonmonotonicity and multivaluedness. Thus we may write that

$$\bar{e}_i = e_i - e_{i0}, \quad s_i \in \partial j_i(\bar{e}_i) + \partial \frac{1}{2} \bar{e}_i^T F_{0i}^{-1} \bar{e}_i, \quad i = 1, \dots, n, \tag{4.6.22}$$

where j_i is a locally Lipshitz superpotential, ∂ is the generalized gradient, e_{0i} is the initial strain vector and F_{0i} is the symmetric natural flexibility matrix. The nonconvex superpotential relation in (4.6.22) may describe the zig-zag behaviour of reinforced concrete in tension (Scanlon's diagram [Flo]) , or of a composite material [Ond, Hult, Schel], the semirigid connections in steel structures etc; In the one-dimensional case the law is as the one of Fig. 1.4.3d. For the whole structure (4.6.22) takes the form

$$s - F_0^{-1}(e - e_0) \in \Lambda(e - e_0), \tag{4.6.23}$$

where $e = [e_1^T, \ldots, e_n^T]$ (T denotes the transpose of a vector or a matrix), $F_0 = \text{diag}[F_{01}, \ldots, F_{0n}]$, and $\Lambda(v) = [\partial j_1(\tilde{e}_1)^T, \partial j_2(\tilde{e}_2)^T, \ldots, \partial j_n(\tilde{e}_n)^T]^T$. In order to complete the equations of the problem we must take into account the conditions of equilibrium and the compatibility conditions

$$Gs = p \tag{4.6.24}$$

and

$$e = G^T u \tag{4.6.25}$$

respectively. From (4.6.24), (4.6.25) we obtain that

$$s^T(e^* - e) = p^T(u^* - u) \quad \forall u^* \in R^l, \tag{4.6.26}$$

where e^* and u^* satisfy (4.6.25) and l is the dimension of the displacement vector u. By the definition of the generalized gradient, (4.6.23) implies that

$$\sum_{i=1}^{n} j_i^0(e_i - e_{0i}, e_i^* - e_i) \geq s^T(e^* - e) - [F_0^{-1}(e - e_0)]^T(e^* - e) \quad \forall e^* \in R^{l'}, \tag{4.6.27}$$

where l' is the dimension of the strain vector e. From (4.6.25)–(4.6.27) we obtain that the structure obeys to the following hemivariational inequality: Find $u \in R^l$ such as to satisfy the inequality

$$\sum_{i=1}^{n} j_i^0([G^T u]_i - e_{0i}, [G^T(u^* - u)]_i) + (u^T G - e_0^T)F_0^{-1}G^T(u^* - u)$$
$$\geq p^T(u^* - u) \quad \forall u^* \in R^l. \tag{4.6.28}$$

Since the lines of matrix G are by definition linearly independent the total stiffness matrix $K = GF_0^{-1}G^T$ is with F_0^{-1} strictly positive definite.

Further we assume that the superpotentials j_i, $i = 1, \ldots, n$, satisfy the relation (4.3.11). Then due to Theorem 4.23 the hemivariational inequality (4.6.28) has at least one solution.

Let us suppose further that nonmonotone, possibly multivalued laws expressed through nonconvex superpotentials hold at the boundary of the structure. We write then

$$p_j = \bar{p}_j + \bar{\bar{p}}_j, \quad -\bar{p}_j \in \partial \varphi_j(u_j) \quad j = 1, \ldots, m, \tag{4.6.29}$$

where $\bar{\bar{p}}_j$ is given, $\bar{p}_j$ is a reaction vector related to the corresponding displacement vector u_j through a nonconvex superpotential relation, ∂ is the generalized gradient and φ_j is a locally Lipschitz function satisfying the hypothesis of Theorem 4.20. From (4.6.29) and (4.6.28) we are led to the following problem: Find $u \in R^l$ such that

$$\sum_{i=1}^{n} j_i^0([G^T u]_i - e_{0i}, [G^T(u^* - u)]_i) + u^T K(u^* - u) + \sum_{j=1}^{m} \varphi_j^0(u_j, u_j^* - u_j)$$
$$\geq (p + GF_0^{-1}e_0)^T(u^* - u) \quad \forall u^* \in R^l. \tag{4.6.30}$$

Combining Theorems 4.23 and 4.20 with Remark 4.21 we obtain that (4.6.30) has at least one solution. Note that (4.6.30) is the most general expression in the Structural Analysis and contains all other expressions concerning linear problem or nonlinear monotone problems as special cases.

4.6.5 Variational-Hemivariational Inequalities in the Theory of Laminated von Kármán Plates

Here we will study the delamination effect for laminated plates undergoing large displacements (von Kármán plates). As in Sect. 3.5 the mechanical behaviour of the interlayer binding material, together with the possibility of debonding is described by a nonmonotone, possibly multivalued law connecting the interlaminar bonding forces with the corresponding relative displacements. In the present problem at the boundary of the plate monotone boundary conditions are assumed to hold, e.g. the Signorini-Fichera boundary condition or the plastic hinge boundary condition (see e.g. [Pan85]). The interlayer law (resp. the boundary law) is expressed through nonconvex (resp. convex) superpotentials leading to hemivariational (resp. variational) inequalities. Thus the whole problem gives rise, as we shall see further, to a variational-hemivariational inequality concerning the bending of the laminae and to a variational equality concerning the stretching of the plate.

Consider a laminated plate consisting of two laminae and the binding material between them (Fig 4.6.7). In the undeformed state the middle surface of lamina j occupies an open, bounded and connected subset Ω_j of R^2, referred to a fixed right-handed Cartesian coordinate system $Ox_1x_2x_3$. Let $\Gamma_j, j = 1, 2$, be the boundary of the j-th lamina. Γ_j is assumed to be appropriately regular (in general, a Lipschitz boundary $C^{0,1}$ is sufficient). Let also the interlaminar binding material occupy a subset Ω' such that $\Omega' \subset \Omega_1 \cap \Omega_2$ and $\bar{\Omega}' \cap \Gamma_1 = \emptyset$, $\bar{\Omega}' \cap \Gamma_2 = \emptyset$. We denote by $\zeta^{(j)}(x)$ the vertical deflection of the point $x \in \Omega_j$ of the j-th lamina, and by $f^{(j)} = (0, 0, f_3^{(j)}(x))$ the distributed vertical load acting on the j-th lamina. Further, let $u^{(j)} = \{u_1^{(j)}, u_2^{(j)}\}$ be the in-plane displacement of the j-th lamina. We assume that the j-th lamina has constant thickness h_j, while the interlaminar binding layer has constant thickness h. Moreover, we assume that each lamina obeys the von Kármán plate theory, i.e. it is a thin plate having large deflections. The following system of differential equations holds for von Kármán plates.

$$K_j \Delta\Delta\zeta^{(j)} - h_j(\sigma_{\alpha\beta}^{(j)}\zeta_{,\beta}^{(j)})_{,\alpha} = f^{(j)} \quad \text{in } \Omega_j, \tag{4.6.31}$$

$$\sigma_{\alpha\beta,\beta}^{(j)} = 0 \quad \text{in } \Omega_j, \tag{4.6.32}$$

and

$$\sigma_{\alpha\beta}^{(j)} = C_{\alpha\beta\gamma\delta}^{(j)}(\varepsilon_{\gamma\delta}^{(j)}(u^{(j)}) + \frac{1}{2}\zeta_{,\gamma}^{(j)}\zeta_{,\delta}^{(j)}) \quad \text{in } \Omega_j. \tag{4.6.33}$$

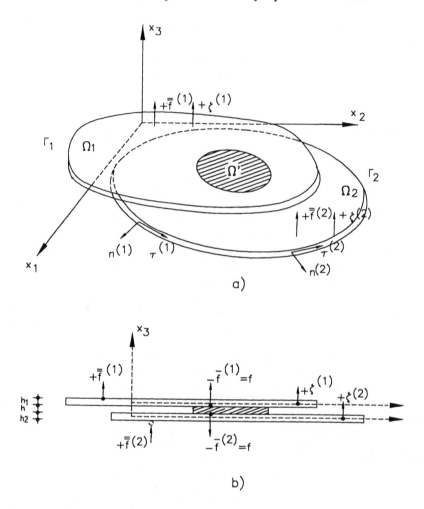

Fig. 4.6.7. The geometry of the laminated plate

Here the subscripts $\alpha, \beta, \gamma, \delta = 1, 2$ correspond to the coordinate directions in the plane of the plates; the superscript or the subscript $j = 1, 2$ refers to the j-th lamina; $\{\sigma_{\alpha\beta}^{(j)}\}, \{\varepsilon_{\alpha\beta}^{(j)}\}$, and $\{C_{\alpha\beta\gamma\delta}^{(j)}\}$ denote the stress, strain and elasticity tensors in the plane of the plate. The components of $C^{(j)}$ are elements of $L^\infty(\Omega_j)$ and have the usual symmetry and ellipticity properties (1.4.42a,b). Moreover, $K_j = Eh_j^3/12(1-\nu^2)$ is the bending rigidity of the j-th plate with E the modulus of elasticity and ν the Poisson ratio. For the sake of simplicity, we consider here isotropic homogeneous plates of constant thickness. In laminated plates, the interlaminar normal stress σ_{33} is one of the main causes for delamination effects (see e.g. [Mos] p.318). Note, however, that this is a simplification of the mechanical problem. In order to reproduce the action of σ_{33}, $f^{(j)}$ is split into $\bar{f}^{(j)}$, which describes the interaction of the two plates, and $\bar{\bar{f}}^{(j)} \in L^2(\Omega_j)$, which represents the external loading applied on the j-th plate:

$$f^{(j)} = \bar{f}^{(j)} + \bar{\bar{f}}^{(j)} \quad \text{in } \Omega_j, \quad j = 1, 2. \tag{4.6.34}$$

If f denotes the stress in the interlaminar binding layer, the following holds

$$f = \bar{f}^{(1)} = -\bar{f}^{(2)} \quad \text{in } \Omega'. \tag{4.6.35}$$

We introduce now a phenomenological law connecting f with the corresponding relative deflection of the plates $[\zeta] = \zeta^{(1)} - \zeta^{(2)}$.

We assume that (cf. also Sect. 3.5)

$$-f \in \partial J(x, [\zeta]) \quad \text{in } \Omega', \tag{4.6.36}$$

where J is locally Lipschitz function satisfying the assumptions (i)–(iv) and (4.3.27) of the Theorem 4.25. Moreover we have that

$$\bar{f}^{(j)} = 0 \quad \text{in } \Omega_j - \Omega', \quad j = 1, 2. \tag{4.6.37}$$

We assume now that the following boundary conditions hold on the subset $\tilde{\Gamma}_j$ of the plate boundaries

$$M_j(\zeta^{(j)}) \in \beta_j\left(\frac{\partial \zeta^{(j)}}{\partial n}\right) \quad \text{on } \tilde{\Gamma}_j, \quad j = 1, 2, \tag{4.6.38a}$$

$$-Q_j(\zeta^{(j)}) \in \beta_j'(\zeta^{(j)}) \quad \text{on } \tilde{\Gamma}_j, \quad j = 1, 2 \tag{4.6.38b}$$

whereas

$$\zeta^{(j)} = \frac{\partial \zeta^{(j)}}{\partial n} = 0 \quad \text{on } \Gamma_j - \tilde{\Gamma}_j, \quad j = 1, 2. \tag{4.6.38c}$$

Here $\beta_j, \beta_j', j = 1, 2$ are possibly multivalued, maximal monotone operators from R into 2^R (cf. Sect. 1.3). Accordingly, convex, l.s.c., proper functionals $\phi_j, \phi_j', j = 1, 2$ can be determined such that (cf. (1.3.4))

$$\beta_j\left(\frac{\partial \zeta^{(j)}}{\partial n}\right) = \partial \phi_j\left(\frac{\partial \zeta^{(j)}}{\partial n}\right), \quad j = 1, 2, \tag{4.6.39a}$$

and

$$\beta_j'(\zeta^{(j)}) = \partial \phi_j'(\zeta^{(j)}), \quad j = 1, 2. \tag{4.6.39b}$$

Let us introduce the Sobolev space $H^2(\Omega_j)$ for the deflections $\zeta^{(j)}$ and define the following convex, l.s.c. and proper functionals on $H^2(\Omega_j), j = 1, 2$:

$$\Phi_j(z^{(j)}) = \begin{cases} \int_{\tilde{\Gamma}_j} \phi_j\left(\frac{\partial z^{(j)}}{\partial n}\right) d\Gamma + \int_{\tilde{\Gamma}_j} \phi_j'(z^{(j)}) d\Gamma, \\ \text{if } \phi_j\left(\frac{\partial z^{(j)}}{\partial n}\right) \in L^1(\tilde{\Gamma}_j) \text{ and } \phi_j'(z^{(j)}) \in L^1(\tilde{\Gamma}_j) \\ \infty \quad \text{otherwise}, \quad j = 1, 2. \end{cases} \tag{4.6.40}$$

For the in-plane displacement we assume the boundary conditions

$$\sigma_{\alpha\beta}^{(j)} n_\beta^{(j)} = 0 \quad \text{on } \Gamma_j. \tag{4.6.41}$$

We now can derive the variational formulation of the problem. From (4.6.31), by assuming sufficiently regular functions, multiplying by $z^{(j)} - \zeta^{(j)}$, integrating and applying the Green-Gauss theorem, we obtain the expressions:

$$a(\zeta^{(j)}, z^{(j)} - \zeta^{(j)}) + \int_{\bar{\Omega}_j} h_j \sigma_{\alpha\beta}^{(j)} \zeta_{,\beta}^{(j)}(z^{(j)} - \zeta^{(j)})_{,\beta} d\Omega \tag{4.6.42}$$

$$= \int_{\Gamma_j} h_j \sigma_{\alpha\beta}^{(j)} \zeta_{,\beta}^{(j)} n_\alpha^{(j)}(z^{(j)} - \zeta^{(j)}) d\Gamma + \int_{\Omega_j} \bar{\bar{f}}^{(j)}(z^{(j)} - \zeta^{(j)}) d\Omega$$

$$+ \int_{\Gamma_j} \bar{Q}_j(\zeta^{(j)})(z^{(j)} - \zeta^{(j)}) d\Gamma - \int_{\Gamma_j} M_j(\zeta^{(j)}) \frac{\partial(z^{(j)} - \zeta^{(j)})}{\partial n^{(j)}} d\Gamma,$$

$$j = 1, 2, \quad \alpha, \beta = 1, 2.$$

Here $n^{(j)}$ denotes the outward normal unit vector to Γ_j and $a(\zeta, z)$, $M(\zeta)$, and $\bar{Q}(\zeta)$ are given by (3.5.12a,b,c). Applying the same technique to (4.6.32), implies the expression

$$\int_{\Omega_j} \sigma_{\alpha\beta}^{(j)} \varepsilon_{\alpha\beta}^{(j)}(v^{(j)} - u^{(j)}) d\Omega \tag{4.6.43}$$

$$= \int_{\Gamma_j} \sigma_{\alpha\beta}^{(j)} n_\beta^{(j)}(v_\alpha^{(j)} - u_\alpha^{(j)}) d\Gamma, \quad j = 1, 2, \ \alpha, \beta = 1, 2.$$

Further, the following notations are introduced:

$$R(m, k) = \int_\Omega C_{\alpha\beta\gamma\delta} m_{\alpha\beta} k_{\gamma\delta} d\Omega, \quad \alpha, \beta, \gamma, \delta = 1, 2 \tag{4.6.44}$$

and

$$P(\zeta, z) = \{\zeta_{,\alpha} z_{,\beta}\}, \quad P(\zeta, \zeta) = P(\zeta), \tag{4.6.45}$$

where $m = \{m_{\alpha\beta}\}$ and $k = \{k_{\alpha\beta}\}, \alpha, \beta = 1, 2$ are 2×2 tensors.

Let us also introduce a functional framework for the B.V.P. We assume that $u^{(j)}, v^{(j)} \in [H^1(\Omega_j)]^2$ and that $\zeta^{(j)}, z^{(j)} \in Z_j$, where

$$Z_j = \left\{ z : z \in H^2(\Omega_j), \quad z = 0 \text{ on } \Gamma_j - \tilde{\Gamma}_j, \quad \frac{\partial \zeta}{\partial n} = 0 \text{ a.e. } \Gamma_j - \tilde{\Gamma}_j \right\}. \tag{4.6.46}$$

Taking into account (4.6.44), (4.6.45) and the variational equalities (4.6.42), (4.6.43), the boundary conditions (4.6.38a,b,c) and (4.6.41), the interface conditions (4.6.36), (4.6.37), and the definition (4.6.40) we obtain the following problem : find $u^{(j)} \in [H^1(\Omega_j)]^2$ and $\zeta^{(j)} \in Z_j, j = 1, 2$, such as to satisfy the variational-hemivariational inequality

$$\sum_{j=1}^{2} a_j(\zeta^{(j)}, z^{(j)} - \zeta^{(j)}) + \sum_{j=1}^{2} h_j R(\varepsilon(u^{(j)}) + \frac{1}{2} P(\zeta^{(j)}), P(\zeta^{(j)}, z^{(j)} - \zeta^{(j)}))$$

$$+ \int_{\Omega'} J^0([\zeta], [z] - [\zeta]) d\Omega + \sum_{j=1}^{2} \{\Phi_j(z^{(j)}) - \Phi_j(\zeta^{(j)})\}$$

$$\geq \sum_{j=1}^{2} \int_{\Omega_j} \overline{\overline{f}}^{(j)} (z^{(j)} - \zeta^{(j)}) d\Omega, \qquad \forall z^{(1)} \in Z_1, \quad \forall z^{(2)} \in Z_2, \qquad (4.6.47)$$

and the variational equalities

$$R(\varepsilon(u^{(j)}) + \frac{1}{2} P(\zeta^{(j)}), \varepsilon(v^{(j)} - u^{(j)})) = 0, \quad \forall v^{(j)} \in [H^1(\Omega_j)]^2, \ j = 1, 2. \quad (4.6.48)$$

Analogously we may derive the variational formulation for r-laminae. Then in (4.6.47) the summation $\sum_{j=1}^{2}$ is replaced by $\sum_{j=1}^{r}$ and the term $\int_{\Omega'} J^0(\cdot, \cdot) d\Omega$ is replaced by the term $\sum_{m=1}^{m'} \int_{\Omega'} J_m^0(\cdot, \cdot) d\Omega$, where m' is the total number of interfaces, i.e. $m' = r - 1$. Then (4.6.47) must hold for every $z^{(j)} \in Z_j, \ j = 1, \ldots, r$. Several other types of boundary conditions may be considered, as it is obvious.

For instance, instead of the homogeneous boundary conditions (4.6.38c) we can have nonhomogeneous ones. They lead to the same variational formulations through appropriate translations. Moreover, instead of (4.6.38a,b) one could define nonmonotone multivalued boundary conditions between M_j and $\frac{\partial \zeta^{(j)}}{\partial n}$, and $-Q_j$ and $J^{(j)}$ expressed by the relations

$$M_j(\zeta^{(j)}) \in \partial \varphi_j(x, \frac{\partial \zeta^{(j)}}{\partial n}) \qquad (4.6.49a)$$

$$-Q_j(\zeta^{(j)}) \in \partial \varphi_j'(x, \zeta^{(j)}) \qquad (4.6.49b)$$

on $\tilde{\Gamma}_j, \ j = 1, 2$, where φ_j and φ_j' are locally Lipschitz function satisfying the assumptions (i)–(iv) and (4.3.27) of the Theorem 4.25.

We assume now that the classical boundary conditions of each plate in bending define the kinematically admissible sets Z_j which are assumed to be closed linear subspaces of $H^2(\Omega_j)$. The following problem is formulated: find $\zeta^{(j)} \in Z_j$ and $u^{(j)} \in [H^1(\Omega_j)]$, $j = 1, \ldots, r$, such as to satisfy the variational-hemivariational inequality

$$\sum_{j=1}^{r} \alpha_j(\zeta^{(j)}, z^{(j)} - \zeta^{(j)}) + \sum_{j=1}^{r} h_j R(\varepsilon(u^{(j)}) + \frac{1}{2} P(\zeta^{(j)}), P(\zeta^{(j)}, z^{(j)} - \zeta^{(j)}))$$

$$+ \sum_{m=1}^{m'} \int_{\Omega_m'} J_m^0([\zeta]^{(m)}, [z]^{(m)} - [\zeta]^{(m)}) d\Omega + \sum_{j=1}^{r} \{\Phi_j(z^{(j)}) - \Phi_j(\zeta^{(j)})\}$$

$$\geq \sum_{j=1}^{r} \int_{\Omega_j} \overline{\overline{f}}^{(j)}(z^{(j)} - \zeta^{(j)}) d\Omega, \quad \forall z^{(j)} \in Z_j, \quad j = 1, \ldots, r, \qquad (4.6.50)$$

and the variational equalities

$$R(\varepsilon(u^{(j)}) + \frac{1}{2} P(\zeta^{(j)}), \varepsilon(v^{(j)} - u^{(j)})) = 0, \qquad (4.6.51)$$

$$\forall v^{(j)} \in [H^1(\Omega_j)]^2, \ j = 1, \ldots, r.$$

Note that if instead of (4.6.38a,b), the nonconvex relations (4.6.49a,b) hold then in (4.6.50) the term $\sum_{j=1}^{r}\{\Phi_j(z^{(j)}) - \Phi_j(\zeta^{(j)})\}$ must be replaced by the expression $\sum_{j=1}^{r} \int_{\tilde{\Gamma}_j} [\varphi_j^0(x, \frac{\partial\zeta^{(j)}}{\partial n}, \frac{\partial(z^{(j)},\zeta^{(j)})}{\partial n}) + \varphi_j^{0'}(x, \zeta^{(j)}, z^{(j)} - \zeta^{(j)})]d\Gamma$.

Further we shall eliminate the in-plane displacements of the plate. To this end we note first that $R(\cdot,\cdot)$ as defined by (4.6.44) is a continuous symmetric,coercive bilinear form on $[L^2(\Omega)]^4$, and that $P : [H^2(\Omega)]^2 \rightarrow [L^2(\Omega)]^4$ of (4.6.45) is a completely continuous operator (cf. e.g. [Pan85] p.219). Thus (4.6.51) and the Lax-Milgram theorem imply that to every deflection $\zeta^{(j)} \in Z_j, j = 1, 2, \ldots, r$, there corresponds a plane displacement $u^{(j)}(\zeta^{(j)}) \in [H^2(\Omega_j)]^2$. Indeed, due to Korn's inequality [Fich72] $R(\varepsilon(u), \varepsilon(v))$ is a bilinear coercive form on the quotient space $[H^1(\Omega)]^2/\bar{R}$,where $\bar{R}$ is the space of in-plane rigid-plate displacements defined by

$$\bar{R} = \{\bar{r} : \bar{r} \in [H^1(\Omega)]^2, \quad \bar{r}_1 = \alpha_1 + bx_2, \bar{r}_2 = \alpha_2 - bx_2, \alpha_1, \alpha_2, b \in R\}. \quad (4.6.52)$$

From (4.6.51) it results (see e.g. [Pan85]) that

$$\varepsilon^{(j)}(u^{(j)}(\zeta^{(j)})) : Z_j \rightarrow [L^2(\Omega_j)]^4 \quad (4.6.53)$$

is uniquely determined and is a completely continuous quadratic function of $\zeta^{(j)}, j = 1, 2, \ldots, r$, since $\varepsilon^{(j)}(u^{(j)}(\zeta^{(j)}))$ is a linear continuous function of $P(\zeta^{(j)})$. We also introduce the completely continuous, quadratic functions $G_j : Z_j \rightarrow [L^2(\Omega_j)]^4$ which are defined by

$$\zeta^{(j)} \rightarrow G_j(\zeta^{(j)}) = \varepsilon^{(j)}(u^{(j)}(\zeta^{(j)})) + \frac{1}{2}P(\zeta^{(j)}) \quad (4.6.54)$$

and due to (4.6.48) satisfy the equations

$$R(G_j(\zeta^{(j)}), \varepsilon^{(j)}(u^{(j)}(\zeta^{(j)}))) = 0. \quad (4.6.55)$$

We now define the operators $A_j : Z_j \rightarrow Z_j^*$ and $C_j : Z_j \rightarrow Z_j^*$ (where Z_j^* is the dual space to Z_j) such that

$$\alpha(\zeta^{(j)}, z^{(j)}) = \langle A_j\zeta^{(j)}, z^{(j)}\rangle. \quad (4.6.56)$$

and

$$h_j R(G_j(\zeta^{(j)}), P(\zeta^{(j)}, z^{(j)})) = \langle C_j(\zeta^{(j)}), z^{(j)}\rangle. \quad (4.6.57)$$

where $\langle\cdot,\cdot\rangle$ denotes here the duality pairing between Z_j and Z_j^*

$$T_j = A_j + C_j. \quad (4.6.58)$$

The A_j's are continuous monotone linear operators and the C_j's are completely continuous operators. By means of (4.2.55) we get the relations

$$\langle C_j(\zeta^{(j)}), \zeta^{(j)}\rangle = h_j R(G_j(\zeta^{(j)}), 2G_j(\zeta^{(j)})) \geq 0,$$

$$\forall\zeta^{(j)} \in Z_j, \quad j = 1, 2, \ldots, r. \quad (4.6.59)$$

Moreover if $\alpha_j(\zeta^{(j)}, \zeta^{(j)})$ are coercive on Z_j, i.e. if the boundary conditions in bending prevent rigid-plate deflections (e.g. a plate is partially clamped, or has a curved boundary partly fixed with free rotation) then

$$\langle (A_j + C_j)(\zeta^{(j)}), \zeta^{(j)} \rangle = \langle T_j(\zeta^{(j)}), \zeta^{(j)} \rangle \geq c||\zeta^{(j)}||^2,$$

$$\forall \zeta^{(j)} \in Z_j, \quad j = 1, 2, \ldots, r, c = \text{const} > 0, \tag{4.6.60}$$

where $||\cdot||$ denotes the Z_j-norm. From (4.6.50), (4.6.51) we obtain the following variational problem: find $\zeta^{(j)} \in Z_j$, $j = 1, 2, \ldots, r$, such as to satisfy the variational-hemivariational inequality

$$\sum_{j=1}^{r} \langle T_j(\zeta^{(j)}), z^{(j)} - \zeta^{(j)} \rangle + \sum_{m=1}^{m'} \int_{\Omega'_m} J_m^0([\zeta]^{(m)}, [z]^{(m)} - [\zeta]^{(m)}) d\Omega \tag{4.6.61}$$

$$+ \sum_{j=1}^{r} \{ \Phi_j(z^{(j)}) - \Phi_j(\zeta^{(j)}) \} \geq \sum_{j=1}^{r} \int_{\Omega_j} \overset{=(j)}{f}(z^{(j)} - \zeta^{(j)}) d\Omega \quad \forall z^{(j)} \in Z_j.$$

Note that if (4.6.49a,b) hold then (4.6.61) is replaced by the following hemivariational inequality: find $\zeta^{(j)} \in Z_j$, $j = 1, 2, \ldots, r$ such that

$$\sum_{j=1}^{r} \langle T_j(\zeta^{(j)}), z^{(j)} - \zeta^{(j)} \rangle + \sum_{m=1}^{m'} \int_{\Omega'_m} J_m^0([\zeta]^{(m)}, [z]^{(m)} - [\zeta]^{(m)}) d\Omega$$

$$+ \sum_{j=1}^{r} \int_{\Gamma_j} [\varphi_j^0(x, \frac{\partial \zeta^{(j)}}{\partial n}, \frac{\partial (z^{(j)} - \zeta^{(j)})}{\partial n}) + \varphi_j'^0(x, \zeta^{(j)}, z^{(j)} - \zeta^{(j)})] d\Gamma$$

$$\geq \sum_{j=1}^{r} \int_{\Omega_j} \overset{=(j)}{f}(z^{(j)} - \zeta^{(j)}) d\Omega \quad \forall z^{(j)} \in Z_j. \tag{4.6.62}$$

As we have mentioned we have assumed that $J_m, m = 1, \ldots, m'$ in (4.6.61) (4.6.62), and φ_j, φ_j' $j = 1, \ldots, r$ in (4.6.62) satisfy the assumptions (i)–(iv) of Theorem 4.26 (with coercivity coefficient $c(r) = r$) and the growth assumption (4.3.27). Moreover the pseudo-monotone operator T in (4.6.61) is bounded and thus quasi-bounded. Thus Theorem 4.25 guarantees the existence of a solution to (4.6.62) and Theorem 4.28 implies the existence of a solution to (4.6.61): one can apply to Theorem 4.28 the same procedure with which Theorem 4.25 is obtained from Theorem 4.23.

4.6.6 Rigid Viscoplastic Flow Problems in Cylindrical Pipes with Adhesion or Nonmonotone Friction

In this Section we will study inequality problems formulated in the framework of rigid-viscoplastic flows, i.e. the flow of Bingham's fluids, which are subjected to general adhesion or nonmonotone friction boundary conditions. First we shall formulate the general problem and then we shall study the stationary laminar flow problem in a cylindrical pipe. The influence of thermal phenomena is neglected; we are interested only in the velocity and stress fields.

We assume, for the sake of simplicity, that the density ρ of the fluid is constant, i.e., that

$$\rho(x, t) = \rho_0 = c^{st}. \tag{4.6.63}$$

Let Ω be an open, bounded, connected subset of R^3 and Γ its boundary. Ω is referred to an orthogonal Cartesian coordinate system $0x_1x_2x_3$ which is fixed. We seek a velocity field $v = \{v_i\}$ and a stress field $\sigma = \{\sigma_{ij}\}$, $i, j = 1, 2, 3$ such as to satisfy in $\Omega \times (0, T)$ (here $(0, T)$ denotes the time interval) the equation of motion

$$\rho\gamma_i = \sigma_{ij,j} + f_i \tag{4.6.64}$$

with $\gamma_i = \partial v_i / \partial t + v_{i,j} v_j$ and $f_i = f_i(x, t)$, the equation of incompressibility

$$\operatorname{div} v = 0 \tag{4.6.65}$$

and the material law, which reads [Duv72, Pan85]

$$D_{ij} = 0 \text{ if } \sigma_{II}^{1/2} < g \tag{4.6.66a}$$

$$D_{ij} = \frac{1}{2\mu}\left(1 - \frac{g}{\sigma_{II}^{1/2}}\right)\sigma_{ij}^{D} \text{ if } \sigma_{II}^{1/2} \geq g. \tag{4.6.66b}$$

Here $D = \{D_{ij}\}$ is the rate of deformation tensor with

$$D_{ij} = \frac{1}{2}(v_{i,j} + v_{j,i}), \tag{4.6.67}$$

where the comma denotes the differentiation, μ is the viscosity coefficient and $\sigma_{II} = \frac{1}{2}\sigma_{ij}^{D}\sigma_{ij}^{D}$, where $\{\sigma_{ij}^{D}\}$ is the Cauchy stress deviator and g is the yield limit. On Γ the boundary condition

$$-S_T \in \partial j_T(x, v_T) \text{ on } \Gamma_1 \subset \Gamma, \quad v_T = 0 \text{ on } \Gamma - \Gamma_1 \tag{4.6.68a}$$

$$v_N = 0 \quad \text{on} \quad \Gamma \tag{4.6.68b}$$

is considered for every $t \in (0, T)$, where v_T (resp. v_N) denote the velocity components tangentially (resp. normally) to the boundary Γ. At $t = 0$ the initial condition

$$v_i(x, 0) = v_{0i}, \quad \text{with} \quad v_{0i} = v_{0i}(x) \tag{4.6.69}$$

holds. The viscosity coefficient μ and the yield limit g are assumed here to be positive constants. (A generalization for $\mu = \mu(x, t)$, $g = g(x, t)$ is also possible). Let us introduce the notation

$$D_{II}(v) = \frac{1}{2}D_{ij}(v)D_{ij}(v), \tag{4.6.70}$$

$$\Phi(v) = 2\int_{\Omega} g(D_{II}(v))^{1/2}d\Omega \tag{4.6.71}$$

and

$$a(v,w) = 2 \int\limits_{\Omega} \mu D_{ij}(v) D_{ij}(w) d\Omega, \quad (f,v) = \int\limits_{\Omega} f_i v_i d\Omega. \tag{4.6.72}$$

Let us denote now with V_{ad} the kinematically admissible set which contains the conditions (4.6.65) (4.6.68b) and that $v_T = 0$ on $\Gamma - \Gamma_1$ and let us assume for the moment that all functions involved in the problem are appropriately smooth. Then from (4.6.64) by multiplying it by $(v^* - v)$, $v^*, v \in V_{ad}$, integrating over Ω and applying the Green-Gauss theorem (cf. also (1.4.44)) we obtain the following variational formulation: find for every $t \in (0,T)$ a $v \in V_{ad}$ such as to satisfy the initial conditions (4.6.69) and the variational-hemivariational inequality

$$a(v, v^* - v) + \Phi(v^*) - \Phi(v) + \int\limits_{\Gamma_1} j_T^0(x, v, v^* - v) d\Gamma \geq (f - \rho\gamma, v^* - v) \quad \forall v^* \in V_{ad}. \tag{4.6.73}$$

We shall not insist here on the study of this dynamic variational-hemivariational inequality but we will apply it to formulate the corresponding stationary laminar flow problem in a cylindrical pipe.

We consider a cylindrical pipe with generators parallel to the $0x_3$-axis of the Cartesian system $0x_1x_2x_3$. The cross section of the cylinder is denoted by Σ and $\Omega = \Sigma \times [0, l]$, where l is a given length. We shall study the flow under a pressure drop, i.e.,

$$p = 0 \quad \text{at} \quad x = 0, \quad p = -p_0 l, \quad p_0 \geq 0 \quad \text{at} \quad x = l, \tag{4.6.74}$$

where p_0 is the pressure drop per unit length. If Γ is the boundary of Σ, we assume that (4.6.68a,b) hold on $\Gamma \times (0, l)$.

Moreover, let (4.6.63)−(4.6.65),(4.6.66a,b) and (4.6.69) hold. We restrict ourselves here to the class of laminar flows. Then the velocity has only one component parallel to $0x_3$, which depends only on x_1 and x_2, i.e. $v = (0, 0, v_3(t, x_1, x_2))$. It is easy to verify that $D_{31} = D_{13} = \frac{1}{2}v_{3,1}$ and $D_{23} = D_{32} = \frac{1}{2}v_{3,2}$ are the only nonzero components of the tensor D. Thus the incompressibility condition (4.6.65) is satisfied. The same also happens for the corresponding components of the stress deviator. On the assumption that $f = 0$, the equation of motion (4.6.64) becomes

$$p_{,1} = 0, \quad p_{,2} = 0, \quad \rho v_3' - (\sigma_{31,1}^D + \sigma_{32,2}^D) = p_{,3}. \tag{4.6.75}$$

But since the right-hand side of (4.6.75) depends only on x_3, whereas the left-hand side depends on x_1 and x_2, it results that $p_{,3}$ is constant. Therefore, from (4.6.74) is follows that

$$p(x_3) = -p_0 x_3. \tag{4.6.76}$$

Moreover the incompressibility condition (4.6.65) and the boundary condition (4.6.68b) are fulfilled identically due to the special form of the velocity field. Further we change the notation and we denote by v the velocity component v_3 which is nonzero. We shall seek a solution v in the subspace $V(\Sigma) = \{v : v \in H^1(\Sigma), v = 0 \text{ on } \Gamma - \Gamma_1\}$ of the Sobolev space $H^1(\Sigma)$.

We thus formulate the following problem: find $v : [0, T] \to V(\Sigma)$ such as to satisfy the dynamic variational-hemivariational inequality

$$\left(\rho v', v^* - v\right) + \mu\bar{a}(v, v^* - v) + g\bar{\bar{\Phi}}(v^*) - g\bar{\bar{\Phi}}(v) + \int_{\Gamma_1} j_T^0(x, v, v^* - v)d\Gamma$$

$$\geq (p_0, v^* - v), \quad \forall v \in V(\Sigma) \tag{4.6.77}$$

and the initial condition

$$v(0, x_1, x_2) = v_0(x_1, x_2). \tag{4.6.78}$$

In (4.6.77) we use the notations

$$\bar{a}(v, v^*) = \int_{\Sigma} v_{,i} v^*_{,i} d\Sigma, \quad (p_0, v) = \int_{\Sigma} p_0 v d\Sigma \tag{4.6.79}$$

and

$$\bar{\bar{\Phi}}(v) = \int_{\Sigma} |\mathrm{grad} v| d\Sigma. \tag{4.6.80}$$

The respective stationary problem has the form: find $u \in V(\Sigma)$ such that

$$\mu\bar{a}(v, v^* - v) + g\bar{\bar{\Phi}}(v^*) - g\bar{\bar{\Phi}}(v) + \int_{\Gamma_1} j_T^0(x, v, v^* - v)d\Gamma \geq (p_0, v^* - v) \quad \forall v \in V(\Sigma). \tag{4.6.81}$$

As in the case of the variational-hemivariational inequality (4.6.61) Theorem 4.28 yields the existence of a solution, if j_T satisfies assumptions (i)–(iv) of Theorem 4.26, the assumption (4.3.33), and since $\bar{a}(\cdot, \cdot)$ is coercive on $V(\Sigma)$.

5. Hemivariational Inequalities for Multidimensional Superpotential Law

In this Chapter we study hemivariational inequalities on vector-valued function spaces, on the assumption that certain new classes of directional growth conditions first introduced by the first author in [Nan93b] hold. These conditions, in the case of scalar-valued functions, generalize the well known sign condition (3.1.14) from Chapter 3 which has also been used for the study of nonlinear elliptic operators with strongly nonlinear perturbations [Rauch, Ken78, Web, Bréz82, Boc, Fig91, Struw, Stu]. Some classes of functions enjoying the mentioned directional growth properties will be given. Our approach is based on the Galerkin approximation combined with the fixed point theorems of Brouwer and Kakutani, the Dunford-Pettis criterion for weakly precompactness in the L^1-space and on the well known topological result concerning the family of compact sets with finite intersection property. We close this Chapter with some applications from Mechanics, Engineering and Economics.

5.1 Formulation of the Problem

Let V be a reflexive Banach space and let $\Omega \subset R^n$ be an open bounded subset of $R^n, n \geq 1$, with sufficiently smooth boundary. Throughout this Chapter we assume that

(a) The injection $V \subset L^p(\Omega; R^N), N \geq 1$, is compact for some $1 < p < \infty$;

(b) $V \cap L^\infty(\Omega; R^N)$ is dense in V.

We use the symbols, $\langle \cdot, \cdot \rangle_V, || \cdot ||_V, || \cdot ||_{L^p(\Omega)}$ to denote the pairing over $V^* \times V$, the norm on V and the norm on $L^p(\Omega; R^N)$, respectively, and we assume that $A : V \to V^*$ is a coercive and weakly continuous operator from V into V^*. For the reader's convenience let us point out the properties of A which hold throughout this section (unless otherwise indicated):

(i) A is defined on the whole space V;

(ii) If $\{u_n\}$ is a sequence converging weakly to u in V, then $\{Au_n\} \subset V^*$ converges weakly to Au in V^*;

(iii)   There exists a real-valued function $c : R^+ \to R^+$, $R^+ = \{x \in R : x \geq 0\}$, with $\lim_{r \to \infty} c(r) = +\infty$, such that for all $v \in V$,

$$\langle Av, v \rangle_V \geq c(\|v\|_V)\|v\|_V. \qquad (5.1.1)$$

Let $j : \Omega \times R^N \to R$ be a function defined for almost all $x \in \Omega$ and all $\xi \in R^N$. We impose on j the following conditions:

(i')   For all $\xi \in R^N$ the function $\Omega \ni x \to j(x, \xi)$ (as a function of the variable x) is measurable on Ω.

(ii')   For almost all $x \in \Omega$ the function $R^N \ni \xi \to j(x, \xi)$ (as a function of the variable ξ) is locally Lipschitz on R^N.

The properties (i') and (ii') ensure that j belongs to the class of functions having the Caratheodory property. In the original definition of the Caratheodory property the condition (ii') requires the corresponding function to be only continuous on R^N. However, for our purposes the continuity must be strengthened and thus here the property "locally Lipschitz" is assumed to hold. This fact permits us to define for almost all $x \in \Omega$ the directional Clarke's differential according to the formula

$$j^0(x; \xi, \eta) = \limsup_{\substack{h \to 0 \\ \lambda \to 0+}} \frac{j(x; \xi + h + \lambda\eta) - j(x; \xi + h)}{\lambda} \quad \xi, \eta \in R^N, \qquad (5.1.2)$$

and the generalized gradient of Clarke

$$\partial j(x; \xi) = \{\eta \in R^N : j^0(x; \xi, \gamma) \geq \eta \cdot \gamma \quad \forall \gamma \in R^N\},[1] \qquad (5.1.3)$$

for almost all $x \in \Omega$ and for all $\xi \in R^N$. It is important in our study to ensure the integrability of $j(\cdot, u(\cdot))$ and $j^0(\cdot; u(\cdot), v(\cdot))$ for any vector-valued functions $u, v \in V \cap L^\infty(\Omega; R^N)$. Therefore it will also be assumed that there exists a function $\beta : \Omega \times R^+ \to R$ having the properties

(i'') $\beta(\cdot, r) \in L^1(\Omega)$ for each $r \geq 0$;   (5.1.4)

(ii'')   If $r' \leq r''$ then for almost all $x \in \Omega, \beta(x, r') \leq \beta(x, r'')$;   (5.1.5)

and such that

$$|j(x, \xi) - j(x, \eta)| \leq \beta(x, r)|\xi - \eta| \quad \forall \xi, \eta \in B(0, r), r \geq 0, \qquad (5.1.6)$$

where $B(0, r)$ stands for the ball in R^N with the origin at 0 and a radius $r \geq 0$, i.e. $B(0, r) = \{\zeta \in R^N : |\zeta| \leq r\}$. Observe that the foregoing conditions are not very restrictive. For instance, any locally Lipschitz function $j : R^N \to R$ which does not depend explicitly on $x \in \Omega$ satisfies (5.1.4)–(5.1.6). Then $\beta(r)$ denotes nothing else than the Lipschitz constant for j on the ball $B(0, r)$.

---

[1] By the symbols $|\cdot|$ and "$\cdot$" we denote the norm and the inner product in the Euclidean space R^N, respectively.

On the assumption that $g \in V^*$ we formulate the following hemivariational inequality.

Problem (P) Find $u \in V$ such that

$$\langle Au - g, v - u \rangle_V + \int_\Omega j^0(u, v - u) d\Omega \geq 0 \quad \forall v \in V. \tag{5.1.7}$$

In Chapter 4 problem (P) has been considered under the hypothesis that $j : \Omega \times R^N \to R$ fulfills the growth conditions (4.3.25) which ensure that the function $J : V \to R$ given by the formula

$$J(v) = \int_\Omega j(v) d\Omega \tag{5.1.8}$$

is locally Lipschitz on V (in fact, Lipschitz continuous on bounded subsets of $L^p(\Omega; R^N)$). The corresponding existence result (Theorem 4.25) has been obtained by making use of the theorem of pseudo-monotone operators. This Chapter is devoted to the study of problem (P) on much weaker restrictions than those of (4.3.25). Namely, in Section 5.2 our considerations will be carried out on the following assumption:

The Directional Growth Condition (G). There exists a function $\alpha : \Omega \times R^+ \to R$ with the properties that

    (a') $\alpha(\cdot, r) \in L^q(\Omega)$ for each $r \geq 0$, where $1/p + 1/q = 1$;   (5.1.9)

    (b') If $r' \leq r''$ then for almost all $x \in \Omega$, $\alpha(x, r') \leq \alpha(x, r'')$;   (5.1.10)

such that the following estimate holds

$$j^0(x; \xi, \eta - \xi) \leq \alpha(x, r)(1 + |\xi|) \tag{5.1.11}$$

for a.e. $x \in \Omega$ and for any $\xi, \eta \in R^N$ with $\eta \in B(0, r), r \geq 0$.

In Section 5.3 we formulate existence results in the case in which A is strongly monotone and (5.1.11) is replaced by

$$j^0(x; \xi, \eta - \xi) \leq \alpha(x, r)(1 + |\xi|^\sigma) \tag{5.1.12}$$

which is valid for a.e. $x \in \Omega$ and all $\xi, \eta \in R^N$ with $\eta \in B(0, r), r \geq 0$, where $1 \leq \sigma < 2$.

The subject of the Section 5.4 is to provide existence results for (P) under the following directional growth conditions:

$$\left.\begin{aligned} j^0(x; \xi, -\xi) &\leq k(x)|\xi| \\ j^0(x; \xi, \eta - \xi) &\leq \alpha(x, r)(1 + |\xi|^s) \end{aligned}\right\} \tag{5.1.13}$$

which is assumed to hold for a.e. $x \in \Omega$ and for any $\xi, \eta \in R^N$ with $\eta \in B(0, r), r \geq 0$, where $s < p$, $k \in L^q(\Omega)$ and $\alpha(\cdot, r) \in L^{q'}(\Omega)$ for any

$r > 0$, $q' = p/(p - s)$. We pay also attention to the fact that there exists an essential difference between the cases $N = 1$ and $N > 1$. Namely, by giving appropriate examples of functions j we show that only when $N = 1$ we can simplify (5.1.11)–(5.1.13). These simplifications lead to the generalization of the sign condition (3.1.14) which for a Caratheodory function $g : \Omega \times R$ takes the form $g(x, \xi) \xi \geq 0$ for each $\xi \in R$ and a.e. $x \in \Omega$.

It is easily seen that conditions (5.1.11)–(5.1.13) are too weak to ensure the finite integrability of $j(\cdot; u(\cdot))$ and $j^0(\cdot; u(\cdot), v(\cdot))$ in Ω for any $u, v \in V$. Moreover, they do not guarantee that the functional (5.1.8) is locally Lipschitz on V and even finite over the whole space V. For this reason we cannot follow the method of Chapter 4. However, as we shall see below, the hypotheses (5.1.11)–(5.1.13) are sufficient to establish the existence of solutions to (P) in the sense of the following definition.

Definition 5.1 An element $u \in V$ is said to be a solution of (P) if there exists $\chi \in L^1(\Omega; R^N)$ such that

$$\langle Au - g, v \rangle_V + \int_\Omega \chi \cdot v d\Omega = 0 \quad \forall v \in V \cap L^\infty(\Omega; R^N) \tag{5.1.14}$$

and

$$\chi(x) \in \partial j(x; u(x)) \quad \text{for a.e. } x \in \Omega. \tag{5.1.15}$$

To justify this definition let us observe that from (5.1.14) it follows that the linear functional $l(\chi)$ given by the formula

$$\langle l(\chi), v \rangle_V = \int_\Omega \chi \cdot v d\Omega \quad \forall v \in V \cap L^\infty(\Omega; R^N), \tag{5.1.16}$$

can be extended to the whole space V as an element $l(\chi) \in V^*$, uniquely, because of the density of $V \cap L^\infty(\Omega; R^N)$ in V. Accordingly, (5.1.14) becomes

$$\langle Au - g, v \rangle_V + \langle l(\chi), v \rangle_V = 0 \quad \forall v \in V. \tag{5.1.17}$$

In particular, if j happens to satisfy the growth condition (4.3.25), then from (5.1.14) it follows that the inequality (5.1.7) is valid for each $v \in V$.

5.2 Hemivariational Inequalities with a Coercive Operator

The aim of this Section is to prove that there exists at least one solution of (P) under the hypotheses (5.1.9)–(5.1.11) provided A is coercive. The idea is to construct the suitable family of regularized solutions, possessing the finite intersection property.

Let us introduce the mollifier $p \in C_0^\infty(R^N)$ with the properties: $p(\xi) \geq 0$, $\text{supp}(p) \in [-1, 1]^N$ and

$$\int_{R^N} p(\xi)d\xi = 1. \tag{5.2.1}$$

Define

$$p_\varepsilon(\xi) = \frac{1}{\varepsilon^N}p(\frac{\xi}{\varepsilon}) \quad \xi \in R^N, \varepsilon > 0, \tag{5.2.2}$$

and

$$j_\varepsilon(x;\xi) = \int_{R^N} p_\varepsilon(\eta)j(x;\xi-\eta)d\eta \quad \xi \in R^N. \tag{5.2.3}$$

Denote by $j'_\varepsilon(x,\xi)$ the derivative of $j_\varepsilon(x;\xi)$ with respect to ξ, for almost all $x \in \Omega$, and all $\xi \in R^N$. Further, let Λ be the family of all finite-dimensional subspaces F of $V \cap L^\infty(\Omega; R^N)$, ordered by inclusion. For any $F \in \Lambda$ and $\varepsilon \in (0,1]$ we formulate the following regularized finite dimensional problem.

Problem $(P_{F\varepsilon})$ Find $u_{F\varepsilon} \in F$ such as to satisfy the variational equality

$$\langle Au_{F\varepsilon} - g, v \rangle_V + \int_\Omega j'_\varepsilon(u_{F\varepsilon}) \cdot vd\Omega = 0 \quad \forall v \in F. \tag{5.2.4}$$

Note that due to $(5.1.4)-(5.1.6)$ the integral above is finite for each $u_{F\varepsilon}, v \in F$. Indeed, from $(5.1.6)$ it follows

$$\left|\frac{j(x;u_{F\varepsilon}(x)+tv(x))-j(x;u_{F\varepsilon}(x))}{t}\right| \leq \beta(x,|u_{F\varepsilon}(x)|+1)|v(x)| \tag{5.2.5}$$

$$\leq \beta(x,\|u_{F\varepsilon}\|_{L^\infty(\Omega)}+1)\|v\|_{L^\infty(\Omega)}$$

for a.e. $x \in \Omega$ and sufficiently small t. Hence,

$$\left|j'_\varepsilon(x;u_{F\varepsilon}(x))\cdot v(x)\right|$$

$$= \left|\lim_{t\to 0}\int_{R^N} p_\varepsilon(\tau)\frac{j(x;u_{F\varepsilon}(x)-\tau+tv(x))-j(x;u_{F\varepsilon}(x)-\tau)}{t}d\tau\right|$$

$$\leq \beta(x,\|u_{F\varepsilon}\|_{L^\infty(\Omega)}+1)\|v\|_{L^\infty(\Omega)} \quad \text{for a.e. } x \in \Omega. \tag{5.2.6}$$

Since by $(5.1.4)$,

$$\beta(\cdot,\|u_{F\varepsilon}\|_{L^\infty(\Omega)}+1) \in L^1(\Omega), \tag{5.2.7}$$

the integrability of $j'_\varepsilon(u_{F\varepsilon}) \cdot v$ in Ω follows immediately for any $v \in F$.

Lemma 5.2 Suppose that $(5.1.11)$ holds. Then the estimate

$$j'_\varepsilon(x;\xi)\cdot(\eta-\xi) \leq \bar\alpha(x,r)(1+|\xi|), \quad 0 < \varepsilon \leq 1, \tag{5.2.8}$$

is valid for any $\xi, \eta \in R^N$ with $|\eta| \leq r, r \geq 0$, and for almost all $x \in \Omega$, where $\bar\alpha(x;r) \equiv 2\alpha(x;r+1)$.

Proof. Choose $\eta \in R^N$ with $|\eta| \leq r$. The assertion $(5.2.8)$ follows directly from the estimates

$$
\begin{aligned}
j_\epsilon'(x;\xi)\cdot(\eta-\xi) &= \lim_{t\to 0}\frac{j_\epsilon(x,\xi+t(\eta-\xi))-j_\epsilon(x;\xi)}{t}d\tau \\
&= \lim_{t\to 0}\int_{R^N} p_\epsilon(\tau)\frac{j(x,\xi-\tau+t(\eta-\xi))-j(x;\xi-\tau)}{t}d\tau \\
&\leq \int_{R^N} p_\epsilon(\tau)\limsup_{t\to 0}\frac{j(x,\xi-\tau+t(\eta-\xi))-j(x;\xi-\tau)}{t}d\tau \\
&\leq \int_{R^N} p_\epsilon(\tau)j^0(x,\xi-\tau,\eta-\tau-(\xi-\tau))d\tau \\
&\leq \int_{R^N} p_\epsilon(\tau)\alpha(x;\epsilon+|\eta|)(1+|\xi-\tau|)d\tau \\
&\leq \int_{R^N} p_\epsilon(\tau)\alpha(x;1+|\eta|)(2+|\xi|)d\tau \\
&\leq \int_{R^N} p_\epsilon(\tau)\alpha(x;1+r)(2+|\xi|)d\tau \\
&\leq 2\alpha(x,r+1)(1+|\xi|)=\bar\alpha(x;r)(1+|\xi|),
\end{aligned}
\tag{5.2.9}
$$

where we have applied Fatou's lemma. The proof is complete.   q.e.d.

To establish the solvability of $(P_{F\epsilon})$ we apply Brouwer's fixed point theorem as shown below.

Lemma 5.3 For any $F\in\Lambda$ and $\epsilon\in(0,1]$ the problem $(P_{F\epsilon})$ has at least one solution $u_{F\epsilon}\in F$. Moreover, the family $\{u_{F\epsilon}\}$ is uniformly bounded in V.

Proof. Let $i_F:F\to V$, $F\in\Lambda$, be the inclusion mapping of F into V, and $i_F^*:V^*\to F^*$ the dual projection mapping of V^* into F^*, F^* being the dual of F. Define $A_F=i_F^*Ai_F$ and $g_F=i_F^*g\in F^*$. Moreover, let $h_{F\epsilon}:F\to F^*$ be given by the formula

$$
\langle h_{F\epsilon}(v),w\rangle_F = \int_\Omega j_\epsilon'(v)\cdot w\,d\Omega \quad \forall w\in F,\ v\in F,^2.
\tag{5.2.10}
$$

Then the equation (5.2.4) can be written equivalently as

$$
\tilde\Gamma_{F\epsilon}(u_{F\epsilon})=0
\tag{5.2.11}
$$

where $\tilde\Gamma_{F\epsilon}\equiv A_F-g_F+h_{F\epsilon}$. Since A is continuous from V to V^* equipped with the weak topology, A_F must be continuous from F into F^*. The continuity of the operator $h_{F\epsilon}$ is a consequence of the well known results of the integral calculus. Accordingly, $\tilde\Gamma_{F\epsilon}$ is a continuous operator from a finite dimensional space F into its dual F^*. To apply Brouwer's fixed point theorem let us show that from the coercivity of A and from (5.2.8) (with $r=0$) the following estimates hold:

---

[2] The symbol $(\cdot,\cdot)_F$ is used to denote the pairing over $F^*\times F$.

$$(\tilde{\Gamma}_{F\epsilon}v, v)_F = \langle A_F v - g_F + h_{F\epsilon}(v), v\rangle_F = \langle Av, v\rangle_V - \langle g, v\rangle_V + \int_\Omega j'_\epsilon(v) \cdot v d\Omega$$

$$\geq c(||v||_V)||v||_V - ||g||_{V*}||v||_V - \int_\Omega \bar{\alpha}(0)(1 + |v|)d\Omega \qquad (5.2.12)$$

$$\geq c(||v||_V)||v||_V - ||g||_{V*}||v||_V - ||\bar{\alpha}(0)||_{L^q(\Omega)}\mathrm{mes}\Omega^{1/p}$$
$$-||\bar{\alpha}(0)||_{L^q(\Omega)}||v||_{L^p(\Omega)}$$

$$\geq c(||v||_V)||v||_V - ||g||_{V*}||v||_V - ||\bar{\alpha}(0)||_{L^q(\Omega)}\mathrm{mes}\Omega^{1/p}$$
$$-||\bar{\alpha}(0)||_{L^q(\Omega)}\gamma||v||_V,$$

where mesΩ denotes the Lebesgue measure of Ω, $1/p + 1/q = 1$, and γ is a positive constant with the property that $||v||_{L^p(\Omega)} \leq \gamma||v||_V$, $\forall v \in V$ (V is compactly imbedded into $L^p(\Omega; R^N)$). Thus, due to the well known property of the coercivity function $c(\cdot)$, for sufficiently large $||v||_V$ we have that

$$(\tilde{\Gamma}_{F\epsilon}v, v)_F \geq 0.$$

Since $\tilde{\Gamma}_{F\epsilon}$ is continuous, by means of Brouwer's fixed point theorem we deduce the existence of at least one solution $u_{F\epsilon} \in F$ to $(P_{F\epsilon})$. Moreover, from (5.2.11) and (5.2.12) it results that

$$c(||u_{F\epsilon}||_V)||u_{F\epsilon}||_V \qquad (5.2.13)$$
$$\leq ||g||_{V*}||u_{F\epsilon}||_V + ||\bar{\alpha}(0)||_{L^q(\Omega)}\mathrm{mes}\Omega^{1/p} + ||\bar{\alpha}(0)||_{L^q(\Omega)}\gamma||u_{F\epsilon}||_V.$$

Thus there exists a constant M not depending on F and ϵ such that

$$||u_{F\epsilon}||_V \leq M, \quad F \in \Lambda, \epsilon \in (0,1]. \qquad (5.2.14)$$

This completes the proof of Lemma 5.3.　　　　　　　　　　q.e.d.

The next lemma corresponds to the compactness property of the set $\{j'_\epsilon(u_{F\epsilon}) : F \in \Lambda, \epsilon \in (0,1]\}$ in $L^1(\Omega; R^N)$.

Lemma 5.4 Let for any $F \in \Lambda$ and $\epsilon \in (0,1]$, $u_{F\epsilon} \in F$ be a solution of $(P_{F\epsilon})$. Then the set $\{j'_\epsilon(u_{F\epsilon}) : F \in \Lambda, \epsilon \in (0,1]\}$ is weakly precompact in $L^1(\Omega; R^N)$.

Proof. According to the Dunford-Pettis theorem it suffices to show that for each $\rho > 0$ a $\delta_\rho > 0$ can be determined such that for any $\omega \subset \Omega$ with mes$\omega < \delta_\rho$,

$$\int_\omega |j'_\epsilon(u_{F\epsilon})|d\Omega < \rho, \quad F \in \Lambda, \quad \epsilon \in (0,1]. \qquad (5.2.15)$$

Fix $r > 0$ and let $\eta \in R^N$ be such that $|\eta| \leq r$. Then from (5.2.8) one obtains

$$j'_\epsilon(x; u_{F\epsilon}(x)) \cdot \eta \leq j'_\epsilon(x, u_{F\epsilon}(x)) \cdot u_{F\epsilon}(x) + \bar{\alpha}(x; r)(1 + |u_{F\epsilon}(x)|). \qquad (5.2.16)$$

Let us denote by $\partial_i j_\epsilon(\cdot), i = 1, 2, \ldots, N$, the partial derivatives of $j_\epsilon(x, \xi)$ with respect to ξ_i, and let us set

$$\eta(x) = \frac{r}{\sqrt{N}}[\mathrm{sgn}\, \partial_1 j_\epsilon(x; u_{F\epsilon}(x)), \ldots, \mathrm{sgn}\, \partial_N j_\epsilon(x; u_{F\epsilon}(x))], \qquad (5.2.17)$$

where

$$\mathrm{sgn}\, y = \begin{cases} 1 & \text{if } y > 0 \\ 0 & \text{if } y = 0. \\ -1 & \text{if } y < 0 \end{cases} \qquad (5.2.18)$$

It is not difficult to verify that $|\eta(x)| \le r$ for almost all $x \in \Omega$ and that

$$j'_\epsilon(x; u_{F\epsilon}(x)) \cdot \eta(x) \ge \frac{r}{\sqrt{N}}|j'_\epsilon(x; u_{F\epsilon}(x))|. \qquad (5.2.19)$$

Hence

$$\frac{r}{\sqrt{N}}|j'_\epsilon(x; u_{F\epsilon}(x))| \le j'_\epsilon(x; u_{F\epsilon}(x)) \cdot u_{F\epsilon}(x) + \bar{\alpha}(x; r)(1 + |u_{F\epsilon}(x)|). \quad (5.2.20)$$

Integrating this inequality over $\omega \subset \Omega$ implies that

$$\int_\omega |j'_\epsilon(u_{F\epsilon})|d\Omega \le \frac{\sqrt{N}}{r}\int_\omega j'_\epsilon(u_{F\epsilon}) \cdot u_{F\epsilon}d\Omega + \frac{\sqrt{N}}{r}||\bar{\alpha}(r)||_{L^q(\Omega)}\mathrm{mes}\omega^{1/p}$$

$$+ \frac{\sqrt{N}}{r}||\bar{\alpha}(r)||_{L^q(\omega)}||u_{F\epsilon}||_{L^p(\Omega)}. \qquad (5.2.21)$$

Thus, by virtue of (5.2.14) we are led to the estimates

$$\int_\omega |j'_\epsilon(u_{F\epsilon})|d\Omega \le \frac{\sqrt{N}}{r}\int_\omega j'_\epsilon(u_{F\epsilon}) \cdot u_{F\epsilon}d\Omega$$

$$+ \frac{\sqrt{N}}{r}||\bar{\alpha}(r)||_{L^q(\Omega)}\mathrm{mes}\omega^{1/p} + \frac{\sqrt{N}}{r}||\bar{\alpha}(r)||_{L^q(\omega)}||u_{F\epsilon}||_{L^p(\Omega)}$$

$$\le \frac{\sqrt{N}}{r}\int_\omega j'_\epsilon(u_{F\epsilon}) \cdot u_{F\epsilon}d\Omega \qquad (5.2.22)$$

$$+ \frac{\sqrt{N}}{r}||\bar{\alpha}(r)||_{L^q(\Omega)}\mathrm{mes}\omega^{1/p} + \frac{\sqrt{N}}{r}||\bar{\alpha}(r)||_{L^q(\omega)}\gamma||u_{F\epsilon}||_V$$

$$\le \frac{\sqrt{N}}{r}\int_\omega j'_\epsilon(u_{F\epsilon}) \cdot u_{F\epsilon}d\Omega$$

$$+ \frac{\sqrt{N}}{r}||\bar{\alpha}(r)||_{L^q(\Omega)}\mathrm{mes}\omega^{1/p} + \frac{\sqrt{N}}{r}||\bar{\alpha}(r)||_{L^q(\omega)}\gamma M.$$

Now we show that

$$\int_\omega j'_\epsilon(u_{F\epsilon}) \cdot u_{F\epsilon}d\Omega \le C \qquad (5.2.23)$$

for some positive constant C not depending on $\omega \subset \Omega$, $F \in \Lambda$ and $\epsilon \in (0,1]$. Indeed, from (5.2.8) it results that

$$j'_\epsilon(x; u_{F\epsilon}(x)) \cdot u_{F\epsilon}(x) + \bar{\alpha}(x; 0)(1 + |u_{F\epsilon}(x)|) \ge 0 \text{ for a.e. } x \in \Omega. \quad (5.2.24)$$

Thus we can write that

$$\int_\omega [j_\varepsilon'(u_{F\varepsilon}) \cdot u_{F\varepsilon} + \bar{\alpha}(0)(1 + |u_{F\varepsilon}|)]d\Omega \leq \int_\Omega [j_\varepsilon'(u_{F\varepsilon}) \cdot u_{F\varepsilon} + \bar{\alpha}(0)(1 + |u_{F\varepsilon}|)]d\Omega$$

(5.2.25)

and consequently

$$\int_\omega j_\varepsilon'(u_{F\varepsilon}) \cdot u_{F\varepsilon} d\Omega \leq \int_\Omega j_\varepsilon'(u_{F\varepsilon}) \cdot u_{F\varepsilon} d\Omega$$

(5.2.26)

$$+ \; ||\bar{\alpha}(0)||_{L^q(\Omega)} \text{mes} \Omega^{1/p} + ||\bar{\alpha}(0)||_{L^q(\Omega)} \gamma ||u_{F\varepsilon}||_V$$

$$\leq \int_\Omega j_\varepsilon'(u_{F\varepsilon}) \cdot u_{F\varepsilon} d\Omega$$

$$+ \; ||\bar{\alpha}(0)||_{L^q(\Omega)} \text{mes} \Omega^{1/p} + ||\bar{\alpha}(0)||_{L^q(\Omega)} \gamma M.$$

But since A is weakly continuous, it maps bounded sets into bounded sets. Hence, by means of (5.2.4) and (5.2.14) we conclude that

$$\int_\Omega j_\varepsilon'(u_{F\varepsilon}) \cdot u_{F\varepsilon} d\Omega = -\langle Au_{F\varepsilon} - g, u_{F\varepsilon} \rangle_V \leq ||Au_{F\varepsilon} - g||_{V*} ||u_{F\varepsilon}||_V \leq \text{ const.}$$

(5.2.27)

Hence, (5.2.26) yields (5.2.23), as desired. Further, by combining (5.2.22) with (5.2.23) we are led directly to the estimate

$$\int_\omega |j_\varepsilon'(u_{F\varepsilon})|d\Omega \leq \frac{\sqrt{N}}{r}C + \frac{\sqrt{N}}{r}||\bar{\alpha}(r)||_{L^q(\Omega)} \text{mes} \omega^{1/p} + \frac{\sqrt{N}}{r}||\bar{\alpha}(r)||_{L^q(\omega)} \gamma M.$$

(5.2.28)

Now, let $\rho > 0$. Fix $r > 0$ such that

$$\frac{\sqrt{N}}{r}C < \frac{\rho}{2}.$$

(5.2.29)

Since $\bar{\alpha}(\cdot, r) \in L^q(\Omega)$, we can determine a $\delta_\rho > 0$ small enough such that

$$\frac{\sqrt{N}}{r}||\bar{\alpha}(r)||_{L^q(\Omega)} \text{mes} \omega^{1/p} + \frac{\sqrt{N}}{r}||\bar{\alpha}(r)||_{L^q(\omega)} \gamma M \leq \frac{\rho}{2}$$

(5.2.30)

whenever $\text{mes} \omega < \delta_\rho$. Finally, (5.2.30) yields

$$\int_\omega |j_\varepsilon'(u_{F\varepsilon})|d\Omega \leq \rho \quad F \in \Lambda, \varepsilon \in (0,1],$$

(5.2.31)

for any $\omega \subset \Omega$ with $\text{mes} \omega < \delta_\rho$. Accordingly, the weak precompactness of $\{j_\varepsilon'(u_{F\varepsilon}) : F \in \Lambda, \varepsilon \in (0,1]\}$ in $L^1(\Omega; R^N)$ has been proved.   q.e.d.

Now we can formulate the following existence result.

Theorem 5.5 Let A be a coercive, weakly continuous operator from V into V^*. Suppose that the injection $V \subset L^p(\Omega; R^N), N \geq 1$, is compact for some $1 <$

$p < \infty$, $V \cap L^\infty(\Omega; R^N)$ is dense in V and that (5.1.4)–(5.1.6), (5.1.9)–(5.1.11) hold. Then problem (P) has at least one solution.

Proof. In order to prove the theorem we have to show that there exist $u \in V$ and $\chi \in L^1(\Omega; R^N)$ such that (5.1.14)–(5.1.15) hold. Let us introduce an order in the Cartesian product $\Lambda \times [0,1]$ by the relation $(F, \varepsilon) \leq (F', \varepsilon') \Longleftrightarrow F \subset F'$ and $\varepsilon' \leq \varepsilon$, and let

$$W_{F\varepsilon} = \bigcup_{\substack{F' \in \Lambda \\ F' \supset F \\ 0 < \varepsilon' \leq \varepsilon}} \{(u_{F'\varepsilon'}, j'_{\varepsilon'}(u_{F'\varepsilon'})) \in V \times L^1(\Omega; R^N) : \qquad (5.2.32)$$

$$(u_{F'\varepsilon'}, j'_{\varepsilon'}(u_{F'\varepsilon'})) \text{ satisfies } (P_{F'\varepsilon'})\}.$$

We use the symbol weakcl $(W_{F\varepsilon})$ to denote the weak closure of $W_{F\varepsilon}$ in $V \times L^1(\Omega; R^N)$. Moreover, let

$$Z = \bigcup_{\substack{F \in \Lambda \\ 0 < \varepsilon \leq 1}} \{j'_{F\varepsilon}(u_{F\varepsilon})\} \subset L^1(\Omega; R^N). \qquad (5.2.33)$$

Denoting by weakcl (Z) the weak closure of Z in $L^1(\Omega; R^N)$ we obtain from (5.2.14),

$$\text{weakcl}\,(W_{F\varepsilon}) \subset B_V(O, M) \times \text{weakcl}\,(Z) \quad \forall (F, \varepsilon) \in \Lambda \times (0, 1]. \qquad (5.2.34)$$

Since $B_V(O, M)$ is weakly compact in V and by Lemma 5.4 weakcl (Z) is weakly compact in $L^1(\Omega; R^N)$, the family $\{\text{weakcl}\,(W_{F\varepsilon}) : F \in \Lambda, \varepsilon \in (0, 1]\}$ is contained in the weakly compact set $B_V(O, M) \times \text{weakcl}(Z)$ in $V \times L^1(\Omega; R^N)$. Now we notice that for any $F_1, \ldots, F_k \in \Lambda$ and any $\varepsilon_1, \ldots, \varepsilon_k > 0$, $k \geq 1$, we have the inclusion $W_{F_1\varepsilon_1} \cap \ldots \cap W_{F_k\varepsilon_k} \supset W_{F\varepsilon}$ with $F = F_1 \cup \ldots \cup F_k$ and $\varepsilon = \min(\varepsilon_1, \ldots, \varepsilon_k)$. Thus, it results by means of Lemma 5.3 that the family $\{\text{weakcl}(W_{F\varepsilon}) : F \in \Lambda, \varepsilon \in (0, 1]\}$ has the finite intersection property. Therefore the intersection

$$\bigcap_{\substack{F \in \Lambda \\ 0 < \varepsilon \leq 1}} \text{weakcl}\,(W_{F\varepsilon}) \qquad (5.2.35)$$

is not empty; let the pair (u, χ) belongs to this intersection. The proof will be complete if we show that (u, χ) satisfies (5.1.14) and (5.1.15). Let $v \in V \cap L^\infty(\Omega; R^N)$ be arbitrary. We choose $F \in \Lambda$ such that $v \in F$. Then there exists a sequence $\{(u_{F_n\varepsilon_n}, j'_{\varepsilon_n}(u_{F_n\varepsilon_n}))\}$ in $W_{F\varepsilon}$, where $\varepsilon_n \to 0$, such that $(u_{F_n\varepsilon_n}, j'_{\varepsilon_n}(u_{F_n\varepsilon_n})) \to (u, \chi)$ weakly in $V \times L^1(\Omega; R^N)$ as $n \to \infty$. In the sequel we shall use the notation $(u_n, j'_n(u_n))$ instead of $(u_{F_n\varepsilon_n}, j'_{\varepsilon_n}(u_{F_n\varepsilon_n}))$, $n \geq 1$, for the sake of simplicity. Accordingly

$$u_n \to u \quad \text{weakly in } V, \qquad (5.2.36)$$

and

$$j'_n(u_n) \to \chi \quad \text{weakly in } L^1(\Omega; R^N). \qquad (5.2.37)$$

Moreover, the following equality holds for $n = 1, 2, \ldots,$

$$\langle Au_n - g, v \rangle_V + \int_\Omega j_n'(u_n) \cdot v d\Omega = 0. \tag{5.2.38}$$

Now, taking into account (5.2.36) and (5.2.37) we can pass to the limit as $n \to \infty$ in (5.2.38). The weak continuity of A and the fact that $v \in V \cap L^\infty(\Omega; R^N)$ is arbitrary permits us to conclude that the equality

$$\langle Au - g, v \rangle_V + \int_\Omega \chi \cdot v d\Omega = 0 \tag{5.2.39}$$

is valid for any $v \in V \cap L^\infty(\Omega; R^N)$. Thus (5.1.14) holds.

In the last step of the proof we will prove (5.1.15). To this end let us notice that due to the compact imbedding $V \subset L^p(\Omega; R^N)$, it results from (5.2.36) that

$$u_n \to u \quad \text{strongly in } L^p(\Omega; R^N). \tag{5.2.40}$$

Hence (by passing to a subsequence, if necessary)

$$u_n \to u \quad \text{a.e. in } \Omega. \tag{5.2.41}$$

Thus we can apply the Egoroff's theorem. According to this theorem for any $\rho > 0$ a subset $\omega \subset \Omega$ with mes$\omega < \rho$ can be determined such that

$$u_n \to u \quad \text{uniformly on } \Omega \setminus \omega \tag{5.2.42}$$

with $u \in L^\infty(\Omega \setminus \omega; R^N)$. Let $v \in L^\infty(\Omega \setminus \omega; R^N)$ be arbitrarily given. Due to Fatou's lemma, for any positive $\mu > 0$ there exists $\delta_\mu > 0$ and N_μ such that

$$\int_{\Omega \setminus \omega} \frac{j(x; u_n(x) - \tau + tv(x)) - j(x; u_n(x) - \tau)}{t} d\Omega$$
$$\leq \int_{\Omega \setminus \omega} j^0(x; u(x), v(x)) d\Omega + \mu \tag{5.2.43}$$

provided $n > N_\mu$, $|\tau| < \delta_\mu$ and $0 < t < \delta_\mu$. This inequality multiplied by p_{ε_n} (with $\varepsilon_n < \delta_\mu$) and integrated over R^N yields

$$\int_{\Omega \setminus \omega} \frac{j_n(x; u_n(x) + tv(x)) - j_n(x; u_n(x))}{t} d\tau$$
$$= \int_{R^N} p_{\varepsilon_n}(\tau)(\int_{\Omega \setminus \omega} \frac{j(x; u_n(x) - \tau + tv(x)) - j(x; u_n(x) - \tau)}{t} d\Omega) d\tau$$
$$\leq \int_{\Omega \setminus \omega} j^0(x; u(x), v(x)) d\Omega + \mu. \tag{5.2.44}$$

But

$$\lim_{t \to 0} \int_{R^N} p_{en}(\tau)(\int_{\Omega \setminus \omega} \frac{j(x; u_n(x) - \tau + tv(x)) - j(x; u_n(x) - \tau)}{t} d\Omega) d\tau$$

$$= \int_{\Omega \setminus \omega} j_n'(x; u_n(x)) \cdot v(x) d\Omega \qquad (5.2.45)$$

and therefore

$$\int_{\Omega \setminus \omega} j_n'(x; u_n(x)) \cdot v(x) d\Omega \leq \int_{\Omega \setminus \omega} j^0(x; u(x), v(x)) d\Omega + \mu, \qquad (5.2.46)$$

which is valid for $n > N_\mu$. Now, letting $n \to \infty$ in (5.2.46) and taking into account (5.2.37) implies that

$$\int_{\Omega \setminus \omega} \chi \cdot v d\Omega \leq \int_{\Omega \setminus \omega} j^0(u, v) d\Omega + \mu. \qquad (5.2.47)$$

Since $\mu > 0$ was chosen arbitrarily,

$$\int_{\Omega \setminus \omega} \chi \cdot v d\Omega \leq \int_{\Omega \setminus \omega} j^0(u, v) d\Omega \quad \forall v \in L^\infty(\Omega \setminus \omega; R^N). \qquad (5.2.48)$$

But the last inequality easily implies that

$$\chi(x) \in \partial j(x; u(x)) \quad \text{for a.e. } x \in \Omega \setminus \omega, \qquad (5.2.49)$$

where mes$\omega < \rho$. Now, since ρ was chosen arbitrarily,

$$\chi(x) \in \partial j(x; u(x)) \quad \text{for a.e. } x \in \Omega \qquad (5.2.50)$$

and the proof of the theorem is complete.   q.e.d.

Remark 5.6 Let us consider the one-dimensional case, i.e. $N = 1$. Then the condition (5.1.11) can be made weaker. Namely, let us suppose that (5.1.4) holds with $\beta(\cdot, r) \in L^q(\Omega)$ for each $r \geq 0$ and that the following estimate holds

$$j^0(x; \xi, -\xi) \leq k(x)|\xi| \quad \forall \xi \in R, \text{ for a.e. } x \in \Omega, \qquad (5.2.51)$$

for some nonnegative function $k \in L^q(\Omega)$. Then here exists $\alpha : \Omega \times R^+ \to R^+$ such that (5.1.9)–(5.1.11) hold. Indeed, to show this let us notice that due to the positive homogeneity of $\eta \to j^0(x; \xi, \eta)$, relation (5.2.51) yields

$$j^0(x; \xi, -1) \leq k(x) \quad \xi \geq 0, \qquad (5.2.52a)$$

$$j^0(x; \xi, 1) \leq k(x) \quad \xi < 0. \qquad (5.2.52b)$$

At the same time, from (5.1.6) we get

$$|j^0(x; \xi, \eta)| \leq \beta(x, |\xi| + 1)|\eta| \quad \xi, \eta \in R. \qquad (5.2.53)$$

Let $\xi, \eta \in B(0, r)$ for some $r \geq 0$. Then from (5.2.53) we obtain the following estimate

$$
\begin{aligned}
j^0(x; \xi, \eta - \xi) &\leq \beta(x; |\xi| + 1)|\eta - \xi| && (5.2.54) \\
&\leq \beta(x; r + 1)(r + |\xi|) \leq \beta(x; r + 1)(r + 1)(1 + |\xi|).
\end{aligned}
$$

If $\xi > r \geq \eta \geq -r$, then by means of the positive homogeneity of $j^0(x; \xi, \cdot)$, relation (5.2.52a) implies that

$$
j^0(x; \xi, \eta - \xi) \leq k(x)(\xi - \eta) = k(x)\xi(1 - \frac{\eta}{\xi}) \leq k(x)\xi \leq 2k(x)(1 + |\xi|). \quad (5.2.55)
$$

Analogously, if $\xi < -r \leq \eta \leq r$, then from (5.2.52b) we deduce the relation

$$
j^0(x; \xi, \eta - \xi) \leq k(x)(\eta - \xi) \leq k(x)(-\xi)(1 - \frac{\eta}{\xi}) \leq k(x)(-\xi) \leq 2k(x)(1 + |\xi|). \tag{5.2.56}
$$

Now it remains to define $\alpha : \Omega \times R^+ \to R$ as

$$
\alpha(x, r) = \max\{2k(x), \beta(x; r + 1)(r + 1)\} \tag{5.2.57}
$$

in order to obtain the desired estimate (5.1.11). By the assumption that both $k(\cdot)$ and $\beta(\cdot, r + 1)$ belong to $L^q(\Omega)$, the same can be said about $\alpha(\cdot, r)$, i.e. $\alpha(\cdot, r) \in L^q(\Omega)$, for any $r \geq 0$.

Corollary 5.7 Let V be a reflexive Banach space and $\Omega \subset R^n$ an open bounded subset of $R^n, n \geq 1$. Suppose that the injection $V \subset L^p(\Omega)$ is compact for some $1 < p < \infty$, and that $V \cap L^\infty(\Omega)$ is dense in V. Moreover, let A be a coercive, weakly continuous operator from V into V^*. Suppose that (5.2.51) and (5.1.4)–(5.1.6), with $\beta(\cdot, r) \in L^q(\Omega)$ for each $r \geq 0$, $1/p + 1/q = 1$, hold. Then problem (P) has at least one solution. If additionally we suppose that for any $v \in V$ there exists a sequence $\{w_n\}$ with the properties that $0 \leq w_n \leq 1$, $(1 - w_n)v \in V \cap L^\infty(\Omega)$ and $(1 - w_n)v \to v$ strongly in V, then $\chi u \in L^1(\Omega)$ and

$$
\langle Au - g, u \rangle_V + \int_\Omega \chi u \, d\Omega = 0, \tag{5.2.58}
$$

where the pair $(u, \chi) \in V \times L^1(\Omega)$ is a solution of (P).

Proof. The existence of solutions to (P) is an immediate consequence of Theorem 5.5 and Remark 5.6. To show that $\chi u \in L^1(\Omega)$ let us choose a function sequence $\{w_n\}$ with the properties that $0 \leq w_n \leq 1$, $(1 - w_n)v \in V \cap L^\infty(\Omega)$ and $(1 - w_n)v \to v$ strongly in V. By the compact imbedding $V \subset L^p(\Omega), (1 - w_n)u \to u$ strongly in $L^p(\Omega)$. Consequently, by passing to a subsequence, if necessary, we may assume that $(1 - w_n(x))u(x) \to u(x)$ for a.e. $x \in \Omega$. On the other hand, from (5.2.51) we get the estimate

$$
\chi(x)(1 - w_n(x))u(x) \geq -k(x)(1 - w_n(x))|u(x)| \geq -k(x)|u(x)| \tag{5.2.59}
$$

for a.e. $x \in \Omega$, with $k|u| \in L^1(\Omega)$. Since

$$\int_\Omega \chi(1 - w_n)u d\Omega = \langle -Au + g, (1 - w_n)u \rangle_V \leq C_0, \quad C_0 = \text{const} > 0, \quad (5.2.60)$$

and $\chi(x)(1 - w_n(x))u(x) \to \chi(x)u(x)$ for a.e. $x \in \Omega$, by applying Fatou's lemma we get $\chi u \in L^1(\Omega)$. To obtain (5.2.58) we pass to the limit as $n \to \infty$ in the equality (5.2.60), where

$$\int_\Omega \chi(1 - w_n)u d\Omega \to \int_\Omega \chi u d\Omega \quad \text{as } n \to \infty, \quad (5.2.61)$$

due to the dominated convergence. The proof is complete.   q.e.d.

The hypothesis on the existence of a sequence $\{w_n\}$ with the properties mentioned in Corollary 5.7 is related to the truncation procedure introduced by Hedberg [Hed] for Sobolev spaces. For the applications of this technique in the study of strongly nonlinear equations we refer to [Web, Brèz82, Boc].

The following question arises when studying the general, N-dimensional case with $N > 1$, namely, whether it is possible to deduce (5.1.11) from the estimate (5.2.51). It turns out that the answer to this question in negative. To show this let us consider the following

Example 5.8 Let $j : R^2 \to R$ be a locally Lipschitz function $j : R^2 \to R$ defined as follows

$$j(\xi) = \begin{cases} \frac{\xi_1^2}{\xi_2^2} & \text{for } |\xi_2| \geq 1 \\ \frac{\xi_2^2}{\xi_1^2} & \text{for } |\xi_2| < 1 \end{cases} \quad \xi = (\xi_1, \xi_2) \in R^2. \quad (5.2.62)$$

Then one can easily find that

$$j^0(\xi, -\xi) \leq 0 \quad \forall \xi \in R^2, \quad (5.2.63)$$

i.e., the condition (5.2.51) holds with $k = 0$. On the other hand, for $\eta = (0, -1)$ and $\xi_2 > 1$ we get

$$j^0(\xi, \eta - \xi) = j'(\xi) \cdot \eta = 2\frac{\xi_1^2}{\xi_2^3}. \quad (5.2.64)$$

Moreover,

$$\sup_{\substack{\xi_1 \in R \\ \xi_2 > 1}} \frac{2\xi_1^2}{\xi_2^3(1 + \sqrt{\xi_1^2 + \xi_2^2})} = +\infty. \quad (5.2.65)$$

Thus there does not exist a constant $\alpha = \alpha(1) \in R$ with the property that $j^0(\xi, \eta - \xi) \leq \alpha(1)(1 + |\xi|)$ for each $\xi \in R^2$. Accordingly, (5.2.51) does not imply (5.1.11) if $N > 1$.

Let $\theta : \Omega \times R \to R$ be a function from $\Omega \times R$ into R such that:

(i) $\theta(x, \cdot) \in L^\infty_{\text{loc}}(R)$ for a.e. $x \in \Omega$;

(ii) The function $\beta : \Omega \times R^+ \to R$ defined by

$$\beta(x,r) = \operatorname*{ess\,sup}_{-r \leq \tau \leq r} |\theta(x,\tau)| \qquad (5.2.66)$$

for a.e. $x \in \Omega$ and each $r \geq 0$, has the property that

$$\beta(\cdot, r) \in L^q(\Omega) \text{ for all } r \geq 0; \qquad (5.2.67)$$

(iii) There exists a nonnegative function $k \in L^q(\Omega)$ such that

$$\theta(x,\tau) \leq k(x) \quad \text{for a.e. } x \in \Omega \text{ and each } \tau < 0, \qquad (5.2.68)$$

$$\theta(x,\tau) \geq -k(x) \quad \text{for a.e. } x \in \Omega \text{ and each } \tau \geq 0. \qquad (5.2.69)$$

Now let us define $j : \Omega \times R \to R$ by the relation

$$j(x,\xi) = \int_0^\xi \theta(x,\tau)d\tau. \qquad (5.2.70)$$

Because of (i)–(iii) one can easily verify that j satisfies all the requirements of Corollary 5.7. Indeed,

$$
\begin{aligned}
|j(x,\xi) - j(x,\eta)| &= \Big| \int_\eta^\xi \theta(x,\tau)d\tau \Big| \leq \operatorname*{ess\,sup}_{-r \leq \tau \leq r} |\theta(x,\tau)||\xi - \eta| \\
&\leq \beta(x,r)|\xi - \eta| \quad \forall \xi, \eta \in B(0,r), r \geq 0. \quad (5.2.71)
\end{aligned}
$$

Hence, the conditions (5.1.4) with $\beta(\cdot, r) \in L^q(\Omega)$ for each $r \geq 0$, (5.1.5) and (5.1.6) hold. Further,

$$
\begin{aligned}
j^0(x; \xi, -\xi) &= \limsup_{\substack{h \to 0 \\ \lambda \to 0+}} \frac{j(x; \xi + h - \lambda\xi) - j(x, \xi + h)}{\lambda} \\
&= \limsup_{\substack{h \to 0 \\ \lambda \to 0+}} \frac{1}{\lambda} \int_{\xi+h}^{\xi+h-\lambda\xi} \theta(x,\tau)d\tau. \quad (5.2.72)
\end{aligned}
$$

This equality, owing to (5.2.68), leads for sufficiently small h and λ to

$$\frac{1}{\lambda} \int_{\xi+h}^{\xi+h-\lambda\xi} \theta(x,\tau)d\tau = \frac{1}{\lambda} \int_{\xi+h-\lambda\xi}^{\xi+h} [-\theta(x,\tau)]d\tau \leq k(x)\xi = k(x)|\xi| \qquad (5.2.73)$$

for $\xi > 0$, and to

$$\frac{1}{\lambda} \int_{\xi+h}^{\xi+h-\lambda\xi} \theta(x,\tau)d\tau \leq k(x)(-\xi) = k(x)|\xi| \qquad (5.2.74)$$

for $\xi < 0$. The foregoing estimates lead to (5.2.51). Now, let us recall (cf. (1.2.45)–(1.2.49)) that for j given by the relation (5.2.70),

$$\partial j(x;\xi) \subset [\underline{\theta}(x,\xi), \bar{\theta}(x,\xi)] \tag{5.2.75}$$

for a.e. $x \in \Omega$ and each $\xi \in R$, where the quantities $\underline{\theta}(x,\xi)$ and $\bar{\theta}(x,\xi)$ are defined by means of

$$\left.\begin{array}{rcl}
\underline{\theta}_\rho(x,\xi) &=& \text{ess inf}_{|\tau-\xi|\le\rho}\, \theta(x,\tau) \\[2mm]
\bar{\theta}_\rho(x,\xi) &=& \text{ess sup}_{|\tau-\xi|\le\rho}\, \theta(x,\tau)
\end{array}\right\} \quad \rho > 0, \tag{5.2.76}$$

according to the relations

$$\left.\begin{array}{rcl}
\underline{\theta}(x,\xi) &=& \lim_{\rho\to 0}\underline{\theta}_\rho(x,\xi) \\[2mm]
\bar{\theta}(x,\xi) &=& \lim_{\rho\to 0}\bar{\theta}_\rho(x,\xi)
\end{array}\right\} \quad \text{for a.e. } x \in \Omega \text{ and each } \xi \in R. \tag{5.2.77}$$

Now we are ready to formulate the following result generalizing Theorem 3.4.

Corollary 5.9 Let V be a reflexive Banach space and $\Omega \subset R^n$ an open bounded subset of $R^n, n \ge 1$. Suppose that the injection $V \subset L^p(\Omega)$ is compact for some $1 < p < \infty$, and that $V \cap L^\infty(\Omega)$ is dense in V. Moreover, let A be a coercive, weakly continuous operator from V into V^*. Suppose that $\theta : \Omega \times R \to R$ possesses all the properties listed in (i)–(iii). Then the problem: find $u \in V$ and $\chi \in L^1(\Omega)$ such that

$$\left.\begin{array}{l}
\langle Au - g, v\rangle_V + \int_\Omega \chi \cdot v\, d\Omega = 0 \quad \forall v \in V \cap L^\infty(\Omega) \\[2mm]
\chi(x) \in [\underline{\theta}(x, u(x)), \bar{\theta}(x, u(x))] \quad \text{for a.e. } x \in \Omega;
\end{array}\right\} \tag{5.2.78}$$

has at least one solution. If additionally we suppose that V has the property that for any $v \in V$ there exists a sequence $\{w_n\}$ such that $0 \le w_n \le 1$, $(1 - w_n)v \in V \cap L^\infty(\Omega)$ and $(1 - w_n)v \to v$ strongly in V, then $\chi u \in L^1(\Omega)$ and

$$\langle Au - g\rangle_V + \int_\Omega \chi u\, d\Omega = 0, \tag{5.2.79}$$

where $(u, \chi) \in V \times L^1(\Omega)$ is a solution of (5.2.78).

Remark 5.10 In a special case when the function θ does not depend on x, i.e. $\theta : R \to R$, the conditions (i)–(iii) reduce to the only two requirements, namely $\theta \in L^\infty_{\text{loc}}(R)$ and

$$\begin{array}{ll}
\theta(\tau) \le k & \text{for a.e. } \tau < 0 \\
\theta(\tau) \ge -k & \text{for a.e. } \tau \ge 0
\end{array} \tag{5.2.80}$$

where k is a nonnegative constant.

Remark 5.11 Let us consider a function $\theta \in L^\infty_{\text{loc}}(R)$ which is supposed to increase ultimately , i.e.

$$\operatorname*{ess\,sup}_{\tau\in(-\infty,-\xi)} \theta(\tau) \leq c \leq \operatorname*{ess\,inf}_{\tau\in(\xi,+\infty)} \theta(\tau) \qquad (5.2.81)$$

for some $\xi \geq 0$, and $c \in R$. Then after easy calculations we are led to the conclusion that

$$\begin{aligned} \theta(\tau) &\leq k \qquad \text{for a.e. } \tau < 0 \\ \theta(\tau) &\geq -k \qquad \text{for a.e. } \tau \geq 0, \end{aligned} \qquad (5.2.82)$$

where

$$k = |c| + \operatorname*{ess\,sup}_{-\xi \leq \tau \leq \xi} |\theta(\tau)|. \qquad (5.2.83)$$

Accordingly, ultimately increasing functions (cf. Chapter 3) belong to the class of functions obeying the directional growth condition (5.2.51).

5.3 Hemivariational Inequalities with Strongly Monotone Operator

This Section is devoted to study the problem (P) under an additional restriction concerning the operator $A : V \to V^*$. Namely, A is supposed to be strongly monotone, i.e.

$$\langle Au - Av, u - v \rangle_V \geq m||u - v||_V^2 \quad \forall u, v \in V, m = \text{ const } > 0. \qquad (5.3.1)$$

It turns out that in such a case it is possible to weaken (5.1.11) and replace it by

$$j^0(x; \xi, \eta - \xi) \leq \alpha(r, x)(1 + |\xi|^s) \qquad (5.3.2)$$

where $\xi, \eta \in R^N$ with $\eta \in B(0, r), r \geq 0$, and $1 \leq s < 2$. However, it needs an additional assumption:

$$\begin{aligned} &p \geq 2 \text{ and } \alpha(r, \cdot) \in L^{\bar{q}}(\Omega) \text{ for each } r \geq 0, \\ &\text{where } \bar{q} = p/(p - s). \end{aligned} \qquad (5.3.3)$$

Under the foregoing hypotheses the following theorem can be formulated.

Theorem 5.12 Let A be a strongly monotone, weakly continuous operator from V into V^* and let V be compactly imbedded into $L^p(\Omega; R^N)$ with $p \geq 2$. Suppose that (5.1.4)–(5.1.6), (5.3.2) and (5.3.3) hold. Then problem (P) has at least one solution.

The proof of this theorem follows along the lines of the proof of Theorem 5.5. First we formulate the regularized finite-dimensional problem $(P_{F\varepsilon})$ given by means of (5.2.4). Then it is shown that statements analogous to those of Lemma 5.2, Lemma 5.3 and Lemma 5.4 hold.

Lemma 5.13 Suppose that (5.1.4)–(5.1.6) and (5.3.2) hold. Then the estimate

$$j'_\varepsilon(x; \xi) \cdot (\eta - \xi) \leq \hat{\alpha}(x, r)(1 + |\xi|^s), \quad 0 < \varepsilon \leq 1, \qquad (5.3.4)$$

holds for almost all $x \in \Omega$ and any $\xi, \eta \in R^N$ with $|\eta| \leq r, r \geq 0$, where $\hat{\alpha}(x;r)$ is defined as $3\alpha(x;r+1)$.

Proof. Let $\eta \in R^N$ with $|\eta| \leq r$. The assertion follows directly from the estimates

$$
\begin{aligned}
j_\epsilon'(x;\xi) \cdot (\eta - \xi) &= \lim_{t \to 0} \frac{j_\epsilon(x, \xi + t(\eta - \xi)) - j_\epsilon(x;\xi)}{t} \\
&\leq \int_{R^N} p_\epsilon(\tau) \limsup_{t \to 0} \frac{j(x;\xi - \tau + t(\eta - \xi)) - j(x;\xi - \tau)}{t} d\tau \\
&\leq \int_{R^N} p_\epsilon(\tau) j^0(x;\xi - \tau, \eta - \tau - (\xi - \tau)) d\tau \\
&\leq \int_{R^N} p_\epsilon(\tau) \alpha(x;\varepsilon + |\eta|)(1 + |\xi - \tau|^s) d\tau \\
&\leq \int_{R^N} p_\epsilon(\tau) \alpha(x;1 + |\eta|)[1 + (1 + |\xi|)^s] d\tau \qquad (5.3.5) \\
&\leq \int_{R^N} p_\epsilon(\tau) \alpha(x;1 + r)[1 + (1 + |\xi|)^s] d\tau \\
&\leq 3\alpha(x, r + 1)(1 + |\xi|^s) = \hat{\alpha}(x;r)(1 + |\xi|^s),
\end{aligned}
$$

where we have applied the Fatou's lemma and the inequality

$$
1 + (t + 1)^s \leq 3(1 + t^s), \quad 1 \leq s < 2, \quad t \geq 0. \qquad (5.3.6)
$$

The proof is complete.   q.e.d.

Lemma 5.14 For any $F \in \Lambda$ and $\varepsilon \in (0, 1]$ the problem $(P_{F\epsilon})$ has at least one solution $u_{F\epsilon} \in F$. Moreover, the family $\{u_{F\epsilon}\}$ is uniformly bounded in the space V.

Proof. The similarity with the previous case allows us to preserve the same notations and to focus on the main part only, which concerns the inequalities

$$
\begin{aligned}
(\tilde{T}_{F\epsilon} v, v)_F &= \langle A_F v - g_F + h_{F\epsilon}(v), v \rangle_F \qquad (5.3.7) \\
&= \langle Av, v \rangle_V - \langle g, v \rangle_V + \int_\Omega j_\epsilon'(v) \cdot v \, d\Omega \\
&\geq m\|v\|_V^2 - \|g\|_{V^*}\|v\|_V - \int_\Omega \hat{\alpha}(0)(1 + |v|^s) d\Omega \\
&\geq m\|v\|_V^2 - \|g\|_{V^*}\|v\|_V \\
&\quad - \|\hat{\alpha}(0)\|_{L^{\bar{q}}(\Omega)} \mathrm{mes}\,\Omega^{1/\bar{p}} - \|\hat{\alpha}(0)\|_{L^{\bar{q}}(\Omega)}\|v\|_{L^p(\Omega)}^s \\
&\geq m\|v\|_V^2 - \|g\|_{V^*}\|v\|_V \\
&\quad - \|\hat{\alpha}(0)\|_{L^{\bar{q}}(\Omega)} \mathrm{mes}\,\Omega^{1/\bar{p}} - \|\hat{\alpha}(0)\|_{L^{\bar{q}}(\Omega)} \gamma'\|v\|_V^s,
\end{aligned}
$$

where $1/\bar{p} + 1/\bar{q} = 1$. Since $s < 2$, we deduce easily that

$$(\tilde{\Gamma}_{F\epsilon}v, v)_F \geq 0 \qquad (5.3.8)$$

for $\|v\|_V$ sufficiently large. Consequently, the existence of a solution to $(P_{F\epsilon})$ follows from Brouwer's fixed point theorem. The uniform boundedness of $\{u_{F\epsilon}\}$ is obtained by taking into account the estimate

$$m\|u_{F\epsilon}\|_V^2 \leq \|g\|_{V^*}\|u_{F\epsilon}\|_V \qquad (5.3.9)$$
$$+\|\hat{\alpha}(0)\|_{L^q(\Omega)}\text{mes}\Omega^{1/\bar{p}} + \|\hat{\alpha}(0)\|_{L^q(\Omega)}\gamma^s\|u_{F\epsilon}\|_V^s.$$

The proof is complete. $\qquad\qquad\qquad\qquad\qquad\qquad\qquad\qquad$ q.e.d.

Lemma 5.15 Let $u_{F\epsilon} \in F$ be a solution to $(P_{F\epsilon})$ for any $F \in \Lambda$ and $\epsilon \in (0,1]$. Then the set $\{j_\epsilon'(u_{F\epsilon}) : F \in \Lambda, \epsilon \in (0,1]\}$ is weakly precompact in $L^1(\Omega; R)$.

Proof. Since we use the same method as in the proof of Lemma 5.3 we preserve all the notations and focus our attention on the main points only. Namely, from the inequality

$$j_\epsilon'(x; u_{F\epsilon}(x)) \cdot \eta \leq j_\epsilon'(x; u_{F\epsilon}(x)) \cdot u_{F\epsilon}(x) + \hat{\alpha}(x; r)(1 + |u_{F\epsilon}(x)|^s), \qquad (5.3.10)$$

when integrated over ω, we obtain

$$\int_\omega |j_\epsilon'(u_{F\epsilon})|d\Omega \leq \frac{\sqrt{N}}{r} \int_\omega j_\epsilon'(u_{F\epsilon}) \cdot u_{F\epsilon}d\Omega \qquad (5.3.11)$$
$$+\frac{\sqrt{N}}{r}\|\hat{\alpha}(r)\|_{L^q(\omega)}\text{mes}\omega^{1/\bar{p}} + \frac{\sqrt{N}}{r}\|\hat{\alpha}(r)\|_{L^q(\omega)}\|u_{F\epsilon}\|_{L^p(\omega)}^s,$$

which due to the boundedness of $\{u_{F\epsilon}\}$ in V yields

$$\int_\omega |j_\epsilon'(u_{F\epsilon})|d\Omega \leq \frac{\sqrt{N}}{r} \int_\omega j_\epsilon'(u_{F\epsilon}) \cdot u_{F\epsilon}d\Omega \qquad (5.3.12)$$
$$+\frac{\sqrt{N}}{r}\|\hat{\alpha}(r)\|_{L^q(\Omega)}\text{mes}\omega^{1/\bar{p}} + \frac{\sqrt{N}}{r}\|\hat{\alpha}(r)\|_{L^q(\omega)}\gamma^s M^s.$$

Now by making use of the estimate

$$j_\epsilon'(x; u_{F\epsilon}(x)) + \hat{\alpha}(x; 0)(1 + |u_{F\epsilon}(x)|^s) \geq 0 \quad \text{for a.e. } x \in \Omega \qquad (5.3.13)$$

we analogously may show that

$$\int_\omega j_\epsilon'(u_{F\epsilon}) \cdot u_{F\epsilon}d\Omega \leq C \qquad (5.3.14)$$

for some positive constant C not depending on $\omega \subset \Omega$, $F \in \Lambda$ and $\epsilon \in (0,1]$. Finally, the inequality results

$$\int_\omega |j_\epsilon'(u_{F\epsilon})|d\Omega \leq \frac{\sqrt{N}}{r}C + \frac{\sqrt{N}}{r}\|\hat{\alpha}(r)\|_{L^q(\Omega)}\text{mes}\omega^{1/\bar{p}} + \frac{\sqrt{N}}{r}\|\hat{\alpha}(r)\|_{L^q(\omega)}\gamma^s M^s$$
$$(5.3.15)$$

from which, by means of the same arguments as previously, we show that for any $\varepsilon > 0$ there exists $\delta > 0$ with the property that

$$\int_\omega |j'_\varepsilon(u_{F\varepsilon})|d\Omega \le \varepsilon \qquad (5.3.16)$$

whenever mes$\omega < \delta$, as desired.   q.e.d.

The remaining parts of the proof of Theorem 5.12 coincide with the corresponding parts of the proof of Theorem 5.5 and therefore the details will be omitted here.

Remark 5.16 In the onedimensional case, i.e. for $N = 1$, the condition (5.3.2) can be relaxed. Namely, if we suppose that $p \ge 2$, $1 \le s < 2$, $\bar{q} = p/(p - s)$, and that (5.1.6) holds with $\beta(\cdot, r) \in L^{\bar{q}}(\Omega)$ for each $r \ge 0$, then the estimate

$$j^0(x; \xi, -\xi) \le k(x)(1 + |\xi|^s) \quad \forall \xi \in R, \text{ for a.e. } x \in \Omega, \qquad (5.3.17)$$

where $k \in L^{\bar{q}}(\Omega)$ is a nonnegative function, implies (5.3.2). To show this, let us notice that (5.3.17) yields

$$j^0(x; \xi, -1) \le k(x)(1 + \xi^s)(1/\xi) \quad \xi \ge 0 \qquad (5.3.18a)$$

$$j^0(x; \xi, 1) \le k(x)(1 + (-\xi)^s)(-1/\xi) \quad \xi < 0. \qquad (5.3.18b)$$

At the same time, eq. (5.1.6) becomes

$$\begin{aligned} j^0(x; \xi, \eta - \xi) &\le \beta(x; |\xi| + 1)|\eta - \xi| \\ &\le \beta(x; r + 1)(r + |\xi|) \le \beta(x; r + 1)(r + 1)(1 + |\xi|) \\ &\le 2\beta(x; r + 1)(r + 1)(1 + |\xi|^s) \end{aligned} \qquad (5.3.19)$$

for any $\xi, \eta \in B(0, r), r \ge 0$. Here we have applied the inequality

$$1 + t \le 2(1 + t^s) \text{ for } t \ge 0 \text{ and } 1 \le s < 2. \qquad (5.3.20)$$

Further, if $\xi > r \ge \eta \ge -r, r \ge 0$, then by the positive homogeneity of $j^0(x; \xi, \cdot)$, Eq. (5.3.18a) implies that

$$j^0(x; \xi, \eta - \xi) \le k(x)(1 + \xi^s)(1 - \frac{\eta}{\xi}) \le 2k(x)(1 + |\xi|^s). \qquad (5.3.21)$$

Analogously, if $\xi < -r \le \eta \le r, r \ge 0$, we obtain that

$$j^0(x; \xi, \eta - \xi) \le k(x)(1 + (-\xi)^s)(1 - \frac{\eta}{\xi}) \le 2k(x)(1 + |\xi|^s). \qquad (5.3.22)$$

Defining $\alpha : \Omega \times R^+ \to R$ as

$$\alpha(x, r) = \max\{2k(x), 2\beta(x; r + 1)(r + 1)\}, \qquad (5.3.23)$$

implies the desired estimate (5.3.2).

Corollary 5.17 Let V be a reflexive Banach space and Ω an open bounded subset of $R^n, n \geq 1$. Suppose that the injection $V \subset L^p(\Omega)$ is compact for some $2 \leq p < \infty$, and that $V \cap L^\infty(\Omega)$ is dense in V. Moreover, let A be a coercive, weakly continuous operator from V into V^*. Suppose that (5.3.17) and (5.1.6) hold with $\beta(\cdot, r) \in L^{\bar{q}}(\Omega)$ for each $r \geq 0$. Then problem (P) has at least one solution $(u, \chi) \in V \times L^1(\Omega)$. If additionally we suppose that for any $v \in V$ there exists a sequence $\{w_n\}$ with the properties that $0 \leq w_n \leq 1$, $(1 - w_n)u \in V \cap L^\infty(\Omega)$ and $(1 - w_n)u \to u$ strongly in V, then $\chi u \in L^1(\Omega)$ and

$$\langle Au - g, u \rangle_V + \int_\Omega \chi u \, d\Omega = 0. \tag{5.3.24}$$

Let $\theta : \Omega \times R \to R$ be a function from $\Omega \times R$ into R satisfying the conditions:

(a) $\theta(x, \cdot) \in L^\infty_{\text{loc}}(R)$ for a.e. $x \in \Omega$;

(b) The function $\beta : \Omega \times R^+ \to R$ defined by

$$\beta(x, r) = \operatorname*{ess\,sup}_{-r \leq \tau \leq r} |\theta(x, \tau)| \tag{5.3.25}$$

for a.e. $x \in \Omega$ and each $r \geq 0$, has the property that

$$\beta(\cdot, r) \in L^{\bar{q}}(\Omega) \text{ for all } r \geq 0; \tag{5.3.26}$$

(c) There exists a nonnegative function $k \in L^{\bar{q}}(\Omega)$ such that

$$\theta(x, \tau) \leq k(x)|\tau|^{s-1} \text{ for a.e. } x \in \Omega \text{ and each } \tau < 0,$$
$$\theta(x, \tau) \geq -k(x)|\tau|^{s-1} \text{ for a.e. } x \in \Omega \text{ and each } \tau \geq 0. \tag{5.3.27}$$

Define $j : \Omega \times R \to R$ by (5.2.70). It is easily verified that if (a)–(c) hold, then the function j fulfills all the hypotheses of Corollary 5.17. Thus we are led to the following result.

Corollary 5.18 Let V be a reflexive Banach space and $\Omega \subset R^n$ an open bounded subset of $R^n, n \geq 1$. Suppose that the injection $V \subset L^p(\Omega)$ is compact for some $2 \leq p < \infty$, and that $V \cap L^\infty(\Omega)$ is dense in V. Moreover, let A be strongly monotone and weakly continuous operator from V into V^*. Suppose that $\theta : \Omega \times R \to R$ satisfies (a)–(c). Then the problem: find $u \in V$ and $\chi \in L^1(\Omega)$ such that

$$\langle Au - g, v \rangle_V + \int_\Omega \chi \cdot v \, d\Omega = 0 \quad \forall v \in V \cap L^\infty(\Omega), \tag{5.3.28}$$

$$\chi(x) \in [\underline{\theta}(x, u(x)), \bar{\theta}(x, u(x))] \quad \text{for a.e. } x \in \Omega; \tag{5.3.29}$$

has at least one solution. If additionally we suppose that for any $v \in V$ there exists a sequence $\{w_n\}$ with the properties that $0 \leq w_n \leq 1$, $(1 - w_n)u \in V \cap L^\infty(\Omega)$ and $(1 - w_n)u \to u$ strongly in $L^p(\Omega)$, then $\chi u \in L^1(\Omega)$ and (5.3.24) holds.

Remark 5.19 In the special case where the function θ does not depend on x, i.e. $\theta : R \to R$, assumptions listed in (a)–(c) reduce only to the two requirements: $\theta \in L^{\infty}_{loc}(R)$ and

$$\theta(\tau) \leq k|\tau|^{s-1} \qquad \text{for a.e. } \tau < 0,$$
$$\theta(\tau) \geq -k|\tau|^{s-1} \qquad \text{for a.e. } \tau \geq 0,$$
(5.3.30)

where k is a nonnegative constant.

Let us consider further the following discontinuous semilinear equation: For $\lambda \in R$, $g \in L^2(\Omega)$ and $\theta : \Omega \times R \to R$ we seek for a $u \in H^1_0(\Omega)$ such that

$$-\Delta u + \lambda u + g \in -[\underline{\theta}(\cdot, u(\cdot)), \bar{\theta}(\cdot, u(\cdot))] \quad \text{in } \Omega. \qquad (5.3.31)$$

Corollary 5.20 Let $\theta : \Omega \times R \to R$ fulfills all the requirements of Corollary 5.18 and let $\lambda > -\lambda_1$, where λ_1 is the smallest eigenvalue of the operator $-\Delta$. Then the discontinuous semilinear equation (5.3.31) with the nonlinearities determined by θ has at least one solution $u \in H^1_0(\Omega)$.

Proof. The inclusion (5.3.31) corresponds to the following hemivariational inequality

$$\langle Au - g, v - u \rangle_{H^1(\Omega)} + \int_{\Omega} j^0(u, v - u)d\Omega \geq 0 \quad \forall v \in H^1_0(\Omega), \qquad (5.3.32)$$

where $A : H^1_0(\Omega) \to H^{-1}(\Omega)$ is a linear operator defined by

$$\langle Au, v \rangle_{H^1(\Omega)} = \int_{\Omega} (\nabla u \, \nabla v + \lambda uv)d\Omega \quad \forall u, v \in H^1_0(\Omega), \qquad (5.3.33)$$

and $j : R \to R$ is represented by the integral (5.2.70). Recalling that the Rayleigh-Ritz characterization of the smallest Dirichlet eigenvalue λ_1 is given by

$$\lambda = \inf_{\substack{u \in H^1(\Omega) \\ u \neq 0}} \frac{\int_{\Omega} |\nabla u|^2 d\Omega}{\int_{\Omega} |u|^2 d\Omega} \qquad (5.3.34)$$

we get the following estimate

$$\langle Av, v \rangle_{H^1(\Omega)} \geq \min\{1, 1 + \frac{\lambda}{\lambda_1}\} \|v\|^2_{H^1(\Omega)}, \qquad (5.3.35)$$

from which the strong monotonicity of A for $\lambda > \lambda_1$ follows. Accordingly, the hypotheses of Corollary 5.18 hold. Consequently, (5.3.31) admits at least one solution $u \in H^1_0(\Omega)$.   q.e.d.

Let us compare the results of this Section with those corresponding to the "smooth" case, i.e., when $j'(x, \xi) = g(x, \xi)$, $x \in \Omega$, $\xi \in R$, where $g : \Omega \times R \to R$ is a Caratheodory function.

As it is well known, the existence result for the problem: find $u \in V$ such that

$$\langle Au - g, v \rangle_V + \int_\Omega g(u)vd\Omega = 0 \quad \forall v \in V \cap L^\infty(\Omega) \tag{5.3.36}$$

has been obtained on the assumption that A is bounded and pseudo-monotone and that g satisfies the sign condition [Web]. In our consideration A has been supposed to be weakly continuous only. Why do we not consider a weaker hypothesis on A assuming that A is a bounded pseudo-monotone operator like in the "smooth" case? The reason is the following: in the approach presented here we cannot deduce that the approximating sequence $\{j_n'(x, u_n(x))\}$ converges to $\chi(x)$ for a.e. $x \in \Omega$, even if $u_n \to u$ a.e. in Ω. This is not the case for a Caratheodory function $g : \Omega \times R \to R$, where, due to the continuity of g with respect to the second variable, the truncation procedure permits to write

$$g^{(n)}(x, u_n(x)) \to g(x, u(x)) \text{ for a.e. } x \in \Omega \tag{5.3.37}$$

provided $u_n(x) \to u(x)$ a.e. in Ω. Accordingly, in the proof of Theorem 5.5 we are not allowed to deduce on the basis of Fatou's lemma that

$$\liminf \int_\Omega j_n'(u_n)u_n d\Omega \geq \int_\Omega \chi u d\Omega, \tag{5.3.38}$$

which would permit us to conclude that

$$\limsup \langle Au_n, u_n - u \rangle_V \leq 0, \tag{5.3.39}$$

and, consequently, to prove the assertion for a pseudo-monotone A. Note also that for the same reason, without an additional hypothesis concerning the existence for each $v \in V$ of a sequence $\{w_n\}$ with the properties $0 \leq w_n \leq 1$, $(1 - w_n)v \in V \cap L^\infty(\Omega)$ and $(1 - w_n)v \to v$ strongly in V, we would not be able to claim that $\chi u \in L^1(\Omega)$, $((u, \chi) \in V \times L^1(\Omega)$ being a solution of $(P))$.

Remark 5.21 To complete this Section we shall characterize a class of locally Lipschitz functions $j : R^N \to R$ satisfying (5.3.2). Let us suppose that the following condition of relaxed monotonicity holds

$$(\xi^* - \eta^*) \cdot (\xi - \eta) \geq -k|\xi - \eta|^s \quad \forall \xi, \eta \in R^N, \tag{5.3.40}$$

for any $\xi^* \in \partial j(\xi)$ and $\eta^* \in \partial j(\eta)$, where $k \geq 0$ is a nonnegative constant. Then by making use of the equality $j^0(\xi, \eta - \xi) = \xi^* \cdot (\eta - \xi)$ for some $\xi^* \in \partial j(\xi), \xi \in R^N$, and of (5.3.40), for $\eta \in B(0, r)$ we obtain

$$\begin{aligned} j^0(\xi, \eta - \xi) &= \xi^* \cdot (\eta - \xi) \leq \eta^* \cdot (\eta - \xi) + k|\xi - \eta|^s \\ &\leq L(r)(r + |\xi|) + k(r + |\xi|)^s \\ &\leq L(r)(r + 1)(1 + |\xi|) + k(r + 1)^s(1 + |\xi|)^s \\ &\leq (L(r)(r + 1) + k(r + 1)^s)(1 + |\xi|)^s, \end{aligned} \tag{5.3.41}$$

where $L(r)$ denotes the Lipschitz constant of j in the ball $B(0, r + 1)$. Thus setting

$$\alpha(r) = L(r)(r + 1) + k(r + 1)^s \qquad (5.3.42)$$

implies directly the estimate (5.3.2). It is possible to weaken the condition of relaxed monotonicity (5.3.40) and to replace it by the following one

$$(\xi^* - \eta^*) \cdot (\xi - \eta) \geq -\pi(r)|\xi - \eta|^s \quad \forall \xi, \eta \in R^N, \ |\eta| \leq r, \ r \geq 0, \quad (5.3.43)$$

where $\pi : R^+ \to R^+$ is a nondecreasing function. Then by making use of the same method as previously, we can prove that the function j has the property (5.3.2) with

$$\alpha(r) = L(r)(r + 1) + \pi(r)(r + 1)^s. \qquad (5.3.44)$$

The above relaxation of the monotonicity condition seems to be an efficient method to construct nonconvex locally Lipschitz functionals satisfying the directional growth condition of the form (5.3.2).

5.4 Hemivariational Inequalities with Relaxed Directional Growth Condition

In this Section we study the problem (P) in the case when A is coercive and $N > 1$, and under a condition weaker than the (5.1.11). We suppose that the following directional growth conditions hold:

$$j^0(x; \xi, -\xi) \leq k(x)|\xi| \quad \forall \xi \in R^N \text{ and a.e. } x \in \Omega; \qquad (5.4.1)$$

and

$$j^0(x; \xi, \eta - \xi) \leq \alpha(x, r)(1 + |x|^s) \qquad \forall \xi, \eta \in R^N \text{ with } \eta \in B(0, r), \quad (5.4.2)$$
$$r > 0, \text{ and a.e. } x \in \Omega,$$

where $1 \leq s < p$ and where functions $k : \Omega \to R^+$ and $\alpha : \Omega \times R^+ \to R^+$ are assumed to have the following properties

$$k(\cdot) \in L^q(\Omega), \text{ where } 1/p + 1/q = 1; \qquad (5.4.3)$$

$$\alpha(\cdot, r) \in L^{q'}(\Omega) \text{ for each } r > 0, \text{ where } q' = p/(p - s); \qquad (5.4.4)$$

$$\text{If } 0 < r' \leq r'', \text{ then for almost all } x \in \Omega, \ \alpha(x, r') \leq \alpha(x, r''). \qquad (5.4.5)$$

The approach presented in Section 5.2 based on the application of a mollifier for the regularization of the corresponding hemivariational inequalities is not suitable for the study of the case (5.4.1)−(5.4.2). The reason is that the regularization of the function j by means of the mollifier does not lead to the same type of directional growth conditions like those of (5.4.1) and (5.4.2). In this

Section we shall use the Galerkin approximation combined with the Kakutani fixed point theorem.

Let Λ be the family of all finite-dimensional subspaces F of $V \cap L^\infty(\Omega; R^N)$, ordered by inclusion. For any $F \in \Lambda$ we formulate the following finite dimensional problem.

Problem (P_F). Find $u_F \in F$ and $\chi_F \in L^1(\Omega; R^N)$ such as to satisfy the following relations

$$\langle Au_F - g, v \rangle_V + \int_\Omega \chi_F \cdot v d\Omega = 0 \quad \forall v \in F, \tag{5.4.6}$$

and

$$\chi_F(x) \in \partial j(x; u_F(x)) \quad \text{for a.e. } x \in \Omega. \tag{5.4.7}$$

Let us recall that if $F \in \Lambda$ and $v_F \in F$, then from (5.1.4) and (5.1.6) it results that

$$|j^0(x; v_F(x), w(x))| \le \beta(x, |v_F(x)| + 1)|w(x)| \le \beta(x, \|v_F\|_{L^\infty(\Omega)} + 1)\|w\|_{L^\infty(\Omega)} \tag{5.4.8}$$

for any $w \in L^\infty(\Omega; R^N)$ and for a.e. $x \in \Omega$. Due to (5.1.4) we have $\beta(\|u_F\|_{L^\infty(\Omega)} + 1) \in L^1(\Omega)$. Thus

$$\int_\Omega |j^0(v_F, w)| d\Omega \le \int_\Omega \beta(\|v_F\|_{L^\infty(\Omega)} + 1) d\Omega \|w\|_{L^\infty(\Omega)} \quad \forall w \in L^\infty(\Omega; R^N) \tag{5.4.9}$$

and the integral on the left hand side of (5.4.9) is finite. It allows us to define the mapping $\Gamma_F : F \to 2^{L^1(\Omega; R^N)}$ by the following relation

$$\Gamma_F(v_F) = \{\psi \in L^1(\Omega; R^N) : \int_\Omega \psi \cdot w d\Omega \le \int_\Omega j^0(v_F, w) d\Omega \quad \forall w \in L^\infty(\Omega; R^N)\}, \tag{5.4.10}$$

for each $v_F \in F$. It follows immediately that, if $\psi \in \Gamma_F(v_F)$, then $\psi(x) \in \partial j(x; v_F(x))$ for a.e. $x \in \Omega$.

Lemma 5.22 Let $v_F \in F$ for some $F \in \Lambda$. Then $\Gamma_F(v_F)$ is a nonempty, convex and weakly compact subset of $L^1(\Omega; R^N)$.

Proof. Since $R^N \ni \xi \to j(x, \xi)$ is locally Lipschitz, $\partial j(x, v_F(x))$ is not empty for a.e. $x \in \Omega$. Setting $\psi(x) \in \partial j(x, v_F(x))$ for a.e. $x \in \Omega$, we obtain the function ψ which, due to (5.4.9) and (5.4.10), is an element of $\Gamma_F(v_F)$. The convexity of $\Gamma_F(v_F)$ follows immediately from the fact that the mapping $L^\infty(\Omega; R^N) \ni w \to j^0(v_F, w)$ is convex. To show the weak precompactness of $\Gamma_F(v_F)$ in $L^1(\Omega; R)$ let us choose a $\psi \in \Gamma_F(v_F)$. Then from (5.4.9) it follows

$$\int_\Omega \psi \cdot w d\Omega \le \int_\Omega j^0(v_F, w) d\Omega \le \int_\Omega \beta(\|v_F\|_{L^\infty(\Omega)} + 1) d\Omega \|w\|_{L^\infty(\Omega)} \tag{5.4.11}$$

for any $w \in L^\infty(\Omega; R^N)$. Let us denote by $\hat{\psi}$ an element from $L^\infty(\Omega; R^N)$ with the properties that $\|\hat{\psi}\|_{L^\infty(\Omega)} \leq \sqrt{N}$ and

$$\int_\omega |\psi| d\Omega \leq \int_\omega \psi \cdot \hat{\psi} d\Omega \tag{5.4.12}$$

for a measurable $\omega \subset \Omega$. This easily implies

$$\int_\omega |\psi| d\Omega \leq \sqrt{N} \int_\omega \beta(\|v_F\|_{L^\infty(\Omega)} + 1) d\Omega. \tag{5.4.13}$$

Since

$$\beta(\|v_F\|_{L^\infty(\Omega)} + 1) \in L^1(\Omega), \tag{5.4.14}$$

for any $\varepsilon > 0$ there exists $\delta > 0$ such that

$$\int_\omega |\psi| d\Omega \leq \sqrt{N} \int_\omega \beta(\|v_F\|_{L^\infty(\Omega)} + 1) d\Omega \leq \varepsilon, \tag{5.4.15}$$

whenever $\mathrm{mes}\,\omega < \delta$. Thus the Dunford-Pettis criterion for the weak precompactness in $L^1(\Omega)$ holds (see cf. [Eke]). Finally, it remains to show that $\Gamma_F(v_F)$ is weakly closed. For this purpose let us suppose that a sequence $\{\psi_n\} \subset \Gamma_F(v_F)$ converges weakly to ψ in $L^1(\Omega; R^N)$. Then we have

$$\int_\Omega \psi_n \cdot w d\Omega \leq \int_\Omega j^0(v_F, w) d\Omega \quad \forall w \in L^\infty(\Omega; R^N), \tag{5.4.16}$$

from which by passing to the limit $n \to \infty$ we obtain that

$$\int_\Omega \psi \cdot w d\Omega \leq \int_\Omega j^0(v_F, w) d\Omega \quad \forall w \in L^\infty(\Omega; R^N). \tag{5.4.17}$$

It means that $\psi \in \Gamma_F(v_F)$ and the proof of Lemma 3.1 is complete.   q.e.d.

For $F \in \Lambda$, let $i_F : F \to V$ denote the inclusion mapping of F into V, and let $i_F^* : V^* \to F^*$ be the transpose of i_F. We denote by $\langle \cdot, \cdot \rangle_F$ the duality pairing over $F^* \times F$. Further, let $\tau_F : L^1(\Omega; R^N) \to F^*$ be an operator which assigns to any $\psi \in L^1(\Omega; R^N)$ the element $\tau_F \psi \in F^*$ defined by

$$\langle \tau_F \psi, v \rangle_F = \int_\Omega \psi \cdot v d\Omega \quad \text{for any } v \in F. \tag{5.4.18}$$

Now let us consider a mapping $T_F : F \to 2^{F^*}$ given by formula

$$T_F(v_F) = \tau_F \Gamma_F(v_F) \quad \text{for } v \in F. \tag{5.4.19}$$

The following properties of T_F can be proved.

Lemma 5.23 T_F is a mapping from F into 2^{F^*} such that for each $v_F \in F$, $T_F(v_F)$ is a nonempty bounded, closed, convex subset of F^*. Moreover, T_F is upper semicontinuous from F into 2^{F^*}.

Proof. Let us note that τ_F defined by (5.4.18) is linear and continuous from the weak topology on $L^1(\Omega; R^N)$ to the (unique) topology on F^*. Thus from Lemma 5.22 it results that for each $v_F \in F$, $T_F(v_F) = \tau_F \Gamma_F(v_F)$ is a nonempty bounded, closed, convex subset of F^*. It remains to show that T_F is upper semicontinuous from F into 2^{F^*}. For this purpose let us suppose that a sequence $\{v_{Fn}\} \subset F$ converges to v_F in F, and that the corresponding sequence $\{\tau_F \psi_n\} \subset F^*$ with $\psi_n \in \Gamma_F(v_{Fn})$, converges to some $\psi^* \in F^*$. We have to prove that there exists $\psi \in \Gamma_F(v_F)$ such that $\psi^* = \tau_F \psi$.

First we notice that the convergence of $\{v_{Fn}\}$ in F implies the existence of a finite upper bound for $\{\|v_{Fn}\|_{L^\infty(\Omega)}\}$, say C, i.e.,

$$\|v_{Fn}\|_{L^\infty(\Omega)} \leq C \quad \text{for } n = 1, 2, \ldots . \tag{5.4.20}$$

Taking into account that

$$\int_\Omega \psi_n \cdot w d\Omega \leq \int_\Omega j^0(v_{Fn}, w) d\Omega \quad \forall w \in L^\infty(\Omega; R^N), \tag{5.4.21}$$

we obtain the estimate

$$\int_\omega |\psi_n| d\Omega \leq \sqrt{N} \int_\omega \beta(C+1) d\Omega, \quad \omega \subset \Omega, \tag{5.4.22}$$

which, due to the integrability of $\beta(C+1)$, allows us to conclude that for any $\varepsilon > 0$ there exists $\delta > 0$ with the property that

$$\int_\omega |\psi_n| d\Omega \leq \sqrt{N} \int_\omega \beta(C+1) d\Omega \leq \varepsilon \quad n = 1, 2, \ldots, \tag{5.4.23}$$

provided mes$\omega < \delta$. Applying the Dunford-Pettis theorem we assert the weak precompactness of $\{\psi_n\}$ in $L^1(\Omega; R^N)$. Accordingly, a subsequence of $\{\psi_n\}$ can be extracted (again denoted by the same symbol) such that $\psi_n \to \psi$ weakly in $L^1(\Omega; R^N)$ for some $\psi \in L^1(\Omega; R^N)$. This fact implies that $\tau_F \psi_n \to \tau_F \psi = \psi^*$ in F^*. We claim that $\psi \in \Gamma_F(v_F)$. Indeed, taking into account the upper semicontinuity of the function

$$F \ni v_{Fn} \to \int_\Omega j^0(v_{Fn}, w) d\Omega, \quad \forall w \in L^\infty(\Omega; R^N) \tag{5.4.24}$$

we are led by means of (5.4.21) directly to the relation

$$\int_\Omega \psi \cdot w d\Omega \leq \int_\Omega j^0(v_F, w) d\Omega \quad \forall w \in L^\infty(\Omega; R^N), \tag{5.4.25}$$

which implies that $\psi \in \Gamma_F(v_F)$. The proof is complete.   q.e.d.

Now we are ready to pass to the existence result for (P_F).

Proposition 5.24 For every $F \in \Lambda$ the problem (P_F) has at least one solution $(u_F, \chi_F) \in F \times L^1(\Omega; R^N)$, i.e., (5.4.6) and (5.4.7) hold. Moreover, the family $\{u_F\}_{F\in\Lambda}$ is uniformly bounded, i.e., there exists a positive constant M independent of $F \in \Lambda$ such that

$$||u_F||_V \leq M, \quad F \in \Lambda. \tag{5.4.26}$$

Proof. Define $A_F = i_F^* A i_F$ and $f_F = i_F^* f$. Notice that in order to obtain the solvability of (P_F) it suffices to show that f_F belongs to the range of the multivalued mapping $A_F + T_F$. We shall show that, in fact, Range $(A_F + T_F) = F^*$. Indeed, due to Lemma 5.23, for each $v \in F, A_F(v) + T_F(v)$ is a nonempty bounded closed convex subset of F^*. Further, the weak continuity of A from V into V^* implies the continuity of A_F from F into F^*. Therefore, due to Lemma 5.23 the mapping $A_F + T_F$ is upper semicontinuous from F into 2^{F^*}. Moreover, using (5.4.1), (5.4.9) and the coercivity of A we obtain that

$$
\begin{aligned}
\langle A_F v, v\rangle_F \;+\; \int_\Omega \psi \cdot v d\Omega &\geq \langle Av, v\rangle_V - \int_\Omega j^0(v, -v) d\Omega \\
&\geq c(||v||_V)||v||_V - \int_\Omega k|v|d\Omega \geq c(||v||_V)||v||_V - ||k||_{L^q(\Omega)}||v||_{L^p(\Omega)} \\
&\geq c(||v||_V)||v||_V - ||k||_{L^q(\Omega)}\gamma||v||_V, \quad v \in F, \; \psi \in \Gamma_F(v),
\end{aligned}
\tag{5.4.27}
$$

where $1/p + 1/q = 1$, and γ is a positive constant, such that $||v||_{L^p(\Omega)} \leq \gamma||v||_V, \; \forall v \in V$ (V is compactly imbedded into $L^p(\Omega; R^N)$). This means that $A_F + T_F$ is also coercive. Now it remains to apply Proposition 2.5 to conclude that Range $(A_F + T_F) = F^*$. Accordingly, for $g \in V^*$ there exist $u_F \in F$ and $\chi_F \in \Gamma_F(u_F)$ such that $g_F = A_F(u_F) + {}_{T_F}\chi_F$, i.e., (5.4.6) and (5.4.7) hold. To show that the family $\{u_F\}$ is uniformly bounded we make use of the estimates

$$
\begin{aligned}
||g||_{V^*}||u_F||_V &\geq \langle g, u_F\rangle_V = \langle g_F, u_F\rangle_F = \langle A_F u_F, u_F\rangle_F + \int_\Omega \chi_F \cdot u_F d\Omega \\
&\geq \langle A u_F, u_F\rangle_V - \int_\Omega j^0(u_F, -u_F) d\Omega \geq c(||u_F||_V)||u_F||_V \\
&\quad - \int_\Omega k|u_F|d\Omega \\
&\geq c(||u_F||_V)||u_F||_V - ||k||_{L^q(\Omega)}||u_F||_{L^p(\Omega)} \\
&\geq c(||u_F||_V)||u_F||_V - ||k||_{L^q(\Omega)}\gamma||u_F||_V,
\end{aligned}
\tag{5.4.28}
$$

which together with the well known properties of the coercive function $c = c(\cdot)$ (cf. (iii) in Sect. 5.1) lead to the result.   q.e.d.

The next lemma concerns the compactness property of the set $\{\chi_F : F \in \Lambda\}$ in $L^1(\Omega; R^N)$.

Lemma 5.25 Let for some $F \in \Lambda$ a pair $(u_F, \chi_F) \in F \times L^1(\Omega; R^N)$ be a solution of (P_F). Then the set $\{\chi_F \in L^1(\Omega; R^N) : (u_F, \chi_F)$ is a solution of (P_F) for some $u_F \in F, F \in \Lambda\}$ is weakly precompact in $L^1(\Omega; R^N)$.

Proof. According to the Dunford-Pettis theorem it suffices to show that for each $\varepsilon > 0$ a $\delta_\varepsilon > 0$ can be determined such that for any $\omega \subset \Omega$ with mes$\omega < \delta_\varepsilon$,

$$\int_\omega |\chi_F| d\Omega < \varepsilon, \quad F \in \Lambda. \tag{5.4.29}$$

Fix $r > 0$ and let $\eta \in R^N$ be such that $|\eta| \leq r$. Then we have

$$\chi_F(x) \cdot (\eta - u_F(x)) \leq j^0(x; u_F(x), \eta - u_F(x)) \tag{5.4.30}$$

from which, by virtue of (5.4.2) it results that

$$\chi_F(x) \cdot \eta \leq \chi_F(x) \cdot u_F(x) + \alpha(x; r)(1 + |u_F(x)|^s) \tag{5.4.31}$$

for a.e. $x \in \Omega$. Denoting by $\chi_{Fi}(x), i = 1, 2, \ldots, N$, the components of $\chi_F(x)$ we set

$$\eta(x) = \frac{r}{\sqrt{N}}(\text{sgn}\chi_{F1}(x), \ldots, \text{sgn}\chi_{FN}(x)), \tag{5.4.32}$$

where

$$\text{sgn}y = \begin{cases} 1 & \text{if } y > 0 \\ 0 & \text{if } y = 0 \\ -1 & \text{if } y < 0 \end{cases}. \tag{5.4.33}$$

It is not difficult to verify that $|\eta(x)| \leq r$ for almost all $x \in \Omega$ and that

$$\chi_F(x) \cdot \eta(x) \geq \frac{r}{\sqrt{N}}|\chi_F(x)|. \tag{5.4.34}$$

Therefore we are led to the estimate

$$\frac{r}{\sqrt{N}}|\chi_F(x))| \leq \chi_F(x) \cdot u_F(x) + \alpha(x; r)(1 + |u_F(x)|^s). \tag{5.4.35}$$

Integrating this inequality over $\omega \subset \Omega$ yields

$$\int_\omega |\chi_F| d\Omega \leq \frac{\sqrt{N}}{r} \int_\omega \chi_F \cdot u_F d\Omega + \frac{\sqrt{N}}{r}||\alpha(r)||_{L^{q'}(\Omega)} \text{mes}\omega^{s/p}$$

$$+ \frac{\sqrt{N}}{r}||\alpha(r)||_{L^{q'}(\omega)}||u_F||_{L^p(\Omega)}^s. \tag{5.4.36}$$

Thus, from (5.4.26) we obtain

$$\int_\omega |\chi_F| d\Omega \leq \frac{\sqrt{N}}{r} \int_\omega \chi_F \cdot u_F d\Omega \tag{5.4.37}$$

$$+ \frac{\sqrt{N}}{r}||\alpha(r)||_{L^{q'}(\Omega)} \text{mes}\omega^{s/p} + \frac{\sqrt{N}}{r}||\alpha(r)||_{L^{q'}(\omega)}\gamma^s||u_F||_V^s$$

$$\leq \frac{\sqrt{N}}{r} \int_\omega \chi_F \cdot u_F d\Omega$$

$$+ \frac{\sqrt{N}}{r}||\alpha(r)||_{L^{q'}(\Omega)} \text{mes}\omega^{s/p} + \frac{\sqrt{N}}{r}||\alpha(r)||_{L^{q'}(\omega)}\gamma^s M^s.$$

Now we show that

$$\int_\omega \chi_F \cdot u_F d\Omega \leq C \qquad (5.4.38)$$

for some positive constant C not depending on $\omega \subset \Omega$ and $F \in \Lambda$. Indeed, from (5.4.1) one can easily deduce that

$$\chi_F(x) \cdot u_F(x) + k(x)(1 + |u_F(x)|) \geq 0 \text{ for a.e. } x \in \Omega. \qquad (5.4.39)$$

Thus we can write that

$$\int_\omega [\chi_F \cdot u_F + k(1 + |u_F|)]d\Omega \leq \int_\Omega [\chi_F \cdot u_F + k(1 + |u_F|)]d\Omega, \qquad (5.4.40)$$

and consequently

$$\begin{aligned}
\int_\omega \chi_F \cdot u_F d\Omega &\leq \int_\Omega \chi_F \cdot u_F d\Omega \\
&\quad + ||k||_{L^q(\Omega)} \operatorname{mes}\Omega^{1/p} + ||k||_{L^q(\Omega)}\gamma ||u_F||_V \qquad (5.4.41) \\
&\leq \int_\Omega \chi_F \cdot u_F d\Omega \\
&\quad + ||k||_{L^q(\Omega)} \operatorname{mes}\Omega^{1/p} + ||k||_{L^q(\Omega)}\gamma M.
\end{aligned}$$

But A maps bounded sets into bounded sets because of its weak continuity. Therefore, by means of (5.4.6) and (5.4.26) we conclude that

$$\int_\Omega \chi_F \cdot u_F d\Omega = -\langle Au_F - g, u_F \rangle_V \leq ||Au_F - g||_{V^*} ||u_F||_V \leq \hat{C}, \quad \hat{C} = \text{const.}$$

$$(5.4.42)$$

From the last two estimates we easily obtain (5.4.38), as desired. From (5.4.38) and (5.4.37) one obtains for $r > 0$ that

$$\int_\omega |\chi_F| d\Omega \leq \frac{\sqrt{N}}{r} C + \frac{\sqrt{N}}{r} ||\alpha(r)||_{L^{q'}(\Omega)} \operatorname{mes}\omega^{s/p} + \frac{\sqrt{N}}{r} ||\alpha(r)||_{L^{q'}(\omega)}\gamma^s M^s.$$

$$(5.4.43)$$

Now, let $\varepsilon > 0$. Fix $r > 0$ with

$$\frac{\sqrt{N}}{r} C < \frac{\varepsilon}{2}. \qquad (5.4.44)$$

Since $\alpha(\cdot, r) \in L^{q'}(\Omega)$, we can determine $\delta_\varepsilon > 0$ small enough such that

$$\frac{\sqrt{N}}{r} ||\alpha(r)||_{L^{q'}(\Omega)} \operatorname{mes}\omega^{s/p} + \frac{\sqrt{N}}{r} ||\alpha(r)||_{L^{q'}(\omega)}\gamma^s M^s \leq \frac{\varepsilon}{2} \qquad (5.4.45)$$

whenever $\operatorname{mes}\omega < \delta_\varepsilon$. Finally, we have from (5.4.43), (5.4.44) and (5.4.45) that

$$\int_\omega |\chi_F| d\Omega \leq \varepsilon \quad F \in \Lambda, \qquad (5.4.46)$$

for any $\omega \subset \Omega$ with mes$\omega < \delta_\epsilon$. Accordingly, the weak precompactness of $\{\chi_F : F \in \Lambda\}$ in $L^1(\Omega; R^N)$ is proved.   q.e.d.

Now we can formulate the main result of this Section.

Theorem 5.26 Let A be a coercive, weakly continuous operator from V into V^*. Suppose that the injection $V \subset L^p(\Omega; R^N)$, $N \geq 1$, is compact for some $1 < p < \infty$, $V \cap L^\infty(\Omega; R^N)$ is dense in V and that (5.1.4)–(5.1.6), (5.4.1)–(5.4.5) hold. Then problem (P) has at least one solution.

Proof. In order to prove the theorem we have to show that there exist $u \in V$ and $\chi \in L^1(\Omega; R^N)$ such that (5.1.14)–(5.1.15) hold. For $F \in \Lambda$ let

$$W_F = \bigcup_{\substack{F' \in \Lambda \\ F' \supset F}} \{(u_{F'}, \chi_{F'}) \in V \times L^1(\Omega; R^N) : (u_{F'}, \chi_{F'}) \text{ satisfies } (P_{F'})\} \quad (5.4.47)$$

We use the symbol weakcl (W_F) to denote the weak closure of W_F in $V \times L^1(\Omega; R^N)$. Moreover, let

$$Z = \bigcup_{F \in \Lambda} \{\chi_F \in L^1(\Omega; R^N) : (u_F, \chi_F) \text{ satisfies } (P_F) \text{ for some } u_F \in F\}. \tag{5.4.48}$$

Denoting by weakcl (Z) the weak closure of Z in $L^1(\Omega; R^N)$, we obtain from (5.4.26)

$$\text{weakcl}\,(W_F) \subset B_V(O, M) \times \text{weakcl}\,(Z) \quad \forall F \in \Lambda. \tag{5.4.49}$$

Since $B_V(O, M)$ is weakly compact in V and, by Lemma 5.25, weakcl (Z) is weakly compact in $L^1(\Omega; R^N)$, the family $\{\text{weakcl}(W_F) : F \in \Lambda\}$ is contained in the weakly compact set $B_V(O, M) \times \text{weakcl}(Z)$ in $V \times L^1(\Omega; R^N)$. Now let us notice that for any $F_1, \ldots, F_k \in \Lambda$, $k = 1, 2, \ldots$, we have the inclusion $W_{F_1} \cap \ldots \cap W_{F_k} \supset W_F$, with $F = F_1 \cup \ldots \cup F_k$, from which it follows by Proposition 5.24 that the family $\{\text{weakcl}(W_F) : F \in \Lambda\}$ has the finite intersection property. Thus the intersection

$$\bigcap_{F \in \Lambda} \text{weakcl}\,(W_F) \tag{5.4.50}$$

is not empty. Let (u, χ) be an element of this intersection.

The proof will be complete if we show that (u, χ) satisfies (5.1.14). Let $v \in V \cap L^\infty(\Omega; R^N)$ be arbitrary. We choose $F \in \Lambda$ such that $v \in F$. There exists a sequence $\{(u_{Fn}, \chi_{Fn})\}$ in W_F, (for simplicity of the notations we denote it by (u_n, χ_n)) converging weakly to (u, χ) in $V \times L^1(\Omega; R^N)$. This fact means that

$$u_n \rightarrow u \quad \text{weakly in } V, \tag{5.4.51}$$

and

$$\chi_n \rightarrow \chi \quad \text{weakly in } L^1(\Omega; R^N). \tag{5.4.52}$$

Moreover, the following equality holds for $n = 1, 2, \ldots$,

$$\langle Au_n - g, v \rangle_V + \int_{\Omega} \chi_n \cdot v d\Omega = 0. \tag{5.4.53}$$

Now, taking into account (5.4.51) and (5.4.52) we can pass to the limit as $n \to \infty$ in (5.4.53). The weak continuity of A and the fact that $v \in V \cap L^\infty(\Omega; R^N)$ is arbitrary permits us to conclude that the equality

$$\langle Au - g, v \rangle_V + \int_{\Omega} \chi \cdot v d\Omega = 0 \tag{5.4.54}$$

is valid for any $v \in V \cap L^\infty(\Omega; R^N)$. But by the density of $V \cap L^\infty(\Omega; R^N)$ in V we finally arrive at (5.1.14).

In the last step of the proof we have to show (5.1.15). For this purpose let us notice that due to the compact imbedding $V \subset L^p(\Omega; R^N)$, it results from (5.4.51) that

$$u_n \to u \quad \text{strongly in } L^p(\Omega; R^N), \tag{5.4.55}$$

which implies that for a subsequence of $\{u_n\}$ (again denoted by the same symbol) we have

$$u_n \to u \quad \text{a.e. in } \Omega. \tag{5.4.56}$$

Thus we can apply Egoroff's theorem. Accordingly, for any $\varepsilon > 0$ a subset $\omega \subset \Omega$ with mes$\omega < \varepsilon$ can be determined such that

$$u_n \to u \quad \text{uniformly on } \Omega \setminus \omega \tag{5.4.57}$$

with $u \in L^\infty(\Omega \setminus \omega; R^N)$. Let $v \in L^\infty(\Omega \setminus \omega; R^N)$ be arbitrary function. From the estimate

$$\int_{\Omega \setminus \omega} \chi_n \cdot v d\Omega \leq \int_{\Omega \setminus \omega} j^0(u_n, v) d\Omega \tag{5.4.58}$$

combined with the weak convergence of χ_n to χ in $L^1(\Omega; R^N)$, (5.4.55) and the upper semicontinuity of

$$L^\infty(\Omega \setminus \omega; R^N) \ni w \to \int_{\Omega \setminus \omega} j^0(w, v) d\Omega \tag{5.4.59}$$

we obtain

$$\int_{\Omega \setminus \omega} \chi \cdot v d\Omega \leq \int_{\Omega \setminus \omega} j^0(u, v) d\Omega \quad \forall v \in L^\infty(\Omega \setminus \omega; R^N). \tag{5.4.60}$$

But the last inequality implies that

$$\chi(x) \in \partial j(x; u(x)) \quad \text{for a.e. } x \in \Omega \setminus \omega, \tag{5.4.61}$$

where mes$\omega < \varepsilon$, therefore, since ω was chosen arbitrarily,

$$\chi(x) \in \partial j(x; u(x)) \quad \text{for a.e. } x \in \Omega, \tag{5.4.62}$$

i.e., (5.1.15) holds. This completes the proof of the theorem.   q.e.d.

To close this Section we make some comments. First of all let us pay attention to the qualitative difference between the cases $N = 1$ and $N > 1$. As it has been shown in Section 5.2, if $N = 1$, then the directional growth condition (5.4.1) implies (5.4.2) with $s = 1$. Thus in such a case (5.4.2) is superfluous. When $N > 1$, then the condition (5.4.2) is necessary. This is shown in the example below.

Example 5.27 Let us consider a locally Lipschitz function $j : R^2 \to R$ defined by the relation

$$j(\xi) = \begin{cases} \dfrac{\xi_1^{2m}}{\xi_2^{2m}} & \text{for } |\xi_2| \geq 1 \\ \xi_1^{2m} & \text{for } |\xi_2| < 1 \end{cases} \quad \xi = (\xi_1, \xi_2) \in R^2, \tag{5.4.63}$$

where m is a natural number. Then one can easily show that

$$j^0(\xi, -\xi) \leq 0 \quad \forall \xi \in R^2, \tag{5.4.64}$$

i.e., the condition (5.4.1) holds with $k = 0$. On the other hand, for $\eta = (0, -1)$ and $\xi_2 > 1$ we have

$$j^0(\xi, \eta - \xi) = 2m \frac{\xi_1^{2m}}{\xi_2^{2m+1}}. \tag{5.4.65}$$

from which it results that we cannot obtain a better estimate than that of

$$j^0(\xi, \eta - \xi) \leq \alpha(1)(1 + |\xi|^{2m}). \tag{5.4.66}$$

with the power no less then $2m$. Since m was arbitrarily chosen , the necessity of introducing (5.4.2) is clear.

It is not difficult to observe that the growth condition (5.4.1) has been used, in fact, to prove the boundedness of the approximating sequence $\{u_F\}$ in V. If more informations on A are available, then this condition can be made weaker. For instance, if A is strongly monotone and $p \geq 2$, then instead of (5.4.1) we can suppose that

$$j^0(x; \xi, -\xi) \leq k(x)(1 + |\xi|^\sigma) \quad \forall \xi \in R^N, \tag{5.4.67}$$

with $1 \leq \sigma < 2$ and $k \in L^{\bar{q}}(\Omega; R^N)$, where $\bar{q} = p/(p - \sigma)$. Moreover, if $p > 2$, then it is possible to consider the estimate

$$j^0(x; \xi, -\xi) \leq h(x) + \varepsilon |\xi|^2 \quad \forall \xi \in R^N, \tag{5.4.68}$$

with $h \in L^1(\Omega)$ and $\varepsilon > 0$ sufficiently small.

To close this Section let us consider the case in which instead of the compact injection $V \subset L^p(\Omega; R^N)$, $1 < p < \infty$, $N \geq 1$, we suppose that

$$V \text{ is compactly imbedded into } L^\infty(\Omega; R^N). \qquad (5.4.69)$$

In such a case Theorems 5.5, 5.12 and 5.26 hold true not only in the sense of (5.1.14) and (5.1.15), but (5.1.7) also holds true for each $v \in V$. Moreover, it is possible to weaken further the weak continuity of A and to prove the existence result for A pseudo-monotone. Indeed, in the proof of Theorem 5.26 we are allowed to additionally assume that $u \in F$. Then

$$\langle Au_n - g, u - u_n \rangle_V + \int_\Omega \chi_n \cdot (u - u_n)d\Omega = 0, \qquad (5.4.70)$$

which, due to the convergence

$$\int_\Omega \chi_n \cdot (u - u_n)d\Omega \to 0 \text{ as } n \to \infty, \qquad (5.4.71)$$

leads to $\limsup \langle Au_n, u_n - u \rangle_V \leq 0$. Thus the result for A pseudo-monotone can be easily achieved.

Now, let $\varphi : V \to \bar{R}$ be a convex lower semicontinuous function with $0 \in D(\partial\varphi)$. The aforementioned remarks ensure the existence of solutions to variational-hemivariational inequalities of the form

$$\langle Au + (\partial\varphi)_\lambda(u) - g, v - u \rangle_V + \int_\Omega j^0(u; v - u)d\Omega \geq 0 \quad \forall v \in V, \qquad (5.4.72)$$

where $(\partial\varphi)_\lambda : V \to 2^{V^*}, \lambda > 0$, is the regularization of $\partial\varphi$ in the sense of Proposition 2.8 (cf. Theorem 2.11). Let $u_\lambda \in V, \lambda > 0$, be a solution of (5.4.72). Hence we obtain

$$Au_\lambda + (\partial\varphi)_\lambda(u_\lambda) + \chi_\lambda = g, \quad \lambda > 0$$
$$\chi_\lambda \in \partial j(u_\lambda) \quad \text{a.e. in } \Omega. \qquad (5.4.73)$$

Since A is assumed to be coercive and $0 \in D(\partial\varphi)$, the coercive function of $A + (\partial\varphi)_\lambda$ does not depend on λ. Therefore, taking into account the boundedness of $\{\chi_\lambda\}$ in $L^1(\Omega; R^N)$ we obtain the boundedness of $\{u_\lambda\}$ in V. Due to (5.4.69) we can also obtain that

$$\int_\Omega \chi_\lambda \cdot (u_\lambda - u)d\Omega \to 0 \text{ as } \lambda \to 0. \qquad (5.4.74)$$

In the next step, as $\lambda \to 0$, we follow the procedure applied in the proof of Theorem 2.11 (with slight modification due to the presence of χ_λ in (5.4.73)) in order to show that there exist $u \in V$, $u^* \in \partial\varphi(u)$ and $\chi \in L^1(\Omega; R^N)$ such that $Au + u^* + \chi = g$; this part of reasoning uses essentially (5.4.74). To show that $\chi \in \partial j(u)$ we make use of the method of Theorem 5.26. Consequently, we establish the existence result for the following variational-hemivariational inequality: Find $u \in V$ such that

$$\langle Au - g, v - u \rangle_V + \varphi(v) - \varphi(u) + \int_\Omega j^0(u; v - u)d\Omega \geq 0 \quad \forall v \in V. \quad (5.4.75)$$

Finally, the following result being a generalization of Theorem 3.11 follows. The details of its proof are left as an exercise for the reader.

Theorem 5.28 Let $A : V \to V^*$ be a pseudo-monotone, coercive, operator from a reflexive Banach space V into V^*. Suppose that the injection $V \subset L^\infty(\Omega; R^N), N \geq 1$, is compact and that (5.1.4)–(5.1.6) and (5.4.1)–(5.4.5) hold. Moreover, let $\varphi : V \to \bar{R}$ be a convex lower semicontinuous function with $0 \in D(\partial\varphi)$, and let either A be quasi-bounded, or $\partial\varphi$ strongly quasi-bounded. Then the variational-hemivariational inequality (5.4.75) has at least one solution.

Let us remark that the solvability of (5.4.75) in the case of the compact injection $V \subset L^p(\Omega; R^N)$ with $1 < p < \infty$, is still an open problem.

5.5 Applications to Mechanics, Engineering and Economics

In the present Section we give some applications concerning the existence of solution of hemivariational inequalities arising in Mechanics, Engineering and Economics. We apply here the results of the previous Sections of this Chapter. As already mentioned, the existence theory developed in this Chapter is based on weaker growth assumptions for the superpotential than those presented in Chapter 4. On the contrary an additional density assumption of $V \cap L^\infty(\Omega; R^N)$ in V like the one of Chapter 3 is assumed to hold. We should note here that all the applications of this Chapter can be also treated with the existence theory developed in the previous Chapter and vice versa, after appropriate modifications of the assumptions been made in each application.

5.5.1 Nonmonotone Skin Friction in Plane Elasticity

The present Subsection deals with BVPs arising in plane deformable bodies when such a body is subjected to nonmonotone friction skin effects.

Such problems arise in many engineering applications, whenever a structure that can globally or locally be idealised as a plane elasticity problem is driven into, or is lying upon, another medium introducing frictional effects. The same problems arise if a plate is adhesively connected in the tangential sense with a support assumed to be rigid. This fits well the case of sheet piles as well as that of bridge decks unilaterally supported at their edge.

Here we shall assume that the frictional effects are of nonmonotone nature in contrast to the monotone friction conditions studied first by Duvaut and Lions [Duv71,80], and giving rise to variational inequalites. Thus here the simplifying assumption of monotonicity is abandoned in favour of more realistic friction laws at the skin of the body that can equally well describe the phenomenon of adhesion. Here the friction laws are two-dimensional extensions in the spirit of Sect. 4.6.1 of the one-dimensional friction laws of Fig. 1.4.3 b,c,d.

Let Ω be an open, bounded connected subset of R^2 occupied by a linear elastic body in its undeformed state. The body is referred to an orthogonal Cartesian coordinate system $0x_1x_2$. The boundary Γ of Ω is assumed to be regular (a Lipschitzian boundary is sufficient).

In the framework of plane elasticity and for small deformations, the relations

$$\sigma_{ij,j} + f_i = 0, \tag{5.5.1}$$

$$\varepsilon_{ij} = 1/2(u_{i,j} + u_{j,i}), \tag{5.5.2}$$

$$\sigma_{ij} = C_{ijhk}\varepsilon_{hk} \tag{5.5.3}$$

hold, where $i, j = 1, 2$ (summation convention), $\sigma = \{\sigma_{ij}\}$ (resp. $\varepsilon = \{\varepsilon_{ij}\}$) , is the stress (resp. strain) tensor and $C = \{C_{ijhk}\}, i, j, h, k = 1, 2$, is the Hookean elasticity tensor fulfilling the well known ellipticity and symmetry conditions.

Moreover, let $u = \{u_i\}$ and $f = \{f_i\}$ be the displacement and volume force, respectively. The comma denotes partial differentiation with respect to x_1 or x_2.

In order to describe skin effects, e.g. skin friction, adhesion, etc., we assume that the body forces f_i consist of two parts: $\bar{\bar{f}}_i$ which is given and $\bar{f}_i$ which is the reaction of the constraint introducing the skin effects. Thus we may write that

$$f_i = \bar{f}_i + \bar{\bar{f}}_i, \quad i = 1, 2, \quad \bar{\bar{f}}_i \in L^2(\Omega). \tag{5.5.4}$$

Here $\bar{\bar{f}}_i$ is the given external loading and $\bar{f}_i$ is a possibly multivalued function of u_i. j is a locally Lipschitz superpotential fulfilling (5.1.4)–(5.1.6), and ∂ is the generalized gradient.

We consider the multivalued reaction-displacement law

$$-\bar{f} \in \partial j(x, u) \quad \text{on } \Omega' \tag{5.5.5}$$

where Ω' is the part of the body where frictional or adhesive effects take place. We assume that $\Omega' \subset \Omega$, that

$$\bar{f}_i = 0 \quad \text{on } \Omega \setminus \Omega' \tag{5.5.6}$$

and that $\bar{\Omega}' \cap \Gamma = \emptyset$.

In order to complete the classical formulation of the considered BVP we have to specify the boundary conditions. Let $\Gamma = \Gamma_U \cup \Gamma_F \cup \Gamma_0$ where Γ_U and Γ_F are open, mutually disjoint sets and mes $\Gamma_0 = 0$.

We assume that on Γ_U (resp. Γ_F) the displacements $u = \{u_i\}$ (resp. the tractions $f = \{f_i\}$) are given i.e.

$$u_i = U_i \quad \text{on } \Gamma_U, \tag{5.5.7}$$

$$S_i = \sigma_{ij}n_j = F_i \quad \text{on } \Gamma_F, \tag{5.5.8}$$

where $n = \{n_i\}$ is the outward unit normal to Γ. Further we obtain (cf. (1.4.44)) for appropriately regular functions the variational form

$$\int_\Omega \sigma_{ij}(\varepsilon_{ij}(v) - \varepsilon_{ij}(u))d\Omega = \int_\Omega \bar{\bar{f}}_i\,(v_i - u_i)d\Omega + \int_{\Omega'} \bar{f}_i(v_i - u_i)d\Omega + \int_{\Gamma_F} F_i(v_i - u_i)d\Gamma \tag{5.5.9}$$

for $u = \{u_i\} \in V_{ad}$ and for every $v = \{v_i\} \in V_{ad}$, where $V_{ad} = \{v : v \in V, v_i = U_i$ on $\Gamma_U\}$ is the kinematically admissible set: V denotes the vector space of the displacements v. We assume that $V = [H^1(\Omega)]^2$, and that $F \in [L^2(\Gamma_F)]^2$. Moreover, we assume that $U_i \in \tilde{H}(\Gamma_U)$ which symbolises a space with the property that $u_i^* \in H^1(\Omega)$ exists such that $u_i^*|_\Gamma = U_i$ on Γ_U (here $u_i^*|_\Gamma$ denotes the trace of u_i^* on Γ which is an element of $H^{1/2}(\Gamma)$). We assume further that $\Gamma_U \neq \emptyset$. For the sake of simplicity let $U_i = 0$ on Γ_U and thus the kinematically admissible set V_{ad} becomes the closed linear subspace

$$V_0 = \{v : v = \{v_i\}, v_i \in H^1(\Omega), v_i = 0 \text{ on } \Gamma_U\}. \tag{5.5.10}$$

If $U_i \neq 0$ on Γ_U then the translation $\tilde{v} = v - u^*$ and $\tilde{u} = u - u^*$ is performed and the problem reduces to the homogeneous one. We assume also that $f_i \in L^2(\Omega)$ and, let $(\cdot, \cdot)$ denote the duality pairing for $[H^1(\Omega)]^2$. We define further the problem: find $u \in V_0$ such as to satisfy the relation

$$a(u, v - u) + \int_{\Omega'} j^0(x, u, v - u)d\Omega \geq (l, v - u) \quad \text{for all } v \in V_0. \tag{5.5.11}$$

Here $l \in V_0^*$ and is defined by

$$l(v) = \int_{\Gamma_F} F_i v_i d\Gamma + \int_\Omega \bar{\bar{f}}_i v_i d\Omega \tag{5.5.12}$$

and $a(\cdot, \cdot)$ is the bilinear form of elasticity (see (1.4.43)) which is symmetric, bounded and coercive due to Korn's inequality. The variational inequality (5.5.11) has at least one solution because we can apply Theorem 5.26 if j satisfies (5.4.1)–(5.4.5). Indeed $H^1(\Omega) \subset L^2(\Omega')$ is compact and $H^1(\Omega) \cap L^\infty(\Omega')$ is dense in $H^1(\Omega)$ for the $H^1(\Omega)$-norm.

5.5.2 The General Problem of Masonry Structures

The structural elements of a masonry structure are stones or bricks which have interfaces between them. There general nonmonotone possibly multivalued adhesive and/or nonmonotone frictional contact boundary conditions derived by nonconvex superpotentials arise in both the normal and in the tangential direction to the interface. The model proposed here considerably improves the no-tension model [Gia] for masonry structures which has been proposed by certain research workers and which actually constitutes only a rough approximation to the real behaviour of such a structure. Indeed, due to the interface mortar, the interfaces can sustain small but nonnegligible tensile forces. The mathematical model proposed here takes into consideration along with the interface tensile forces the possibly sawtooth stress-strain behaviour of the interface in the tangential direction. This describes local cracking and crushing effects, local debonding and sliding. All these facts show the possibilities offered by the theory of hemivariational inequalities.

Let $\Omega^{(m)}$, $m = 1, 2, \ldots, l$, be a set of deformable bodies, possibly with different elasticity properties, with the boundaries $\Gamma^{(m)}$, $m = 1, 2, \ldots, l$, assumed to be appropriately regular (a Lipschitzian boundary is sufficient). Let $x = \{x_i\}$, $i = 1, 2, 3$, be a point of R^3 and let $\sigma^{(m)} = \sigma_{ij}^{(m)}$ and $\varepsilon^{(m)} = \varepsilon_{ij}^{(m)}$, $i, j = 1, 2, 3$, be the stress and strain tensors of the m-body. We denote by $f^{(m)} = \{f_i^{(m)}\}$ and $u^{(m)} = \{u_i^{(m)}\}$ the volume force and the displacement vector in each body. If $n^{(m)} = \{n_i^{(m)}\}$ is the outward unit normal vector to $\Gamma^{(m)}$, the boundary force on $\Gamma^{(m)}$ is $S_i^{(m)} = \sigma_{ij}^{(m)} n_j^{(m)}$. Let $S_N^{(m)}$ and $S_T^{(m)}$ be the normal and tangential components of it respectively. The corresponding displacement components are $u_N^{(m)}$ and $u_T^{(m)}$. The boundary $\Gamma^{(m)}$ is divided into three non-overlapping parts $\Gamma_U^{(m)}, \Gamma_F^{(m)}$ and $\Gamma_S^{(m)}$. On $\Gamma_U^{(m)}$ the displacements are given; let us take for simplicity that

$$u_i^{(m)} = 0 \quad \text{on} \quad \Gamma_U^{(m)}. \tag{5.5.13}$$

On $\Gamma_F^{(m)}$ the forces are prescribed, i.e.,

$$S_i^{(m)} = F_i^{(m)} \quad \text{on} \quad \Gamma_F^{(m)} \tag{5.5.14}$$

and on $\Gamma_S^{(m)}$, which corresponds to the interface of structure m with other substructures, nonmonotone interface conditions hold describing slip and delamination effects. We write in the general case $\Omega^{(m)} \subset R^3$ the interface conditions in the form

$$-S_N^{(m)} \in \partial j_{N(m)}(x, [u_N^{(m)}]) \tag{5.5.15a}$$

$$-S_T^{(m)} \in \partial j_{T(m)}(x, [u_T^{(m)}]) \tag{5.5.15b}$$

in the normal and in the tangential direction to the interface. The superpotentials j_N and j_T are assumed to be functions of the interlayer gap $[u_N]$ and slip $[u_T]$ (locally Lipschitz continuous) respectively and ∂ denotes the generalized gradient. Here $[u_N^{(m)}]$ and $[u_T^{(m)}]$ are the relative normal and tangential displacements between $\Gamma^{(m)}$ and the adjacent boundary, say $\Gamma^{(m-1)}$. For instance, $[u_N^{(m)}] = u_N^{(m)} + u_N^{(m-1)}$ because $u_N^{(m)}$ (resp. $u_N^{(m-1)}$) is assumed as positive if it is directed outwards of $\Gamma^{(m)}$ (resp. of $\Gamma^{(m-1)}$).

In the framework of small deformations and linear elastic behaviour for $\Omega^{(m)}$ $m = 1, 2, \ldots, l$, we can write the relations

$$\sigma_{ij,j}^{(m)} + f_i^{(m)} = 0, \tag{5.5.16}$$

$$\varepsilon_{ij}^{(m)} = \frac{1}{2}(u_{i,j}^{(m)} + u_{j,i}^{(m)}) = \varepsilon_{ij}(u^{(m)}), \tag{5.5.17}$$

$$\sigma_{ij}^{(m)} = C_{ijhk}^{(m)}\varepsilon_{hk}^{(m)}. \tag{5.5.18}$$

Hooke's tensor $C^{(m)} = \{C_{ijhk}^{(m)}\}$ satisfies the well-known symmetry and ellipticity conditions (1.4.42a,b). For every body $\Omega^{(m)}$ and for sufficiently regular functions we obtain (cf. (1.4.44)) the relation

$$\int_{\Omega^{(m)}} \sigma_{ij}^{(m)}\varepsilon_{ij}^{(m)}(v^{(m)} - u^{(m)})d\Omega = \int_{\Omega^{(m)}} f_i^{(m)}(v_i^{(m)} - u_i^{(m)})d\Omega \tag{5.5.19}$$

$$+ \int_{\Gamma_F^{(m)}} F_i^{(m)}(v_i^{(m)} - u_i^{(m)})d\Gamma$$

$$+ \int_{\Gamma_S^{(m)}} \left[S_N^{(m)}(v_N^{(m)} - u_N^{(m)}) + S_{T_i}^{(m)}(v_{T_i}^{(m)} - u_{T_i}^{(m)}) \right] d\Gamma \quad \forall v \in V_{ad}^{(m)},$$

where $V_{ad}^{(m)}$ is the kinematically admissible set of $\Omega^{(m)}$, i.e.

$$V_{ad}^{(m)} = \{v^{(m)} : v^{(m)} = \{v_i^{(m)}\}, \, v_i^{(m)} \in V(\Omega^{(m)}), \, v_i^{(m)} = 0 \quad \text{on } \Gamma_U^{(m)}\}. \tag{5.5.20}$$

Here $V(\Omega^{(m)})$ denotes a space of functions defined on $\Omega^{(m)}$. Adding with respect to m all the expressions (5.5.19) and taking into account the interconnection of the bodies yields a relation of the form

$$\sum_{m=1}^{l} \int_{\Omega^{(m)}} \sigma_{ij}^{(m)}\varepsilon_{ij}^{(m)}(v^{(m)} - u^{(m)})d\Omega \tag{5.5.21}$$

$$= \sum_{m=1}^{l} \left[\int_{\Omega^{(m)}} f_i^{(m)}(v_i^{(m)} - u_i^{(m)})d\Omega + \int_{\Gamma_F^{(m)}} F_i^{(m)}(v_i^{(m)} - u_i^{(m)})d\Gamma \right]$$

$$+ \sum_{q=1}^{k} \left[\int_{\Gamma^{(q)}} S_N^{(q)}([v_N^{(q)}] - [u_N^{(q)}])d\Gamma + \int_{\Gamma^{(q)}} S_{T_i}^{(q)}([v_{T_i}^{(q)}] - [u_{T_i}^{(q)}])d\Gamma \right] \quad \forall v \in V_{ad},$$

where $V_{ad} = \bigcup_{m=1}^{l} V_{ad}^{(m)}$.

In (5.5.21) the integrals along the joints Γ_q, $q = 1, \ldots, k$, have been introduced. The new numbering of the $\Gamma_S^{(m)}$-boundaries has the advantage that finally the energy of each joint appears. Further we introduce the elastic energy of the m-structure

$$a(u^{(m)}, v^{(m)}) = \int_{\Omega^{(m)}} C_{ijhk}^{(m)}\varepsilon_{ij}(u^{(m)})\varepsilon_{hk}(v^{(m)})d\Omega \tag{5.5.21a}$$

and by taking into account (5.5.15a,b), we get from (5.5.21) the following hemivariational inequality: find $u \in V_{ad}$ such as to satisfy

$$\sum_{m=1}^{l} a(u^{(m)}, v^{(m)} - u^{(m)}) + \sum_{q=1}^{k} \left[\int_{\Gamma_q} \left[j^0_{N(q)}(x, [u_N^{(q)}], [v_N^{(q)}] - [u_N^{(q)}]) \right. \right.$$

$$\left. \left. + j^0_{T(q)}(x, [u_T^{(q)}], [v_T^{(q)}] - [u_T^{(q)}]) \right] d\Gamma \right] \geq \sum_{m=1}^{l} \left[\int_{\Omega^{(m)}} f_i^{(m)}(v_i^{(m)} - u_i^{(m)}) d\Omega \right.$$

$$\left. + \int_{\Gamma_F^{(m)}} F_i^{(m)}(v_i^{(m)} - u_i^{(m)}) d\Gamma \right] \quad \forall v \in V_{\mathrm{ad}}. \tag{5.5.22}$$

To check in which sense a solution of (5.5.22) satisfies (5.5.16), the boundary conditions on $\Gamma_F^{(m)}$ $m = 1, \ldots, l$ and the interface relations (5.5.15a,b) we must make the functional setting of the problem more precise. So we assume that $f_i^{(m)} \in L^2(\Omega^{(m)})$, $F_i^{(m)} \in L^2(\Gamma_F^{(m)})$, $C_{ijhk}^{(m)} \in L^\infty(\Omega^{(m)})$, $u_i^{(m)}, v_i^{(m)} \in H^1(\Omega^{(m)})$. Then $u_N^{(m)}, u_{T_i}^{(m)} \in H^{1/2}(\Gamma^{(m)})$ and $S_N^{(m)}, S_{T_i}^{(m)} \in H^{-1/2}(\Gamma^{(m)})$ and (5.5.19) must be written as (1.4.44) with the integrals replaced by $\langle, \rangle_{1/2}$ and $\langle, \rangle_{H_T}$. We set in (5.5.22) $v_i^{(m)} - u_i^{(m)} = \pm\phi_i^{(m)}$ where $\phi_i^{(m)}$ belongs to the space $C_0^\infty(\Omega^{(m)})$ of infinitely differentiable functions with compact support in $\Omega^{(m)}$. Then from (5.5.22) by setting $v_i^{(m)} - u_i^{(m)} = \pm\phi_i^{(m)}$ for $m = n$ and $v_i^{(m)} - u_i^{(m)} = 0$ for $m \neq n$ we obtain

$$a(u^{(n)}, \phi^{(n)}) = \int_{\Omega^{(n)}} f_i^{(n)} \phi_i^{(n)} d\Omega \tag{5.5.23}$$

since $\phi_i^{(n)} = 0$ on $\Gamma^{(n)}$. Relation (5.5.23) implies that (5.5.16) holds on $\Omega^{(n)}$ in the sense of distributions over $\Omega^{(n)}$. This procedure is repeated for $n = 1, 2, \ldots, l$. Now applying the Green-Gauss theorem to each body we obtain the equality

$$a(u^{(m)}, v^{(m)} - u^{(m)}) = \int_{\Omega^{(m)}} f_i^{(m)}(v_i^{(m)} - u_i^{(m)}) d\Omega \tag{5.5.24}$$

$$+ \int_{\Gamma_F^{(m)}} S_i^{(m)}(v_i^{(m)} - u_i^{(m)}) d\Gamma$$

$$+ \langle S_N^{(m)}, v_N^{(m)} - u_N^{(m)} \rangle_{\frac{1}{2}, \Gamma_S^{(m)}} + \langle S_T^{(m)}, v_T^{(m)} - u_T^{(m)} \rangle_{H_T, \Gamma_S^{(m)}}.$$

From (5.5.24) and (5.5.22) we obtain the inequality

$$\sum_{q=1}^{k} \left[\int_{\Gamma_q} [j^0_{N(q)}(x, [u_N^{(q)}], [v_N^{(q)}] - [u_N^{(q)}]) + j^0_{T(q)}(x, [u_T^{(q)}], [v_T^{(q)}] - [u_T^{(q)}])] d\Gamma \right]$$

$$+ \sum_{m=1}^{l} \langle S_i^{(m)} - F_i^{(m)}, v_i^{(m)} - u_i^{(m)} \rangle_{\Gamma_F^{(m)}} + \sum_{q=1}^{k} \{ \langle S_N^{(q)}, [v_N^{(q)}] - [u_N^{(q)}] \rangle_{\frac{1}{2}, \Gamma_q}$$

$$+ \langle S_T^{(q)}, [v_T^{(q)}] - [u_T^{(q)}] \rangle_{H_T, \Gamma_q} \} \geq 0 \quad \forall v \in V_{\mathrm{ad}}. \tag{5.5.25}$$

If in (5.5.25) we consider that on $\Gamma_F^{(m)}$, $v_i^{(m)} - u_i^{(m)} = \pm r_i^{(m)} \in H^{1/2}(\Gamma^{(m)})$ for $m = n$, and that $v_i^{(m)} - u_i^{(m)} = 0$ for $m \neq n$ on $\Gamma_F^{(m)}$ and on Γ_q for every q, we obtain $S_i^{(n)} = F_i^{(n)}$ as an equality in $H^{-1/2}(\Gamma^{(n)})$; this can be shown for every

n. From (5.5.25) by setting $[v_N^{(q)}] - [u_N^{(q)}] = r_N^{(q)}$ on Γ_q for $q = n$ and the same difference is zero for $q \neq n$, and setting $[v_T^{(q)}] - [u_T^{(q)}] = 0$ on Γ_q for every q we obtain

$$\int_{\Gamma_n} j^0_{N(n)}(x, [u_N^{(n)}], r_N^{(n)}) d\Gamma \geq -\langle S_N^{(n)}, r_N^{(n)} \rangle_{\frac{1}{2}, \Gamma_n} \quad \forall r_N^{(n)} \in H^{1/2}(\Gamma) \qquad (5.5.26)$$

which is a weak formulation of (5.5.15a) on $H^{-1/2}(\Gamma) \times H^{1/2}(\Gamma)$. Analogously we obtain from (5.5.25) a weak form of (5.5.15b). Due to the zero displacements on $\Gamma_U^{(m)}$ the bilinear form $a(u^{(m)}, v^{(m)})$ is coercive due to Korn's inequality and thus also strongly monotone (relation (5.3.1) is satisfied). The corresponding linear operator to the bilinear form is bounded and thus continuous and thus weakly continuous. We assume further that $j_{N(q)}(x, \cdot)$ and $j_{T(q)}(x, \cdot)$ in (5.5.22) are locally Lipschitz and satisfy (5.1.4)–(5.1.6), (5.3.2) and (5.3.3). Moreover the continuous linear trace mapping $H^1(\Omega^{(m)}) \subset L^2(\Gamma^{(m)})$ for each m is compact. Taking into account the definitions of $[u_N^{(q)}]$ and $[u_T^{(q)}]$(cf. also the functional framework of (1.4.37)) we apply Theorem 5.12, as it is slightly modified by replacing in it $L^p(\Omega; R^N)$ by $L^p(\Gamma; R^{N-1})$. Here we take $p = 2$ and thus the existence of the solution results. Note that if (5.3.3) does not hold and only (5.4.1)–(5.4.5) hold then we can apply Theorem 5.26 as it is modified for hemivariational inequalities involving boundary integrals $\int_\Gamma j^0(\cdot, \cdot) d\Gamma$ instead of $\int_\Omega j^0(\cdot, \cdot) d\Omega$. As one can easily verify we replace in Theorem 5.26 $L^p(\Omega, R^N)$ by $L^p(\Gamma; R^{N-1})$ and $V \cap L^\infty(\Omega; R^N)$ by $V \cap \{v : v|_\Gamma \in L^\infty(\Gamma)\}$; here Γ may be replaced by $\Gamma_1 \subset \Gamma$. For $V = H^1(\Omega, R^3)$ the above condition are verified and thus (modified) Theorem 5.26 implies the existence of a solution of the hemivariational inequality (5.5.22). Closing this Section we would like to note that instead of (5.5.15a,b) we can consider an interface relation of the form

$$-S^{(m)} \in \partial j(x, u^{(m)}) \quad x \in \Gamma_S^{(m)}, \qquad (5.5.27)$$

where the superpotential j may result as in Sect. 4.6.1.

5.5.3 On the Nonconvex Semipermeability Problem

Semipermeability problems were first studied in [Duv72] in the context of convexity, i.e. for monotone semipermeability relations. These relation lead to variational inequalities connected with the Δ-operator and arise in heat conduction, in electrostatics and in flow problems through porous media.

To analogous variational inequalities lead certain categories of temperature control problems in heat conduction, or potential control and pressure control in electrostatics and hydraulics, respectively. Here, we shall study the same problems but without assuming monotonicity. The arising BVPs lead to hemivariational inequalities instead of variational inequalities since the potentials are nonconvex. After the formulation of the BVP, the corresponding hemivariational inequality is studied concerning the existence of the solution. The semipermeability conditions studied here are realized by various types of membranes, natural or artificial ones. We consider further an open, bounded, connected subset Ω

of R^3 referred to a fixed Cartesian coordinate system $0x_1x_2x_3$ and we formulate the equation

$$-\Delta u = f \quad \text{in } \Omega \tag{5.5.28}$$

for stationary problems.

Here u represents the temperature in the case of heat conduction problems, whereas in problems of hydraulics and electrostatics the pressure and the electric potential are represented, respectively. We denote further by Γ the boundary of Ω and we assume that Γ is sufficiently regular ($C^{1,1}$-boundary is sufficient). If $n = \{n_i\}$ denotes the outward unit normal to Γ then $\partial u/\partial n$ is the flux of heat, fluid or electricity through Γ for the aforementioned classes of problems, respectively. Further, we shall use the language of heat conduction problems.

We may consider two main classes of semipermeability problems: the interior and the boundary semipermeability problems.

In the first class of problems the classical boundary conditions

$$u = 0 \quad \text{on } \Gamma \tag{5.5.29}$$

are assumed to hold, whereas in the second class the boundary conditions are defined as a relation between $\partial u/\partial n$ and u. In the first class the semipermeability conditons are obtained by assuming that $f = \bar{f} + \bar{\bar{f}}$ where $\bar{f}$ is given and $\bar{\bar{f}}$ is a known function of u. Here, we consider (5.5.29) for the sake of simplicity. All these problems may be put in the following general framework. For the first class we seek a function u such as to satisfy (5.5.28), (5.5.29) with

$$f = \bar{f} + \bar{\bar{f}}, \quad -\bar{\bar{f}} \in \partial j_1(x, u) \text{ in } \Omega. \tag{5.5.30}$$

For the second class we seek a function u such that (5.5.28) is satisfied together with the boundary condition

$$-\frac{\partial u}{\partial n} \in \partial j_2(x, u) \text{ on } \Gamma_1 \subset \Gamma \quad \text{and} \quad u = 0 \text{ on } \Gamma - \Gamma_1. \tag{5.5.31}$$

Both $j_1(x, \cdot)$ and $j_2(x, \cdot)$ are locally Lipschitz function and ∂ denotes the generalized gradient. Note, that if $q = \{q_i\}$ denotes the heat flux vector and $k > 0$ is the coefficient of thermal conductivity of the material we may write by Fourier's law that $q_i n_i = -k\partial u/\partial n$.

For instance, the graph of Fig. 5.5.1a corresponds to the behaviour of a semipermeable membrane (e.g. a wall) of finite thickness: If $u < h$ the heat flux tends to leave Ω but the semipermeable membrane does not permit it, i.e. $\frac{\partial u}{\partial n} = 0$. If $u = h$ heat flux enters into Ω through Γ for constant temperature until a given value a is reached and then we may assume that $-\frac{\partial u}{\partial n} = \alpha(u - h_1), \alpha > 0, h_1 > h$ for $u > h$. Analogous relations may be considered in the case of internal semipermeability. The graph of Fig. 5.5.1b corresponds to a temperature control problem in which we regulate the temperature to deviate as little as possible from the interval $[h_1, h_2]$.

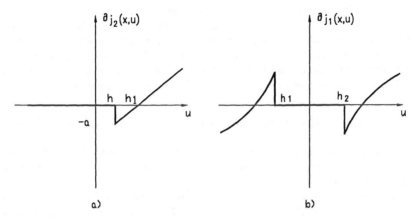

Fig. 5.5.1. Semipermeability relations

From (5.5.28) we obtain by multiplying by $v - u$, integrating over Ω and applying for appropriately regular functions the Green-Gauss theorem that

$$\alpha(u, v - u) = \int_{\Gamma} \frac{\partial u}{\partial n}(v - u)d\Gamma + (f, v - u), \qquad (5.5.32)$$

where

$$\alpha(u, v) = \int_{\Omega} \mathrm{grad}u \cdot \mathrm{grad}v d\Omega \qquad (5.5.33)$$

and

$$(f, u) = \int_{\Omega} fu d\Omega. \qquad (5.5.34)$$

Now we can formulate the hemivariational inequalities corresponding to the two classes of problems. Let for the first class $V = \overset{0}{H}^1(\Omega)$ and $\overline{\overline{f}} \in L^2(\Omega)$; for the second class $V = \{v : v \in H^1(\Omega), \ v = 0 \text{ on } \Gamma - \Gamma_1\}$ and $f \in L^2(\Omega)$. Then combining (5.5.32) with the definitions of (5.5.30) and (5.5.31) we are led to the following two hemivariational inequalities for the first and for the second class of semipermeability problems respectively
(i) Find $u \in V$ such that

$$\alpha(u, v - u) + \int_{\Omega} j_1^0(x, u, v - u)d\Omega \geq (\overline{\overline{f}}, v - u) \quad \forall v \in V. \qquad (5.5.35)$$

(ii) Find $u \in V$ such that

$$\alpha(u, v - u) + \int_{\Gamma_1} j_2^0(x, u, v - u)d\Gamma \geq (f, v - u) \quad \forall v \in V. \qquad (5.5.36)$$

Since $\alpha(\cdot, \cdot)$ is strongly monotone on V both in (i) and (ii) and the imbeddings $V \subset L^2(\Omega)$ and $V \subset L^2(\Gamma_1)$ are compact we can prove the existence of solutions of (i) and of (ii) by applying Theorem 5.12, if j_1 and j_2 satisfy conditions (5.1.4)–(5.1.6) and (5.3.2), (5.3.3). As in Sect. 5.5.2 we have to slightly modify Theorem 5.12 (replace Ω by Γ etc.) before applying it to (5.5.36). Also Theorem 5.26 can be applied because the corresponding density requirements hold.

5.5.4 Hemivariational Inequalities in Nonlinear Elasticity Problems

Let Ω be an open, bounded, connected subset of R^3 occupied by a deformable body in its undeformed state. We denote by Γ the boundary of Ω which is assumed to be Lipschitzian.

We assume further that the boundary is divided into three disjoint open subsets Γ_U, Γ_F, and Γ_S, i.e. $\Gamma = \bar{\Gamma}_U \cup \bar{\Gamma}_F \cup \bar{\Gamma}_S$. On Γ_U the displacements are given

$$u_i = U_i, \quad U_i = U_i(x) \text{ on } \Gamma_U, \tag{5.5.37}$$

on Γ_F the forces are prescribed, i.e.,

$$S_i = F_i, \quad F_i = F_i(x) \text{ on } \Gamma_F, \tag{5.5.38}$$

and on Γ_S nonmonotone boundary conditions hold, causing, as we shall see further, the formulation of the problem as a hemivariational inequality. We consider the following general boundary condition

$$-S \in \partial j(x, u) \text{ on } \Gamma_S, \tag{5.5.39}$$

where $S(x) \in R^3$, $u(x) \in R^3$, $j(x, \cdot)$ is locally Lipschitz and ∂ denotes the generalized gradient. In the framework of small deformations and nonlinear elastic behaviour we write the equation of equilibrium

$$\sigma_{ij,j} + f_i = 0, \tag{5.5.40}$$

the strain-displacement relations

$$\varepsilon_{ij} = \frac{1}{2}(u_{i,j} + u_{j,i}), \tag{5.5.41}$$

and the nonlinear stress-strain relation

$$\sigma_{ij} = [\mathrm{grad}\, w(\varepsilon)]_{ij} \tag{5.5.42}$$

where w is a convex Gateaux differentiable function of ε. With an appropriate choice of w (5.5.42) describes the polygonal stress-strain laws (cf. e.g. [Pan 85] p. 98), the law of the deformation theory of plasticity e.t.c. For instance for the polygonal stress-strain laws case

$$w(\varepsilon) = \frac{1}{2}C_{ijhk}\varepsilon_{ij}\varepsilon_{hk} + \frac{1}{2\mu}(\varepsilon_{ij} - (P_K\varepsilon)_{ij})(\varepsilon_{ij} - (P_K\varepsilon)_{ij}), \tag{5.5.43}$$

where $C = \{C_{ijhk}\}$ is Hooke' tensor, $\mu > 0$, K is a closed convex subset of R^6 and P_K is the projection operator; for the deformation theory of plasticity

$$w(\varepsilon) = \frac{E}{6(1 - 2\nu)}\varepsilon_{ij}^2 + \int_0^\gamma g(\gamma)d\gamma, \qquad (5.5.44)$$

where the stress and strain deviators σ_{ij}^D and ε_{ij}^D satisfy the relation

$$\sigma_{ij}^D = \alpha(\gamma)\varepsilon_{ij}^D \qquad (5.5.45)$$

with

$$g^2 = \sigma_{ij}^D \sigma_{ij}^D, \quad \gamma^2 = \varepsilon_{ij}^D \varepsilon_{ij}^D, \quad \frac{g(\gamma)}{\gamma} = \alpha(\gamma) \geq c_1, \quad \frac{dg(\gamma)}{d\gamma} = \beta(\gamma) \geq c_2. \quad (5.5.46)$$

Here $\alpha(\gamma)$ is continuously differentiable function of γ and c_1, c_2 are positive constants.

One can easily verify that $w(\cdot)$ is in both cases strictly convex and that $\mathrm{grad}\,w(\cdot)$ exists everywhere.

We further assume that $u_i, v_i \in W^{1,p}(\Omega)$ with $p \geq 2$ (the well-known Sobolev space) and that $F_i \in L^{q'}(\Gamma_F)$. Here $1/q + 1/q' = 1$ and q is determined by the theorem ([Kuf] p.319) stating that the mapping $u \in W^{1,p}(\Omega) \to u|_\Gamma \in L^q(\Gamma)$ is linear and continuous for $p \geq n$, $q \geq 1$ arbitrary finite as well as for $1 \leq p < n$, $q = \frac{np-p}{n-p}$, i.e. for $\Omega \subset R^3$ $n = 3$ and for $p = 2$, $q = 4$ etc. Moreover we assume that $U_i \in V(\Gamma_U)$ which is a space with the property that there exists $u_i^* \in W^{1,p}(\Omega)$ such that $u_i^*|_\Gamma = U_i$ on $\Gamma_U(u_i^*|_\Gamma$ is the trace of u_i^* on Γ which is an element on $W^{1-1/p,p}(\Gamma)$). We further assume that Γ_U is nonempty. For the sake of simplicity let $U_i = 0$ on Γ_U and thus $U_{\mathrm{ad}} = \{v : v_i \in W^{1,p}(\Omega), v_i = 0 \text{ on } \Gamma_U\}$. (If $U_i \neq 0$ on Γ_U we perform the translation $\tilde{v} = v - u^*$ and $\tilde{u} = u - u^*$). We also assume that $f_i \in L^{p'}(\Omega)$ $(1/p + 1/p' = 1)$ and let $(\cdot, \cdot)$ denote the duality pairing on $L^p(\Omega) \times L^{p'}(\Omega)$. As in the previous sections we can show by means of the theorem of Green-Gauss that the problem leads to the hemivariational inequality: Find $u \in V_{\mathrm{ad}}$ such that

$$\int_\Omega \left[\frac{\partial w(\varepsilon(u))}{\partial \varepsilon}\right]_{ij} \varepsilon_{ij}(v - u)\partial\Omega + \int_{\Gamma_S} j^0(x, u, v - u)d\Gamma$$

$$\geq \int_\Omega f_i(v_i - u_i)d\Omega + \int_{\Gamma_F} F_i(v_i - u_i)d\Gamma \quad \forall v \in U_{\mathrm{ad}}. \quad (5.5.47)$$

We shall give now two equivalent formulations of (5.5.47) for w convex: find $u \in U_{\mathrm{ad}}$ such that

$$\int_\Omega [w(\varepsilon(v)) - w(\varepsilon(u))]d\Omega + \int_{\Gamma_S} j^0(x, u, v - u)d\Gamma$$

$$\geq \int_\Omega f_i(v_i - u_i)d\Omega + \int_{\Gamma_F} F_i(v_i - u_i)d\Gamma \quad \forall v \in U_{\mathrm{ad}}. \quad (5.5.48)$$

To show this equivalence let us set first in (5.5.48) $v = u + \lambda(v^* - u)$, $\lambda > 0$. Then take $\lambda \to 0_+$ and note that $j^0(x, u, \cdot)$ is positively homogeneous. Conversely, from (5.5.47) and the obvious inequality

$$w(\varepsilon(v)) - w(\varepsilon(u)) \geq \left[\frac{\partial w\varepsilon(u)}{\partial \varepsilon}\right]_{ij} \varepsilon_{ij}(v - u) \qquad (5.5.49)$$

holding for every $\varepsilon(v) \in R^6$ we get (5.5.48). It is also easy to verify that (5.5.48) or (5.5.47) is equivalent to the hemivariational inequality: find $u \in U_{\mathrm{ad}}$ such that

$$\int_\Omega \left[\frac{\partial w(\varepsilon(v))}{\partial \varepsilon}\right]_{ij} \varepsilon_{ij}(v - u)d\Omega + \int_{\Gamma_S} j^0(x, u, v - u)d\Gamma$$

$$\geq \int_\Omega f_i(v_i - u_i)d\Omega + \int_{\Gamma_F} F_i(v_i - u_i)d\Gamma \quad \forall v \in U_{\mathrm{ad}}. \quad (5.5.50)$$

Indeed (5.5.47) together with the monotonicity inequality ($w(\cdot)$ is convex)

$$\left[\frac{\partial w(\varepsilon(v))}{\partial \varepsilon} - \frac{\partial w(\varepsilon(u))}{\partial \varepsilon}\right]_{ij} \varepsilon_{ij}(v - u) \geq 0 \quad \forall \varepsilon(v), \varepsilon(u) \in R^6 \qquad (5.5.51)$$

implies (5.5.47). Conversely , in (5.5.51) we put $\bar{v} = u + \lambda(v - u), 0 < \lambda < 1$ and due to the monotonicity of $\lambda \to [(\mathrm{grad}w(\varepsilon(u + \lambda(v - u)))_{ij}\varepsilon_{ij}(v - u)]$ we get the inequality (5.5.47) at the limit for $\lambda \to 0_+$.

Let us set $v_i - u_i = \pm\varphi_i \in C_0^\infty(\Omega)$ in (5.5.47). This implies

$$[\mathrm{grad}w(\varepsilon(u))]_{ij,j} + f_i = 0 \qquad (5.5.52)$$

in the sense of distributions on Ω. But due to $f_i \in L^{p'}(\Omega)$ we may apply the Green-Gauss theorem and write (1.4.38) in the given functional framework in this subsection, for $\sigma_{ij} = [\mathrm{grad}w(\varepsilon(u))]_{ij}$. The resulting expression together with (5.5.47) implies first the boundary condition (5.5.38) as equality in the space $[W^{1-1/p,p}(\Gamma)]^*$, and secondly the boundary condition (5.5.39) in the weak form,

$$\int_{\Gamma_S} j^0(x, u, v - u)d\Gamma \geq -\langle S, v - u\rangle \quad \forall v \in [W^{1-1/p,p}(\Gamma)]^3, \qquad (5.5.53)$$

where here $\langle \cdot, \cdot \rangle$ denotes the duality pairing on $[W^{1-1/p,p}(\Gamma)]^3 \times [(W^{1-1/p,p}(\Gamma)^*]^3$.

In order to study the hemivariational inequality (5.5.47) we make the following assumption on the nonlinear elastic energy $w(\cdot)$: For every $u = \{u_i\} \in [W^{1,p}(\Omega)]^3$ there exists a constant $c > 0$ such that for $p \geq 2$

$$\int_\Omega [\mathrm{grad}w(\varepsilon(u))]_{ij}\varepsilon_{ij}(v)d\Omega \geq c \int_\Omega [\varepsilon_{ij}(u)\varepsilon_{ij}(v_i)]^{p/2}d\Omega. \qquad (5.5.54)$$

Korn's first inequality in $W^{1,p}$-spaces holds for $p > 1$ (see [Orn, Léné73,74]) and thus we have that

$$\int_{\Omega} [\varepsilon_{ij}(u)\varepsilon_{ij}(u)]^{p/2} d\Omega \geq c||u||^p_{W^{1,p}} \quad \forall u \in U_{ad}, \tag{5.5.55}$$

where c is a const > 0 and $p > 1$.

The theory of the present Chapter cannot be applied to the hemivariational inequality (5.5.47) because the weak continuity of the nonlinear mapping grad $w(\varepsilon(\cdot))$ cannot be verified for large classes of nonlinear elastic materials. Therefore we will assume that j fulfills the assumptions (i)–(iv) and (4..3.27) of Theorem 4.25. Indeed this Theorem can be applied (modified, Ω is replaced by Γ) because due to the convexity of $w(\cdot)$ and the continuity of the linear mapping $[W^{1,p}(\Omega)]^3 \ni u \to \varepsilon(u) \in [L^p(\Omega)]^6$ the nonlinear mapping grad $w(\varepsilon(\cdot))$ is maximal monotone and thus pseudomonotone (Prop. 2.3). Moreover it is coercive because of (5.5.54), (5.5.55) and the compactness of $W^{1,p}(\Omega) \subset L^p(\Gamma)$ (see [Kuf] p.344)

5.5.5 Nonmonotone Laws in Networks

We shall give now an application in Economics concerning a network flow problem. The present Section follows the basic ideas of W. Prager [Prag, Oet] and for the consideration of the nonlinearities combines them with the notion of nonconvex superpotential.

The generally nonmonotone nonlinearity is caused by the law relating the two branch variables of the network, the "flow intensity" and the "price differential" which here can also be vectors. The problem is formulated as a hemivariational inequality and the existence of its solution is discussed by the methods of the present Chapter.

Networks with directed branches are considered. The nodes are denoted by Latin and the branches by Greek letters. We suppose that we have m modes and ν branches. We take as branch variables the "flow intensity" S_γ and the "price differential" e_γ. As node variables the "amount of flow" p_k and the "shadow price" u_k are considered. The terminology has been taken from [Prag]. Moreover each branch may have an "initial price differential" vector e^0_γ. The above given quantities are assembled in vectors e, e^0, u, s, p. The node-branch incidence matrix G permits us to write the following relations

$$Gs = p, \tag{5.5.56}$$

$$e = G^T u, \tag{5.5.57}$$

where the lines of G are linearly independent. Upper index T denotes the transpose of a matrix or a vector. The network analysis is completed physically by defining an algebraic structure on the network, consisting of the relation between the "flow intensity" s_γ and the "price differential" e_γ. We accept that s_γ is a nonmonotone function of the e_γ expressed by the relation

$$e_\gamma - e^0_\gamma \in \partial j_\gamma(s_\gamma) + \frac{1}{2}\partial s^T_\gamma k_\gamma s_\gamma \tag{5.5.58}$$

where k_γ is a positive definite symmetric matrix and ∂ is the generalized gradient. The graph of the $s_\gamma - e_\gamma$ law is called γ-characteristic.

The problem to be solved consists in the determination for the whole network of the vectors s, e, u, for given vectors p and e_0. The problem can be given in a different form as well. Suppose that in some nodes belonging to a set denoted by S_p the amount of flow p_k is given and the shadow prices are unknown, while in the remaining nodes, which form the set S_u, the shadow prices u_k are given and the amounts of flow are unknown. We add a hypothetical "ground" node O, and we join this node with all the nodes of the set S_u with branches which have initial price differentials equal to the given shadow prices and "rigid" γ-characteristics, i.e. $e_\gamma = 0$ and s_γ may have any value. Then this form of the problem is transformed into the preceding one.

From (5.5.57) we obtain that

$$e^T(s^* - s) = 0 \quad \forall s^* \in S_{ad} \tag{5.5.59}$$

where

$$S_{ad} = \{s : s \in R^n, \quad Gs = p\}. \tag{5.5.60}$$

From (5.5.59) and (5.5.58) we get the following formulation of the problem: find $s \in S_{ad}$ such as to satisfy

$$\sum_\gamma j_\gamma^0(s_\gamma, s_\gamma^* - s_\gamma) + s^T K(s^* - s) + e^{0T}(s^* - s) \geq 0 \quad \forall s^* \in S_{ad}. \tag{5.5.61}$$

Here $K = \text{diag}[K_1, \ldots, K_\gamma, \ldots]$ and the summation $\sum_\gamma$ is extended over all branches. Now we will transform the hemivariational inequality by using some elements of graph theory. Let us consider the graph which corresponds to the network and a corresponding tree. The tree results from the initial graph by cuting all the branches creating the closed loops. Let us denote by s_T (resp. s_M) the part of the vector s corresponding to the tree branches (resp. to the cut branches giving rise to closed loops). Then (5.5.56) becomes

$$G_T s_T + G_M s_M = p. \tag{5.5.62}$$

Here G_T is nonsingular and thus we may write that

$$s = \begin{bmatrix} s_T \\ s_M \end{bmatrix} = \begin{bmatrix} G_T^{-1} \\ 0 \end{bmatrix} p + \begin{bmatrix} -G_T^{-1}G_M \\ I \end{bmatrix} s_M = s_0 + B s_M, \tag{5.5.63}$$

where I denotes the unit matrix. From (5.5.63) and (5.5.61) we obtain a hemivariational inequality with respect to s_M which reads: find $s_M \in R^{n_1}$ (n_1 is the dimension of s_M) such that

$$\sum_\gamma j_\gamma^0((s_0 + B s_M)_\gamma, (B s_M^* - B s_M)_\gamma) + s_M^T B^T K B(s_M^* - s_M)$$

$$+ s_0^T K B(s_M^* - s_M) + e^{0T} B(s_M^* - s_M) \geq 0 \quad \forall s_M^* \in R^{n_1}. \tag{5.5.64}$$

Since the rank of B is equal to the number of its columns and K is symmetric and positive definite the same happens for $B^T K B$. In the finite dimensional case treated here, one can easily verify that Theorem 5.12 holds without the weak continuity and the compactness assumption, if $j_\gamma(\cdot, \cdot)$ satisfies (5.1.4)–(5.1.6) and (5.3.2), (5.3.3). Thus (5.5.64) has at least one solution.

6. Noncoercive Hemivariational Inequalities Related to Free Boundary Problems

The aim of this Chapter is the mathematical study of a system of variational-hemivariational inequalities related to a large class of free boundary problems. As a pilot problem we study a system of a variational and a hemivariational inequalities related to the continuous model of the delamination process in multi-layered structures. This model is characterized by the introduction of an "auxiliary" function, the so-called delamination density scalar field δ, which takes values in $[0,1]$; $\delta = 0$ corresponds to the case of lack of delamination, $\delta = 1$, to total delamination and $0 < \delta < 1$ to partial delamination. The characterization of the free boundary by means of such "auxiliary" functions is common in free boundary problems. We refer in this context, besides to the delamination model of Woźniak [Woź90] considered here , also to the adhesive contact and the fissuration models due to Frémond (see e.g. [Frém82a,b,83a,84,87a,b,c,88]) as well as to an analogous model of Frémond concerning the shape memory alloys [Frém87b,90], and numerous models on the frost action in soils [Frém82c,83b, Niez]. For other analogous models concerning general phase transition problems, the Stefan problem and other free boundary problems we also refer to [Frém85, Vis, Rod, Hoff90a,b, Neit] as well as to the theory of coexistent phases by Dunn and Fosdick as presented in [Pan85](Sect.4.2).

Here we have chosen to study the delamination model of Woźniak because it leads to a general mathematical problem, i.e. a system of noncoercive variational and hemivariational inequalities, in contrast to many of the aforementioned free boundary problems, which due to either their inherent nature or, to certain simplifications of the physical models lead simply to coercive variational inequalities. Thus the methodology developed in this Chapter may be applied to the study of many other free boundary problems of more advanced physical and mechanical structures. In the pilot problem of this Chapter, due to unilateral-bilateral internal stress constraints, the corresponding potential energy is not coercive even with respect to the L^1-norm of the strain tensor. Moreover we cannot work along the lines of plasticity theory, where only unilateral restrictions occur. The existence of displacements is proved in a space consisting of vectors with certain of their strain components missing in the totally delaminated parts of the composite structure. Such a space was initially introduced by the first author in [Nan92b,93a].

6.1 A System of a Variational and a Hemivariational Inequality Related to Delamination Problems

The contact between adjacent laminae of multilayered composite structures cannot be fully described by means of displacements and stresses only, when the possibility of debonding has to be taken into account. This is the main idea of the approach presented in [Woź90], where a continuum model for the delamination has been presented. The leading notion there is the delamination density scalar field $\delta \in [0, 1]$ defined in the region occupied by the composite; the cases $\delta = 0, \delta = 1, \delta \in (0, 1)$ correspond to the nondelaminated, totally delaminated and partly delaminated parts of the composite, respectively.

Our aim is to establish the existence of solution in the quasistatic case in which the delamination is total in some part of the composite and some nonmonotone subdifferential boundary conditions are prescribed. We shall show that although certain components of the strain field do not exist because of interlaminar slides, both the displacement field and the stress field are well defined in the whole composite provided the external loading satisfies some compatibility conditions. This purely mathematical result is consistent with the experimental observations. Indeed, it is well known that the delamination effects do not lead to the destruction of a multilayered composite structure if the loading is "appropriately" safe.

The problem which is studied in the present Section differs from those studied in the theory of plastic type materials (plastic materials, textile-type materials, materials not sustaining tractions) first of all in the character of the stress constraints imposed. In the present approach the total delamination is described by mixed unilateral-bilateral constraints, whereas in the theory of plastic-type materials we deal with unilateral ones only. Due to the bilateral restrictions and the prescribed nonmonotone subdifferential boundary conditions we face new serious mathematical difficulties: the corresponding potential energy is nonconvex and noncoercive, even with respect to the L^1-norm of the strain tensor. Therefore we cannot pose our problem in the framework of the BD-space which has proved to be suitable for the aforementioned ideal plasticity problems. The mathematical model is formulated in the displacement space which here consists of vectors such that certain strain components do not exist in the totally delaminated part of the composite.

Let Ω be an open, bounded and connected subset of the three-dimensional Euclidean space R^3, with a Lipschitz continuous boundary Γ. Ω is occupied by the composite structure in its undeformed state and is referred to the Cartesian coordinate system $0x_1x_2x_3$. We decompose Γ into three mutually disjoint parts Γ_U, Γ_S and Γ_F assuming that $\mathrm{mes}\Gamma_U > 0$. We denote as in Sect. 1.4 by $n : \Gamma \to R^3$ the outward unit normal vector field to Γ. For the space of 3×3 symmetric real valued matrices we use here the symbol S^3. The inner product in R^3 and S^3 will be denoted by the same symbol ".", if no ambiguity occurs. As usual, the well known notations $u : \Omega \to R^3$ and $\sigma : \Omega \to S^3$ will be introduced to denote the displacement vector field and the stress vector field, respectively.

The layering of the composite will be determined by the unit vector field n', $n' : \Omega \to R^3$, which determines the directions normal to the lamina interfaces. Following [Woź90] we introduce the delamination density δ, $\delta : \Omega \to [0,1]$, to characterize the intensity of micro-delaminated interfaces of the adjoining laminae. The cases $\delta = 0$, $\delta = 1$ and $\delta \in (0,1)$ correspond to the nondelaminated, totally delaminated and partly delaminated parts of the composite, respectively.

Our considerations will be carried out within the small deformation theory. Consequently, the interrelation between u and the corresponding strain tensor field $\varepsilon(u)$, $\varepsilon(u) : \Omega \to S^3$, is given by

$$\varepsilon_{ij}(u) = \frac{1}{2}(u_{i,j} + u_{j,i}), \quad i,j = 1,2,3. \tag{6.1.1}$$

The material properties of the composite are determined by the effective tensor of elasticity $C = \{C_{ijhk}\}$ (actually the elasticity tensor resulting by means of homogenization), corresponding to the properties of the laminae, and by the longitudinal and shear moduli γ_N and γ_T characterizing the properties of the interlaminar bonding material. Although the quantities C, γ_N and γ_T, as obtained in the original model by Woźniak [Woź90] (see also [Woź87,91]) using the homogenization procedure are constants, in this Chapter they are supposed to belong to $L^\infty(\Omega)$. As usual, it will be assumed that C has the symmetry property

$$C_{ijkl} = C_{jikl} = C_{klij} \quad i,j,k,l = 1,2,3, \tag{6.1.2}$$

and that the following ellipticity property holds

$$C_{ijkl}\varepsilon_{ij}\varepsilon_{kl} \geq \alpha_0 \varepsilon_{ij}\varepsilon_{ij} \quad \text{in } \Omega, \quad \alpha_0 = \text{const} > 0, \tag{6.1.3}$$

for any $\varepsilon \in S^3$. At the same time we assume that

$$\begin{aligned} \gamma_N &\geq \gamma_0 \\ \gamma_T &\geq \gamma_0 \end{aligned} \quad \text{in } \Omega, \quad \gamma_0 = \text{const} > 0. \tag{6.1.4}$$

For any vector $v : \Omega \to R^3$ and tensor $\tau : \Omega \to S^3$ the following notations will be introduced:

$$\begin{aligned} v_N &= v_i n_i' \\ v_{Ti} &= v_i - v_N n_i' \\ \tau_N &= \tau_{ij} n_i' n_j' \\ \tau_{Ti} &= \tau_{ij} n_j' - \tau_N n_i'. \end{aligned} \tag{6.1.5}$$

Denote by $D = C^{-1}$ the inverse operator to the elasticity tensor C. It is clear that $D = \{D_{ijkl}\}$ satisfies symmetry and ellipticity conditions analogous to that of (6.1.2) and (6.1.3). Further, we introduce the bonding energy function $\psi_\delta : S^3 \to \bar{R}$ by the relation [Nan93a]

$$\psi_\delta(\sigma) = \begin{cases} \dfrac{1}{1-\delta}[\dfrac{1}{2\gamma_N}(\sigma_{N+})^2 + \dfrac{1}{2\gamma_T}\sigma_T^2] & \text{if } \delta < 1 \\[2mm] I_\Xi(\sigma) & \text{if } \delta = 1, \end{cases} \tag{6.1.6}$$

where

$$\Xi = \{\tau \in S^3 : \tau_N \leq 0 \text{ and } \tau_T = 0\} \qquad (6.1.7)$$

is the set of admissible stresses in the totally delaminated part of the composite, $\sigma_{N+} = \max(\sigma_N, 0)$ and I_Ξ denotes the indicator of Ξ. The function ψ_δ can be interpreted as the energy of bonding material. Suppose now that the composite structure Ω with the boundary Γ is submitted to external forces consisting of body forces $f : \Omega \to R^3$ and surface tractions $F : \Gamma_F \to R^3$, on $\Gamma_F \subset \Gamma$. On another part of the boundary Γ_U the displacements $u_0 : \Gamma \to R^3$ are prescribed while on the remaining part of the boundary Γ_S nonmonotone boundary conditions are assumed to hold, that is $-S \in \partial j(u)$, where $j : R^3 \to R$ is assumed to be a locally Lipschitz function and $S = \{S_i\} = \{\sigma_{ij}n_j\}$ denotes the stress vector on Γ.

Now we can give the classical formulation of the problem.

Problem (P). Given f, F, u_0, j and δ. Find $u = \{u_j\}$ and $\sigma = \{\sigma_{ij}\}, i, j = 1, 2, 3$, satisfying

(i) The equilibrium equations

$$\sigma_{ij,j} + f_i = 0 \quad \text{in } \Omega, \qquad (6.1.8)$$

(ii) The displacement boundary conditions

$$u_i = u_{0i} \quad \text{on } \Gamma_U, \qquad (6.1.9)$$

(iii) The surface traction conditions

$$S_i = F_i \quad \text{on } \Gamma_F, \qquad (6.1.10)$$

(iv) The nonmonotone subdifferential boundary conditions .

$$-S \in \partial j(u) \quad \text{on } \Gamma_S, \qquad (6.1.11)$$

(v) The constitutive law which reads

$$\varepsilon(u) - D\sigma \in \partial\psi_\delta(\sigma) \quad \text{in } \Omega. \qquad (6.1.12)$$

In the relations (6.1.8)–(6.1.12) all the functions are assumed to be appropriately smooth. These relations constitute the classical formulation of the problem under consideration and formally, by applying the Green-Gauss theorem, they give rise to system of a variational and a hemivariational inequality

$$\int_\Omega \sigma_{ij}[\varepsilon_{ij}(v) - \varepsilon_{ij}(u)]d\Omega + \int_{\Gamma_S} j^0(u, v - u)d\Gamma$$

$$- \int_\Omega f_i(v_i - u_i)d\Omega - \int_{\Gamma_F} F_i(v_i - u_i)d\Gamma \geq 0 \quad \forall v \in V, \qquad (6.1.13)$$

$$\int_\Omega [D_{ijkl}\sigma_{kl} - \varepsilon_{ij}(u)](\tau_{ij} - \sigma_{ij})d\Omega + \int_\Omega \psi_\delta(\tau)d\Omega - \int_\Omega \psi_\delta(\sigma)d\Omega \geq 0 \quad \forall \tau \in Y.$$

Here V is the space of kinematically admissible displacements, while Y is the space for stresses; both spaces will be defined in the sequel.

The constitutive law (6.1.12) can be written equivalently as $\sigma \in (D + \partial\psi_\delta)^{-1}(\varepsilon(u))$ which gives rise to the mapping $G^\delta = (D + \partial\psi_\delta)^{-1}$. We shall give an explicit formula for G^δ under the hypotheses that

$$n' = (0,0,1) \quad \text{in } \Omega, \tag{6.1.14}$$

and that C has a plane of elastic symmetry orthogonal to n'. In such a case the layers of the composite coincide with the planes $x_3 = \text{const}$ and

$$C_{\alpha\beta\gamma 3} = C_{333\mu} = 0. \tag{6.1.15}$$

The Latin indices $k, l, \ldots$ run over 1,2,3, while the Greek ones $\alpha, \beta, \ldots$ run over 1,2. The following formulas for $\partial\psi_\delta$ can now be obtained:

$$\partial\psi_\delta(\sigma) = \frac{1}{1-\delta}\begin{bmatrix} 0 & 0 & \dfrac{1}{2\gamma_T}\sigma_{13} \\ 0 & 0 & \dfrac{1}{2\gamma_T}\sigma_{23} \\ \dfrac{1}{2\gamma_T}\sigma_{13} & \dfrac{1}{2\gamma_T}\sigma_{23} & \dfrac{1}{\gamma_N}\sigma_{33+} \end{bmatrix} \quad \text{for } \delta < 1. \tag{6.1.16}$$

If $\delta = 1$, then we have: $\partial\psi_\delta(\sigma) = \emptyset$ for $\sigma \notin \Xi$, and

$$\partial\psi_\delta(\sigma) = \frac{1}{1-\delta}\left\{\begin{bmatrix} 0 & 0 & \tau_{13} \\ 0 & 0 & \tau_{23} \\ \tau_{13} & \tau_{23} & \tau_{33} \end{bmatrix} : \tau_{13}, \tau_{23} \in R, \tau_{33} \geq 0, \sigma_{33}\tau_{33} = 0\right\} \tag{6.1.17}$$

for $\sigma \in \Xi$, where $\Xi = \{\eta \in S^3 : \eta_{13} = \eta_{23} = 0, \eta_{33} \leq 0\}$. The above relations yield the following equivalent representation of (6.1.12):

$$\left.\begin{array}{l} \varepsilon_{\alpha\beta} - D_{\alpha\beta kl}\sigma_{kl} = 0 \quad \delta \in [0,1] \\[2ex] \left.\begin{array}{l} \varepsilon_{\alpha 3} - D_{\alpha 3\beta 3}\sigma_{\beta 3} = \dfrac{1}{2\gamma_T(1-\delta)}\sigma_{\alpha 3} \\[2ex] \varepsilon_{33} - D_{33kl}\sigma_{kl} = \dfrac{1}{\gamma_N(1-\delta)}\sigma_{33+} \end{array}\right\} \quad \text{if } \delta < 1 \\[4ex] \left.\begin{array}{l} \sigma_{33} \leq 0 \\[1ex] \sigma_{\alpha 3} = 0 \\[1ex] \varepsilon_{33} - D_{33kl}\sigma_{kl} \geq 0 \\[1ex] (\varepsilon_{33} - D_{33kl}\sigma_{kl})\sigma_{33} = 0 \end{array}\right\} \quad \text{if } \delta = 1 \quad \alpha,\beta = 1,2 \ k,l = 1,2,3 \end{array}\right\} \tag{6.1.18}$$

for any $\varepsilon, \sigma \in S^3$. From these expressions we can deduce the following form of the operator G^δ:

$$
\begin{aligned}
G^\delta_{\alpha\beta}(\varepsilon) &= C_{\alpha\beta kl}\varepsilon_{kl} - \frac{C_{\alpha\beta 33}}{(1-\delta)\gamma_N + C_{3333}}(C_{33kl}\varepsilon_{kl})_+, \\
G^\delta_{\alpha 3}(\varepsilon) &= \frac{2(1-\delta)\gamma_T}{k(\delta)}[k(1)\varepsilon_{\alpha 3} + 2(1-\delta)\gamma_T C_{\alpha 3\beta 3}\varepsilon_{\beta 3}], \qquad (6.1.19) \\
G^\delta_{33}(\varepsilon) &= C_{33kl}\varepsilon_{kl} - \frac{C_{3333}}{(1-\delta)\gamma_N + C_{3333}}(C_{33kl}\varepsilon_{kl})_+,
\end{aligned}
$$

which is valid for any $\delta \in [0,1]$. Here we have used the notation

$$
k(\delta) = 4(1-\delta)\gamma_T^2 + 2(1-\delta)\gamma_T(C_{1313}+C_{2323}) + C_{1313}C_{2323} - C_{1323}C_{1323}. \quad (6.1.20)
$$

The foregoing expressions show that if $\delta = 1$ then G^δ does not depend on $\varepsilon_{\alpha 3}$. Consequently, this operator cannot be coercive with respect to the strain tensor. To overcome this difficulty we shall extend G^δ to displacement fields whose $\varepsilon_{\alpha 3}$-strains are not defined in the region where $\delta = 1$.

6.2 The Strain Energy Function

This section is devoted to the study of the properties of the strain energy function. Our considerations will be carried out under the assumptions (6.1.14) and (6.1.15).

It is clear that $\partial\varphi_\delta = D + \partial\psi_\delta$, where

$$
\varphi_\delta(\tau) = \frac{1}{2}D\tau \cdot \tau + \psi_\delta(\tau), \quad \tau \in S^3. \qquad (6.2.1)
$$

Accordingly, $G^\delta = \partial E^\delta$, where E^δ is the conjugate of φ_δ (cf. Sect. 1.1). Thus taking into account the positive homogeneity of G^δ we arrive at

$$
E^\delta(\varepsilon) = \int_0^1 G^\delta(t\varepsilon) \cdot \varepsilon \, dt = \frac{1}{2}G^\delta(\varepsilon) \cdot \varepsilon, \quad \varepsilon \in S^3. \qquad (6.2.2)
$$

E^δ is the stored energy functional. By virtue of (6.1.19) one gets

$$
\begin{aligned}
2E^\delta(\varepsilon) &= C_{3333}\varepsilon_{33}\varepsilon_{33} + 2C_{33\alpha\beta}\varepsilon_{33}\varepsilon_{\alpha\beta} + C_{\alpha\beta\gamma\mu}\varepsilon_{\alpha\beta}\varepsilon_{\gamma\mu} \\
&\quad - \frac{1}{(1-\delta)\gamma_N + C_{3333}}[(C_{33kl}\varepsilon_{kl})_+]^2 \qquad (6.2.3) \\
&\quad + \frac{4(1-\delta)\gamma_T}{k(\delta)}[k(1)\varepsilon_{\alpha 3}\varepsilon_{\alpha 3} + 2(1-\delta)\gamma_T C_{\alpha 3\mu 3}\varepsilon_{\alpha 3}\varepsilon_{\mu 3}], \quad \varepsilon \in S^3.
\end{aligned}
$$

After introducing the notations

$$
g^\delta(\varepsilon) = \begin{cases} 1 & \text{if } C_{33kl}\varepsilon_{kl} \le 0 \\ \dfrac{(1-\delta)\gamma_N}{(1-\delta)\gamma_N + C_{3333}} & \text{if } C_{33kl}\varepsilon_{kl} > 0 \end{cases}
$$

and $\hspace{9cm}$ (6.2.4)

$$h^\delta(\varepsilon) = \begin{cases} 0 & \text{if } C_{33kl}\varepsilon_{kl} \leq 0 \\[2mm] \dfrac{(1-\delta)\gamma_N}{(1-\delta)\gamma_N + C_{3333}} & \text{if } C_{33kl}\varepsilon_{kl} > 0 \end{cases}$$

it results that

$$\begin{aligned} E^\delta(\varepsilon) &= \frac{1}{2}g^\delta(\varepsilon)[C_{3333}\varepsilon_{33}\varepsilon_{33} + 2C_{33\alpha\beta}\varepsilon_{33}\varepsilon_{\alpha\beta} + C_{\alpha\beta\gamma\mu}\varepsilon_{\alpha\beta}\varepsilon_{\gamma\mu}] \hspace{1.5cm} (6.2.5)\\ &+ \frac{1}{2}h^\delta(\varepsilon)[C_{3333}C_{\alpha\beta\gamma\mu} - C_{33\alpha\beta}C_{33\gamma\mu}]\varepsilon_{\alpha\beta}\varepsilon_{\gamma\mu}\\ &+ \frac{2(1-\delta)\gamma_T}{k(\delta)}[k(1)\varepsilon_{\alpha3}\varepsilon_{\alpha3} + 2(1-\delta)\gamma_T C_{\alpha3\mu3}\varepsilon_{\alpha3}\varepsilon_{\mu3}], \quad \varepsilon \in S^3. \end{aligned}$$

The representation of E^δ obtained is a convenient starting point for the further study.

Proposition 6.1 Let (6.1.3) be valid. Then the following estimate holds

$$E^\delta(\varepsilon) \geq \frac{\alpha_0}{2}\varepsilon_{\alpha\beta}\varepsilon_{\alpha\beta} + c(1-\delta)(\varepsilon_{33}\varepsilon_{33} + \varepsilon_{\alpha3}\varepsilon_{\alpha3}) \quad \forall \varepsilon \in S^3, \hspace{1cm} (6.2.6)$$

for some positive constant c.

Proof. The desired inequality will be deduced from the following ones:

$$C_{3333}\varepsilon_{33}\varepsilon_{33} + 2C_{33\alpha\beta}\varepsilon_{33}\varepsilon_{\alpha\beta} + C_{\alpha\beta\gamma\mu}\varepsilon_{\alpha\beta}\varepsilon_{\gamma\mu} \geq \alpha_0(\varepsilon_{33}\varepsilon_{33} + \varepsilon_{\alpha\beta}\varepsilon_{\alpha\beta}), \hspace{0.5cm} (6.2.7a)$$

$$(C_{3333}C_{\alpha\beta\gamma\mu} - C_{33\alpha\beta}C_{33\gamma\mu})\varepsilon_{\alpha\beta}\varepsilon_{\gamma\mu} \geq \alpha_0 C_{3333}\varepsilon_{\alpha\beta}\varepsilon_{\alpha\beta}, \quad \varepsilon \in S^3, \hspace{0.5cm} (6.2.7b)$$

$$C_{1313}C_{2323} - C_{1323}C_{1323} \geq \alpha_0^2. \hspace{2.5cm} (6.2.7c)$$

To show (6.2.7a) it suffices to substitute $\varepsilon \in S^3$ with $\varepsilon_{\alpha3} = 0$ into (6.1.3). For (6.2.7b) we argue as follows. Setting $\varepsilon \in S^3$ with $\varepsilon_{\alpha3} = 0$ and $\varepsilon_{33} = s$, $s \in R$, into (6.1.3) one obtains

$$(C_{3333} - \alpha_0)s^2 + 2C_{33\alpha\beta}\varepsilon_{\alpha\beta}s + C_{\alpha\beta\gamma\mu}\varepsilon_{\alpha\beta}\varepsilon_{\gamma\mu} - \alpha_0\varepsilon_{\alpha\beta}\varepsilon_{\alpha\beta} \geq 0. \hspace{0.8cm} (6.2.8)$$

Since s is arbitrary, the discriminant $\Delta \leq 0$, i.e.

$$\Delta = 4(C_{33\alpha\beta}\varepsilon_{\alpha\beta})^2 - 4(C_{3333} - \alpha_0)(C_{\alpha\beta\gamma\mu}\varepsilon_{\alpha\beta}\varepsilon_{\gamma\mu} - \alpha_0\varepsilon_{\alpha\beta}\varepsilon_{\alpha\beta}) \leq 0. \hspace{0.5cm} (6.2.9)$$

which leads easily to (6.2.7b). The inequality (6.2.7c) follows from the obvious inequality

$$C_{\alpha3\gamma3}\varepsilon_{\alpha3}\varepsilon_{\gamma3} \geq \alpha_0\varepsilon_{\alpha3}\varepsilon_{\alpha3}, \quad \varepsilon \in S^3, \hspace{1.5cm} (6.2.10)$$

when one applies the same method as that applied to (6.2.7b). Now we can complete the proof of (6.2.6). Combining (6.2.7) and (6.1.4) with (6.2.5) yields

$$E^\delta(\varepsilon) \geq \frac{\alpha_0}{2}(g^\delta(\varepsilon) + C_{3333}h^\delta(\varepsilon))\varepsilon_{\alpha\beta}\varepsilon_{\alpha\beta} \tag{6.2.11}$$

$$+\frac{\alpha_0}{2}g^\delta(\varepsilon)\varepsilon_{33}\varepsilon_{33} + (1-\delta)\frac{2\gamma_T k(1)}{k(\delta)}\varepsilon_{\alpha3}\varepsilon_{\alpha3}$$

$$\geq \frac{\alpha_0}{2}\varepsilon_{\alpha\beta}\varepsilon_{\alpha\beta} + (1-\delta)[\frac{\alpha_0\gamma_0}{2(\gamma_N + C_{3333})}\varepsilon_{33}\varepsilon_{33}$$

$$+\frac{2\gamma_0\alpha_0^2}{4\gamma_T^2 + 2\gamma_T(C_{1313} + C_{2323}) + k(1)}\varepsilon_{\alpha3}\varepsilon_{\alpha3}] \quad \varepsilon \in S^3,$$

where the following identity has been used

$$g^\delta(\varepsilon) + C_{3333}h^\delta(\varepsilon) \equiv 1, \quad \varepsilon \in S^3. \tag{6.2.12}$$

Finally, because C, γ_N, and γ_T are bounded, we easily obtain from (6.2.11) the estimate (6.2.6). This completes the proof of Proposition 6.1.   q.e.d.

Remark 6.2 The strain energy functional E^δ does not depend on $\varepsilon_{\alpha3}$ in the region where the delamination is total ($\delta = 1$). For this reason the potential energy

$$\mathcal{E}^\delta(v) = \int_\Omega E^\delta(\varepsilon(v))d\Omega + \int_{\Gamma_S} j(v)d\Gamma - \int_\Omega f_iv_id\Omega - \int_{\Gamma_F} F_iv_id\Gamma \tag{6.2.13}$$

cannot be coercive with respect to any norm of the strain tensor.

6.3 Study of the Case of Partial Delamination

In this Section we shall prove the existence of solutions to (P) assuming that partial delamination occurs in the whole composite. We use the theory of pseudo-monotone mappings in Hilbert spaces. We start with the functional formulation of (P) by introducing spaces

$$V = \{v = \{v_k\} : v_k \in H^1(\Omega) \text{ and } v_{k|\Gamma_U} = 0\} \tag{6.3.1}$$
$$Y = \{\tau = \{\tau_{ij}\} : \tau_{ij} = \tau_{ji}, \tau_{ij} \in L^2(\Omega)\}.$$

The scalar product and the norm in Y will be denoted by

$$(\cdot, \cdot)_{L^2(\Omega)} \quad \text{and} \quad ||\cdot||_{L^2(\Omega)}, \tag{6.3.2}$$

respectively. Let V^* be the dual of V. For the norm in V and the pairing over $V^* \times V$ we use the symbols

$$||\cdot||_{H^1(\Omega)} \quad \text{and} \quad \langle\cdot, \cdot\rangle_{H^1(\Omega)}, \tag{6.3.3}$$

respectively. Let us define $g \in V^*, L : V \to Y$ and $\Psi_\delta : Y \to \bar{R}$ according to the formulas:

$$\langle g, v \rangle_{H^1(\Omega)} = \int_\Omega f_iv_id\Omega + \int_{\Gamma_F} F_iv_id\Gamma, \quad v \in V, \tag{6.3.4}$$

$$Lv = \varepsilon(v), \quad v \in V, \tag{6.3.5}$$

$$\Psi_\delta(\tau) = \begin{cases} \int_\Omega \psi_\delta(\tau)d\Gamma & \text{if } \psi_\delta(\tau) \in L^1(\Omega) \\ +\infty & \text{otherwise,} \end{cases} \quad \tau \in Y, \tag{6.3.6}$$

where f and F are assumed to be elements of $[L^2(\Omega)]^3$ and $[L^2(\Gamma_F)]^3$, respectively. The transpose operator $L^* : Y \to V^*$ is given by

$$\langle L^*\tau, v \rangle_{H^1(\Omega)} = (\tau, Lv)_{L^2(\Omega)}, \quad \forall v \in V, \quad \forall \tau \in Y. \tag{6.3.7}$$

Concerning the functional framework of the nonmonotone boundary conditions (6.1.11) on Γ_S we assume that $j : \Gamma_S \times R^3 \to R$ satisfies the following conditions:

(i)   For all $\xi \in R^3$ the function

$$\Omega \ni x \to j(x, \xi)$$

(as a function of the variable x) is measurable on Γ_S.

(ii) For almost all $x \in \Omega$, $\hspace{5cm}$ (6.3.8)

$$|j(x, \xi) - j(x, \eta)| \le k(x)|\xi - \eta| \quad \forall \xi, \eta \in R^3,$$

with some $k \in L^2(\Gamma_S)$.

(iii) The function $j(\cdot, 0)$ is finitely integrable in Γ_S, i.e.

$$j(\cdot, 0) \in L^1(\Gamma_S).$$

As it is well known the functional $J : L^2(\Gamma_S; R^3) \to R$ defined by the relation

$$J(v) = \int_{\Gamma_S} j(v)d\Gamma \quad v \in [L^2(\Gamma_S)]^3, \tag{6.3.9}$$

is then Lipschitz continuous, i.e.

$$|J(v) - J(u)| \le \int_{\Gamma_S} |j(v) - j(u)|d\Gamma \le ||k||_{L^2(\Gamma_S)}||v - u||_{L^2(\Gamma_S)}, \quad \forall u, v \in [L^2(\Gamma_S)]^3, \tag{6.3.10}$$

and the condition $\chi \in \partial J(u)$ implies that $\chi(x) \in \partial j(x, u(x))$ for almost all $x \in \Omega$ [Aub79, Clar83].

For the sake of simplicity and without any loss of generality we may assume that $u_0 = 0$ on Γ_U. Then the system (6.1.13) becomes

$$\langle L^*\sigma - g, v - u \rangle_{H^1(\Omega)} + \int_\Omega j^0(u; v - u)d\Gamma \ge 0 \quad \forall v \in V, \tag{6.3.11a}$$

$$(D\sigma - Lu, \tau - \sigma)_{L^2(\Omega)} + \Psi^\delta(\tau) + \Psi^\delta(\sigma) \ge 0 \quad \forall \tau \in Y. \tag{6.3.11b}$$

Accordingly, we are led to the following

Problem $(\tilde{P})$. Find $u \in V$ and $\sigma \in Y$ such that (6.3.11a) and (6.3.11b) hold.

The inequality (6.3.11a) is related to the equilibrium condition while (6.3.11b) represents the functional extension of the constitutive law (6.1.13) to Y. It is well known ([Bréz72, Pan85, p.104]) that from (6.3.11b) it follows that (6.1.12) holds for almost all $x \in \Omega$.

The following noncoercive hemivariational inequality results by combining (6.3.11) with $G^\delta = (D + \partial\psi_\delta)^{-1}$:

Problem $(\tilde{P}_u)$. Find $u \in V$ such that

$$\langle L^\star G^\delta L u - g, v - u \rangle_{H^1(\Omega)} + \int_{\Gamma_S} j^0(u; v - u)d\Gamma \geq 0 \quad \forall v \in V. \qquad (6.3.12)$$

Let us define

$$\Xi_u = \{\tau \in Y : \langle L^\star\tau - g, v \rangle_{H^1(\Omega)} + \int_{\Gamma_S} j^0(u; v)d\Gamma \geq 0 \quad \forall v \in V\}. \qquad (6.3.13)$$

Then by making use of the well known results of the subdifferential calculus one can easily deduce that σ is a solution of the inclusion

$$0 \in D\sigma + \partial I_{\Xi_u}(\sigma) + \partial\Psi_\delta(\sigma). \qquad (6.3.14)$$

Here I_{Ξ_u} denotes the indicator of the set Ξ_u.

Let us consider the case in which delamination is partial in the composite. To be more precise, we shall suppose that

$$\delta \leq \delta_0 = \text{const} < 1 \quad \text{in } \Omega. \qquad (6.3.15)$$

This condition together with (6.1.19) allows us to conclude that G^δ is a maximal monotone operator from Y into Y, the effective domain of which coincides with whole space Y. By the assumption that the measure of Γ_U is positive, Korn's inequality is valid on V. Thus due to (6.2.6) the mapping $L^\star G^\delta L$ is not only maximal monotone but also coercive. Because of the compact imbedding $V \subset [L^2(\Gamma_S)]^3$ using Theorem 4.25 we can establish the existence of a solution to $(\tilde{P}_u)$. Hence problem $(\tilde{P})$ admits a solution. Accordingly, for the partial delamination case we obtain the following existence result.

Theorem 6.3 Let (6.3.15) hold. Then there exists a pair $(u, \sigma) \in V \times Y$ fulfilling (6.3.11), i.e. the problem $(\tilde{P})$ has at least one solution. Moreover, (6.3.12) and (6.3.14) hold.

6.4 Study of the General Case

In this section we shall examine the solvability of (P) on the additional assumption that in a certain part of the composite structure the delamination is total. Namely, we will suppose that there exists an open subset Ω_0 of Ω such that

$$\delta(x) \le \delta_0 = \text{const} < 1 \quad \text{for a.e. } x \in \Omega \setminus \Omega_0,$$

$$\delta(x) = 1 \qquad\qquad \text{for a.e. } x \in \Omega_0. \tag{6.4.1}$$

Moreover, the following conditions are assumed to hold:

$$\bar{\Omega}_0 \cap (\bar{\Gamma}_U \cup \bar{\Gamma}_F) = \emptyset,$$

$$F|_{\Omega_0 \cap \Gamma_F} = 0. \tag{6.4.2}$$

The foregoing conditions mean that the totally delaminated part of the composite $\bar{\Omega}_0$ can reach only a traction free part of the boundary. We exclude the case that Γ_S or Γ_U has common points with $\bar{\Omega}_0$. This is physically motivated by the fact that in a composite structure nontrivial boundary conditions cannot occur on a debonded part of the structure.

We start with the regularized problem. Its form results by the following substitutions in (6.3.11a,b). The delamination density δ is replaced by δ_λ, where

$$\delta_\lambda(x) = \begin{cases} \delta(x) & \text{if } x \in \Omega \setminus \Omega_0 \\ \\ 1 - \lambda & \text{if } x \in \Omega_0 \end{cases} \qquad \lambda > 0 \tag{6.4.3}$$

and J by J_λ, where

$$J_\lambda(v) = \int_{\Gamma_S} j_\lambda(v) d\Gamma, \quad v \in V. \tag{6.4.4}$$

Here we have denoted by $j_\lambda : \Gamma_S \times R^3 \to R$ the regularized function given by the relation

$$j_\lambda(x; \xi) = \int_{R^3} \omega_\lambda(\eta) j(x; \xi - \eta) d\eta \quad \xi \in R^3, \tag{6.4.5}$$

where $\omega \in C_0^\infty(R^3)$ is a mollifier fulfilling the properties: $\omega(\xi) \ge 0$, $\text{supp}(\omega) \subset [-1,1]^3$ and

$$\int_{R^3} \omega(\xi) d\xi = 1, \quad \omega_\lambda(\xi) = \frac{1}{\lambda^3} \omega\left(\frac{\xi}{\lambda}\right) \quad x \in R^3, \lambda > 0. \tag{6.4.6}$$

Under the hypotheses (i)–(iii) of Section 6.3 the function J_λ is differentiable with the derivative J_λ' satisfying the estimates

$$|\langle J_\lambda'(v), w\rangle_V| = \left|\int_{\Gamma_S} j_\lambda'(v) \cdot w d\Gamma\right| \le ||k||_{L^2(\Gamma_S)} ||w||_{L^2(\Gamma_S)} \tag{6.4.7}$$

$$\le c||k||_{L^2(\Gamma_S)} ||w||_{H^1(\Omega)} \quad \forall w \in V,$$

for some positive constant c. Hence

$$||J_\lambda'(v)||_{V^*} \le c||k||_{L^2(\Gamma_S)} \tag{6.4.8}$$

for any $v \in V$. Now the following regularized problem is considered.

Problem (P_λ). Find $u^\lambda \in V$ and $\sigma^\lambda \in Y$ satisfying the conditions

$$L^*\sigma^\lambda + J'_\lambda(u^\lambda) - g = 0 \tag{6.4.9a}$$

$$Lu^\lambda - D\sigma^\lambda = \partial\Psi_\lambda(\sigma^\lambda). \tag{6.4.9b}$$

Here, for the sake of simplicity we have denoted by Ψ_λ the regularized function Ψ_{δ_λ}.

Under the foregoing assumptions the existence of solutions to (P_λ) follows directly from Theorem 6.3. At the same time, from (6.3.12) and (6.3.13) it results that

$$L^*G^\lambda Lu^\lambda + J'_\lambda(u^\lambda) - g = 0 \tag{6.4.10}$$

and

$$0 \in D\sigma^\lambda + \partial I_{\Xi_\lambda}(\sigma^\lambda) + \partial\Psi_\lambda(\sigma^\lambda), \tag{6.4.11}$$

where $\Xi_\lambda = \{\tau \in Y : L^*\tau + J'_\lambda(u^\lambda) - g = 0\}$ and we write G^λ instead of G^{δ_λ}.

Now we examine the behavior of $(u^\lambda, \sigma^\lambda)$ when $\lambda \to 0$ and we construct a solution of the original problem by applying the compactness method in appropriately chosen function spaces.

It is evident, due to the mixed unilateral-bilateral constraints in Ω_0 and the nature of these constraints (i.e., no constraints concerning displacements on Ω_0), that the problem (P) cannot have solutions for arbitary loads g. Therefore, further study will be carried out under additional restrictions concerning the body forces f_i and surface tractions F_i. The first condition in this direction reads:

Compatibility Conditions (C.1). There exists $\sigma^0 \in Y$ such that

(i) $L^*\sigma^0 - g = 0$ $\qquad\qquad$ (6.4.12a)

(ii) $\sigma^0_{\alpha 3} = 0$ a.e. in Ω_0 $\qquad\qquad$ (6.4.12b)

(iii) $\sigma^0_{33} \leq -\mu_0$, $\mu_0 = \text{const} > 0$ a.e. in Ω_0. $\qquad\qquad$ (6.4.12c)

The hypothesis (C.1) is analogous to the safe load condition [Suq] in the plasticity theory. It allows us to derive some a priori estimates for σ^λ and $\varepsilon(u^\lambda)$ which contrary to the plasticity theory, do not permit the complete solution of (P).

Proposition 6.4 Let u^λ and σ^λ satisfy (6.4.9). If (C.1) holds then the following a priori estimates are satisfied (c is a constant > 0)

$$\|\sigma^\lambda_{ij}\|_{L^2(\Omega)} \leq c, \qquad i,j = 1,2,3 \tag{6.4.13a}$$

$$\|\sigma^\lambda_{33+}\|_{L^2(\Omega_0)} \leq c\sqrt{\lambda}, \tag{6.4.13b}$$

$$\|\sigma^\lambda_{\alpha 3}\|_{L^2(\Omega_0)} \leq c\sqrt{\lambda}, \qquad \alpha = 1,2 \tag{6.4.13c}$$

$$\|\sigma^\lambda_{33+}\|_{L^1(\Omega_0)} \leq c\lambda. \tag{6.4.13d}$$

Proof. For δ^λ given by (6.4.3) the subdifferential $\partial\psi_\lambda$ takes the form

$$\partial\psi_\delta(\sigma^\lambda) = \frac{1}{1-\delta_\lambda}\begin{bmatrix} 0 & 0 & \dfrac{1}{2\gamma_T}\sigma_{13}^\lambda \\[2mm] 0 & 0 & \dfrac{1}{2\gamma_T}\sigma_{23}^\lambda \\[2mm] \dfrac{1}{2\gamma_T}\sigma_{13}^\lambda & \dfrac{1}{2\gamma_T}\sigma_{23}^\lambda & \dfrac{1}{\gamma_N}\sigma_{33+}^\lambda \end{bmatrix}. \tag{6.4.14}$$

Further, (6.4.11) implies that

$$(D\sigma^\lambda + \partial\Psi_\lambda(\sigma^\lambda), \tau - \sigma^\lambda)_{L^2(\Omega)} \geq 0, \quad \forall \tau \in \Xi_\lambda. \tag{6.4.15}$$

Recalling that $\Xi_\lambda = \{\tau \in Y : L^*\tau + J_\lambda'(u^\lambda) - g = 0\}$ we introduce a sequence $\{\hat{\sigma}^\lambda\} \subset Y$ with the properties that

$$\left.\begin{array}{c} L^*\hat{\sigma}^\lambda = J_\lambda'(u^\lambda) \\[2mm] \mathrm{supp}\,(\hat{\sigma}^\lambda) \cap \bar{\Omega}_0 = \emptyset \\[2mm] \|\hat{\sigma}^\lambda\|_{L^2(\Omega)} \leq c. \end{array}\right\} \tag{6.4.16}$$

The existence of $\{\hat{\sigma}^\lambda\}$ with the properties mentioned above can be easily deduced by making use of the fact that $\mathrm{supp}\, J_\lambda'(u^\lambda) \subset \bar{\Gamma}_F$, $\bar{\Omega}_0 \cap \bar{\Gamma}_F = \emptyset$ and that $J_\lambda'(u^\lambda)$ is uniformly bounded (see (6.4.8)). Indeed, let $\Delta \subset \Omega$ be an open subregion of Ω with sufficiently smooth boundary $\partial\Delta$ such that $\bar{\Delta} \cap \bar{\Omega}_0 = \emptyset$ and $\partial\Delta \supset \Gamma_S$. The existence of such Δ follows from the first condition in (6.4.2). We introduce

$$\tilde{V} = \{w \in [H^1(\Delta)]^3 : w|_{\partial(\Delta\backslash\Gamma_S)} = 0\} \tag{6.4.17}$$

and denote by $\bar{w}^\lambda \in \tilde{V}$ an element from $\tilde{V}$ satisfying the variational equality

$$\int_\Delta C\varepsilon(\bar{w}^\lambda) \cdot \varepsilon(v) d\Omega = \int_{\Gamma_S} j_\lambda'(u^\lambda) \cdot v d\Gamma \quad v \in \tilde{V}, \tag{6.4.18}$$

where $C = \{C_{ijkl}\}$ has the properties (6.1.2) and (6.1.3). The classical Lax-Milgram theorem ensures the existence of such a $\bar{w}^\lambda$. Then by setting

$$\hat{w}^\lambda = \begin{cases} \bar{w}^\lambda & \text{in } \Delta \\[2mm] 0 & \text{in } \Omega \setminus \Delta \end{cases} \tag{6.4.19}$$

we obtain $\hat{w}^\lambda \in V$ with $\mathrm{supp}(\hat{w}^\lambda) \subset \bar{\Delta}$. Now it is enough to define $\hat{\sigma}^\lambda = C\varepsilon(\hat{w}^\lambda)$ in order to obtain an element from Y with the desired properties (6.4.16). Taking into account (C.1) it can be verified that $\sigma^0 + \hat{\sigma}^\lambda \in \Xi_\lambda$. Hence, by substituting $\tau = \sigma^0 + \hat{\sigma}^\lambda$ into (6.4.15) and taking into account (6.4.16) we obtain

$$(D\sigma^\lambda, \sigma^0 + \hat\sigma^\lambda)_{L^2(\Omega)} + \int_{\Omega\backslash\Omega_0} \frac{1}{1-\delta}[\frac{1}{\gamma_T}\sigma_{\alpha 3}^\lambda(\sigma_{\alpha 3}^0 + \hat\sigma_{\alpha 3}^\lambda) + \frac{1}{\gamma_N}\sigma_{33}^\lambda(\sigma_{33}^0 + \hat\sigma_{33}^\lambda)]d\Omega$$

$$\geq (D\sigma^\lambda, \sigma^\lambda)_{L^2(\Omega)} + \frac{\mu_0}{\lambda}\int_{\Omega_0} \frac{1}{\gamma_N}\sigma_{33+}^\lambda d\Omega \qquad (6.4.20)$$

$$+ \frac{1}{\lambda}\int_{\Omega_0}[\frac{1}{\gamma_N}(\sigma_{33+}^\lambda)^2 + \frac{1}{\gamma_T}(\sigma_{13}^\lambda)^2 + \frac{1}{\gamma_T}(\sigma_{23}^\lambda)^2]d\Omega.$$

Therfore the left-hand side of the foregoing inequality is bounded from above by

$$c||\sigma^\lambda||_{L^2(\Omega)}, \quad c = \text{const} > 0, \qquad (6.4.21)$$

while due to the strong ellipticity of D, to (6.1.4) and to (6.4.1), the right-hand side is bounded from below by

$$c||\sigma^\lambda||_{L^2(\Omega)}^2 + \frac{c}{\lambda}(||\sigma_{33+}^\lambda||_{L^1(\Omega_0)} + \sum_i ||\sigma_{i3}^\lambda||_{L^2(\Omega_0)}^2), \quad c = \text{const} > 0. \qquad (6.4.22)$$

Hence (6.4.13) can be easily deduced. \hfill q.e.d.

Further *a priori* estimates are obtained for the strains $\varepsilon(u^\lambda)$.

Proposition 6.5 Under the assumptions of Proposition 6.4 the estimates (c is a constant > 0)

$$||\varepsilon_{\alpha\beta}(u^\lambda)||_{L^2(\Omega)} \leq c, \quad \alpha, \beta = 1, 2 \qquad (6.4.23a)$$

$$||\varepsilon_{i3}(u^\lambda)||_{L^2(\Omega\backslash\Omega_0)} \leq c, \quad i = 1, 2, 3 \qquad (6.4.23b)$$

$$||\varepsilon_{i3}(u^\lambda)||_{L^2(\Omega_0)} \leq \frac{c}{\sqrt\lambda}, \quad i = 1, 2, 3 \qquad (6.4.23c)$$

$$||\varepsilon_{33}(u^\lambda)||_{L^1(\Omega_0)} \leq c, \qquad (6.4.23d)$$

hold.

Proof. Step 1. First we shall show that

$$\langle g - J_\lambda'(u^\lambda), u^\lambda\rangle_{H^1(\Omega)} \leq c. \qquad (6.4.24)$$

Indeed, from (6.4.12a) one gets

$$\langle g - J_\lambda'(u^\lambda), u^\lambda\rangle_{H^1(\Omega)} = \langle L^*(\sigma^0 + \hat\sigma^\lambda), u^\lambda\rangle_{H^1(\Omega)} = ((\sigma^0 + \hat\sigma^\lambda), \varepsilon(u^\lambda))_{L^2(\Omega)}$$

$$= \int_{\Omega\backslash\Omega_0} (\sigma_{ij}^0 + \hat\sigma_{ij}^\lambda)\varepsilon_{ij}(u^\lambda))d\Omega \qquad (6.4.25)$$

$$+ \int_{\Omega_0}[\sigma_{\alpha\beta}^0\varepsilon_{\alpha\beta}(u^\lambda) + \sigma_{33}^0\varepsilon_{33}(u^\lambda)]d\Omega.$$

Hence, because of

$$\varepsilon(u^\lambda) = D\sigma^\lambda + \partial\Psi_\lambda(\sigma^\lambda), \qquad (6.4.26)$$

it results that

$$\langle g - J'_\lambda(u_\lambda), u^\lambda \rangle_{H^1(\Omega)} = \int_{\Omega \backslash \Omega_0} [(\sigma^0_{\alpha\beta} + \hat{\sigma}^\lambda_{\alpha\beta})D_{\alpha\beta kl}\sigma^\lambda_{kl} \tag{6.4.27}$$

$$+ (\sigma^0_{\alpha 3} + \hat{\sigma}^\lambda_{\alpha 3})(D_{\alpha 3\mu 3}\sigma^\lambda_{\mu 3} + \frac{1}{2\gamma_T(1 - \delta)}\sigma^\lambda_{\alpha 3})$$

$$+ (\sigma^0_{33} + \hat{\sigma}^\lambda_{33})(D_{33kl}\sigma^\lambda_{kl} + \frac{1}{\gamma_N(1 - \delta)}\sigma^\lambda_{33+})]d\Omega$$

$$+ \int_{\Omega_0} [\sigma^0_{33}D_{33kl}\sigma^\lambda_{kl} + \sigma^0_{\alpha\beta}D_{\alpha\beta kl}\sigma^\lambda_{kl}]d\Omega$$

$$+ \int_{\Omega_0} \frac{1}{\lambda\gamma_N}\sigma^0_{33}\sigma^\lambda_{33+}d\Omega.$$

From (6.4.13), (6.4.1) and (6.1.4) the boundedness of all integrals except for that of

$$\int_{\Omega_0} \frac{1}{\lambda\gamma_N}\sigma^0_{33}\sigma^\lambda_{33+}d\Omega \tag{6.4.28}$$

easily follows. But the last integral, due to (6.4.12c), is nonpositive. Thus (6.4.24) has been established.

Step 2. For (6.4.23a)–(6.4.23c) we argue as follows. The first equation in (6.4.9) yields that

$$\int_\Omega G^\lambda(\varepsilon(u^\lambda)) \cdot \varepsilon(u^\lambda)d\Omega = \langle g - J'_\lambda(u_\lambda), u^\lambda \rangle_{H^1(\Omega)}. \tag{6.4.29}$$

This equality combined with (6.4.24) and (6.2.6) leads to the estimate

$$\alpha_0 \sum_{\alpha,\beta} ||\varepsilon_{\alpha\beta}(u^\lambda)||^2_{L^2(\Omega)} + 2c\lambda \sum_i ||\varepsilon_{3i}(u^\lambda)||^2_{L^2(\Omega_0)} \tag{6.4.30}$$

$$+ 2c(1 - \delta_0) \sum_i ||\varepsilon_{i3}(u^\lambda)||^2_{L^2(\Omega\backslash\Omega_0)} \leq c$$

from which (6.4.23a)–(6.4.23c) follow.

Step 3. In order to show (6.4.23d) we decompose Ω_0 into $\Omega_0^{\lambda+}$ and $\Omega_0^{\lambda-}$ according to the formulas

$$x \in \Omega_0^{\lambda+} \iff \sigma^\lambda_{33}(x) > 0,$$

$$x \in \Omega_0^{\lambda-} \iff \sigma^\lambda_{33}(x) \leq 0. \tag{6.4.31}$$

Recalling that

$$\sigma^\lambda_{33} = \begin{cases} C_{33kl}\varepsilon_{kl}(u^\lambda) & \text{in } \Omega_0^{\lambda-} \\ \dfrac{\lambda\gamma_N}{\lambda\gamma_N + C_{3333}}C_{33kl}\varepsilon_{kl}(u^\lambda) & \text{in } \Omega_0^{\lambda+}, \end{cases} \tag{6.4.32}$$

we consider the cases of $\Omega_0^{\lambda+}$ and $\Omega_0^{\lambda-}$ separately.

Applying (6.4.13d) with respect to $\Omega^{\lambda+}$ yields the estimate

$$\frac{1}{\lambda}\|\frac{\lambda\gamma_N}{\lambda\gamma_N + C_{3333}}[C_{3333}\varepsilon_{33}(u^\lambda) + C_{33\alpha\beta}\varepsilon_{\alpha\beta}(u^\lambda)]\|_{L^1(\Omega_0^{\lambda+})}$$
$$= \frac{1}{\lambda}\|\sigma_{33}^\lambda\|_{L^1(\Omega_0^{\lambda+})} \le \frac{1}{\lambda}\|\sigma_{33+}^\lambda\|_{L^1(\Omega_0)} \le c. \qquad (6.4.33)$$

Hence

$$\|\frac{\gamma_N C_{3333}}{\lambda\gamma_N + C_{3333}}\varepsilon_{33}(u^\lambda)\|_{L^1(\Omega_0^{\lambda+})} \le \|\frac{\gamma_N}{\lambda\gamma_N + C_{3333}}C_{33\alpha\beta}\varepsilon_{\alpha\beta}(u^\lambda)\|_{L^1(\Omega^{\lambda+})} \le c \qquad (6.4.34)$$

and thus from Hölder's inequality, (6.1.4) and (6.4.23a) we obtain the estimate

$$\|\varepsilon_{33}(u^\lambda)\|_{L^1(\Omega^{\lambda+})} \le c. \qquad (6.4.35)$$

Further, from (6.4.13a) it results that

$$\|C_{3333}\varepsilon_{33}(u^\lambda) + C_{33\alpha\beta}\varepsilon_{\alpha\beta}(u^\lambda)\|_{L^2(\Omega^{\lambda-})} = \|\sigma_{33}^\lambda\|_{L^2(\Omega_0^{\lambda-})} \le \|\sigma_{33}^\lambda\|_{L^2(\Omega_0)} \le c. \qquad (6.4.36)$$

This relation combined with (6.4.23a) leads to

$$\|\varepsilon_{33}(u^\lambda)\|_{L^2(\Omega^{\lambda-})} \le c. \qquad (6.4.37)$$

Therefore, due to Hölder's inequality

$$\|\varepsilon_{33}(u^\lambda)\|_{L^1(\Omega^{\lambda-})} \le c. \qquad (6.4.38)$$

But (6.4.35) and (6.4.38) imply (6.4.23d); thus the proof of Proposition 6.5 is complete.   q.e.d.

Another a priori estimate results from (6.4.13a), (6.4.23b), (6.4.1) and the relation

$$\varepsilon_{33}(u^\lambda) - D_{33kl}\sigma_{kl}^\lambda = \frac{1}{\gamma_N(1-\delta)}\sigma_{33+}^\lambda \quad \text{in } \Omega \setminus \Omega_0. \qquad (6.4.39)$$

This estimate leads to the following result.

Proposition 6.6 Let all the hypotheses of Proposition 6.4 hold. Then

$$\|\sigma_{33+}^\lambda\|_{L^2(\Omega\setminus\Omega_0)} \le c. \qquad (6.4.40)$$

Having obtained (6.4.13), (6.4.23) and (6.4.40) let us make some comments. In the theory of plastic type materials (plastic materials, textile-type materials, materials not sustaining tractions, etc.) the safe load condition of (C.1)-type permits us to show the existence of displacements in the BD-space. This is not the case for the present approach because the a priori estimates (6.4.13), (6.4.23) and (6.4.40) resulting from (C.1) do not allow us to solve the corresponding variational problem in the BD-space; indeed they do not yield the uniform boundedness of $\varepsilon_{\alpha 3}(u^\lambda)$ in Ω_0. This is justified by the fact that due to interlaminar slip the strains $\varepsilon_{\alpha 3}$ do not exist in Ω_0. Thus the suggestion is that we should search for the displacements in a space whose elements have their $\varepsilon_{\alpha 3}$-type strains well defined only out of Ω_0.

To realize this idea it is necessary to obtain appropriate *a priori* estimates for u^λ in Ω_0, independently of (6.4.13), (6.4.23) and (6.4.40). It turns out that such estimates can be constructed under some more restrictive assumptions concerning the load acting on the composite structure and under stronger regularity conditions related to $C_{\alpha 3\beta 3}$, γ_N and γ_T in Ω_0.

Now we will find the lacking *a priori* estimates for u^λ in Ω_0. Let us denote by $H_0^1(\Omega_0)$ the subspace of $H^1(\Omega_0)$ whose elements vanish on the boundary $\partial\Omega_0$ (in the sense of traces). From the equilibrium condition (6.4.9a) one gets

$$\int_{\Omega_0} G_\lambda(\varepsilon(u^\lambda)) \cdot \varepsilon(v)d\Omega = \int_{\Omega_0} f_i v_i d\Omega \quad \forall v \in [H_0^1(\Omega_0)]^3. \tag{6.4.41}$$

Further, let us introduce the following bilinear forms on $[H_0^1(\Omega_0)]^3 \times [H_0^1(\Omega_0)]^3$:

$$
\begin{aligned}
a_\lambda(v, w) &= \int_{\Omega_0} \{g^\lambda(\varepsilon(u^\lambda))[C_{3333}\varepsilon_{33}(v)\varepsilon_{33}(w) + C_{33\alpha\beta}\varepsilon_{33}(v)\varepsilon_{\alpha\beta}(w) \\
&+ C_{33\alpha\beta}\varepsilon_{\alpha\beta}(v)\varepsilon_{33}(w) + C_{\alpha\beta\gamma\mu}\varepsilon_{\alpha\beta}(v)\varepsilon_{\gamma\mu}(w)] \\
&+ h^\lambda(\varepsilon(u^\lambda))(C_{3333}C_{\alpha\beta\gamma\mu} - C_{33\alpha\beta}C_{33\gamma\mu})\varepsilon_{\alpha\beta}(v)\varepsilon_{\gamma\mu}(w) \quad (6.4.42) \\
&+ \frac{\lambda\gamma_T}{k(1-\lambda)}[k(1)(v_{\alpha,3}w_{\alpha,3} + v_{3,\alpha}w_{3,\alpha}) \\
&+ 2\lambda\gamma_T C_{\alpha 3\mu 3}(v_{\alpha,3}w_{\mu,3} + v_{3,\alpha}w_{3,\mu})]\}d\Omega
\end{aligned}
$$

and

$$
\begin{aligned}
d_\lambda(v, w) &= \int_{\Omega_0} \{\frac{\lambda\gamma_T}{k(1-\lambda)}[k(1)(v_{\alpha,3}w_{3,\alpha} + v_{3,\alpha}w_{\alpha,3}) \quad (6.4.43) \\
&+ 2\lambda\gamma_T C_{\alpha 3\mu 3}(v_{\alpha,3}w_{3,\mu} + v_{3,\alpha}w_{\mu,3})]\}d\Omega.
\end{aligned}
$$

The foregoing forms have been constructed in such a way so as to satisfy the condition

$$\int_{\Omega_0} G_\lambda(\varepsilon(u^\lambda)) \cdot \varepsilon(v)d\Omega = a_\lambda(u^\lambda, v) + d_\lambda(u^\lambda, v) \quad \forall v \in H_0^1(\Omega_0)^3. \tag{6.4.44}$$

Accordingly, the equilibrium relation (6.4.9a) implies that the equality

$$a_\lambda(u^\lambda, v) + d_\lambda(u^\lambda, v) = \int_{\Omega_0} f_i v_i d\Omega \tag{6.4.45}$$

is valid for any $v \in [H_0^1(\Omega_0)]^3$.

Proposition 6.7 The bilinear form $a_\lambda(\cdot, \cdot)$ is symmetric, bounded and satisfies the coercivity condition

$$a_\lambda(v,v) \geq \frac{\alpha_0}{2} \sum_{\alpha,\beta} ||\varepsilon_{\alpha\beta}(v)||^2_{L^2(\Omega)} + \bar{c}\lambda||v||^2_{H^1(\Omega_0)} \quad \forall v \in H^1_0(\Omega_0)^3, \qquad (6.4.46)$$

with $\bar{c}$ some positive constant.

Proof. The first two properties are obvious. For the coercivity condition we argue by making use of (6.2.7). It yields

$$\begin{aligned}
a_\lambda(v,v) \;\geq\; & \alpha_0 \int_{\Omega_0} \varepsilon_{\alpha\beta}(v)\varepsilon_{\alpha\beta}(v)d\Omega \\
& + c\lambda \int_{\Omega_0} [2\varepsilon_{33}(v)\varepsilon_{33}(v) + v_{3,\alpha}v_{3,\alpha} + v_{\alpha,3}v_{\alpha,3}]d\Omega \\
\;\geq\; & \alpha_0 \int_{\Omega_0} \varepsilon_{\alpha\beta}(v)\varepsilon_{\alpha\beta}(v)d\Omega + 2c\lambda \int_{\Omega_0} [\varepsilon_{33}(v)\varepsilon_{33}(v) + \varepsilon_{\alpha3}(v)\varepsilon_{\alpha3}(v)]d\Omega \\
\;\geq\; & \frac{\alpha_0}{2} \int_{\Omega_0} \varepsilon_{\alpha\beta}(v)\varepsilon_{\alpha\beta}(v)d\Omega + 2\hat{c}\lambda \int_{\Omega_0} \varepsilon_{ij}(v)\varepsilon_{ij}(v)d\Omega, \qquad (6.4.47)
\end{aligned}$$

where $\hat{c}$ denotes a positive constant. Now Korn's inequality leads to the result. q.e.d.

Let us fix $\lambda > 0$ and choose $w^\lambda \in [C_0^\infty(\Omega_0)]^3$, with

$$||w^\lambda||_{L^2(\Omega_0)} \leq ||u^\lambda||_{L^2(\Omega_0)}. \qquad (6.4.48)$$

There exists an open region Ω'_0 such that $\mathrm{supp}(w) \subset \Omega'_0$ and $\bar{\Omega}'_0 \subset \Omega_0$. Since $a_\lambda(\cdot,\cdot)$ is bilinear bounded and positive definite, the existence of $\bar{u}^\lambda \in [H^1_0(\Omega_0)]^3$ follows with

$$a_\lambda(\bar{u}^\lambda,v) = \int_{\Omega_0} w^\lambda \cdot v d\Omega \quad \forall v \in [H^1_0(\Omega_0)]^3. \qquad (6.4.49)$$

It is clear that

$$\mathrm{supp}(\bar{u}^\lambda) \subset \bar{\Omega}'_0 \subset \Omega_0. \qquad (6.4.50)$$

By substituting $v = \bar{u}^\lambda$ into (6.4.44) we have that

$$a_\lambda(u^\lambda,\bar{u}^\lambda) + d_\lambda(u^\lambda,\bar{u}^\lambda) - \int_{\Omega_0} f_i\bar{u}^\lambda_i d\Omega = 0. \qquad (6.4.51)$$

On the other hand, choosing $\varphi \in C_0^\infty(\Omega_0)$ with $\varphi(x) = 1$ in $\bar{\Omega}'_0$ and applying (6.4.49) implies

$$a_\lambda(\bar{u}^\lambda,\varphi u^\lambda) = \int_{\Omega_0} w^\lambda \cdot \varphi u^\lambda d\Omega = \int_{\Omega_0} w^\lambda \cdot u^\lambda d\Omega. \qquad (6.4.52)$$

Now, the symmetry of $a_\lambda(\cdot,\cdot)$ and (6.4.50) yield the relation

$$a_\lambda(\bar{u}^\lambda,\varphi u^\lambda) = a_\lambda(u^\lambda,\bar{u}^\lambda). \qquad (6.4.53)$$

Hence, (6.4.51) becomes

$$\int_{\Omega_0} w^\lambda \cdot u^\lambda d\Omega + d_\lambda(u^\lambda, \bar{u}^\lambda) - \int_{\Omega_0} f_i \bar{u}_i^\lambda d\Omega = 0. \tag{6.4.54}$$

The foregoing equality is a convenient starting point for obtaining *a priori* estimates for u^λ in Ω_0. We will show that both

$$d_\lambda(u^\lambda, \bar{u}^\lambda) \quad \text{and} \quad \int_{\Omega_0} f_i \bar{u}_i^\lambda d\Omega \tag{6.4.55}$$

admit bounds of the type

$$c||u^\lambda||_{L^2(\Omega_0)} \tag{6.4.56}$$

where c is a constant. The forthcoming proofs will be carried out under some additional restrictions concerning $f_i, C_{\alpha 3 \beta 3}, \gamma_N$, and γ_T.

Lemma 6.8 Suppose that

$$\left. \begin{array}{l} \gamma_N \\ \gamma_T \\ C_{\alpha 3 \beta 3} \end{array} \right\} = \text{constants in } \Omega_0. \tag{6.4.57}$$

Then

$$|d_\lambda(u^\lambda, \bar{u}^\lambda)| \le c||u^\lambda||_{L^2(\Omega_0)}. \tag{6.4.58}$$

Proof. Step 1. Replacing v by $\bar{u}$ in (6.4.49) and applying (6.4.46) yields the relation

$$\bar{c}\lambda||\bar{u}^\lambda||^2_{H^1(\Omega_0)} \le ||w^\lambda||_{L^2(\Omega_0)}||\bar{u}^\lambda||_{L^2(\Omega_0)} \le c||w^\lambda||_{L^2(\Omega_0)}||\bar{u}^\lambda||_{H^1(\Omega_0)}. \tag{6.4.59}$$

Thus due to (6.4.48),

$$||\bar{u}^\lambda||_{H^1(\Omega_0)} \le \frac{c}{\lambda}||u^\lambda||_{L^2(\Omega_0)}. \tag{6.4.60}$$

Step 2. By the substitution of $v = \bar{u}$ into (6.4.49) and the application of (6.4.46) one gets

$$\sum_{\alpha,\beta} ||\varepsilon_{\alpha\beta}(\bar{u}^\lambda)||^2_{L^2(\Omega_0)} \le ||w^\lambda||_{L^2(\Omega_0)}||\bar{u}^\lambda||_{L^2(\Omega_0)} \le c||w^\lambda||_{L^2(\Omega_0)}||\bar{u}^\lambda||_{H^1(\Omega_0)}. \tag{6.4.61}$$

Taking into account (6.4.48) and (6.4.60) we conclude that

$$\sum_{\alpha,\beta} ||\varepsilon_{\alpha\beta}(\bar{u}^\lambda)||_{L^2(\Omega_0)} \le \frac{c}{\sqrt{\lambda}}||u^\lambda||_{L^2(\Omega_0)}. \tag{6.4.62}$$

Step 3. We rearrange $d_\lambda(u^\lambda, \bar{u}^\lambda)$ in a form more convenient for the further study. It is easily seen that $d_\lambda(u^\lambda, \bar{u}^\lambda)$ is composed of two kinds of integrals. These are

$$\int_{\Omega_0} \lambda r_{\alpha\beta} u^\lambda_{\alpha,3} \bar{u}^\lambda_{3,\beta} d\Omega \text{ and } \int_{\Omega_0} \lambda r_{\alpha\beta} u^\lambda_{3,\alpha} \bar{u}^\lambda_{\beta,3} d\Omega, \tag{6.4.63}$$

where

$$r_{\alpha\beta} = \frac{\gamma_T k(1)}{3k(1-\lambda)} \delta_{\alpha\beta} \text{ or } r_{\alpha\beta} = \gamma_T C_{\alpha3\beta3}. \qquad (6.4.64)$$

Here $\delta = \{\delta_{\alpha\beta}\}$ denotes the Kronecker delta. It is well known that $\bar{u}^\lambda \in [H_0^1(\Omega_0)]^3$ can be approximated by elements of $[C_0^\infty(\Omega_0)]^3$. This fact together with the integration by parts permits the following representations:

$$\left. \begin{array}{l} \displaystyle\int_{\Omega_0} \lambda r_{\alpha\beta} u_{\alpha,3}^\lambda \bar{u}_{3,\beta}^\lambda d\Omega = \int_{\Omega_0} \lambda r_{\alpha\beta} u_{\alpha,\beta}^\lambda \bar{u}_{3,3}^\lambda d\Omega = \int_{\Omega_0} \lambda r_{\alpha\beta} \varepsilon_{\alpha\beta}(u^\lambda) \varepsilon_{33}(\bar{u}^\lambda) d\Omega \\[3mm] \displaystyle\int_{\Omega_0} \lambda r_{\alpha\beta} u_{3,\alpha}^\lambda \bar{u}_{\beta,3}^\lambda d\Omega = \int_{\Omega_0} \lambda r_{\alpha\beta} u_{3,3}^\lambda \bar{u}_{\beta,\alpha}^\lambda d\Omega = \int_{\Omega_0} \lambda r_{\alpha\beta} \varepsilon_{33}(u^\lambda) \varepsilon_{\alpha\beta}(\bar{u}^\lambda) d\Omega. \end{array} \right\} $$

$$(6.4.65)$$

Hence we obtain the estimates

$$\left. \begin{array}{l} \displaystyle\left| \int_{\Omega_0} \lambda r_{\alpha\beta} u_{\alpha,3}^\lambda \bar{u}_{3,\beta}^\lambda d\Omega \right| \leq c\lambda \|\varepsilon_{33}(\bar{u}^\lambda)\|_{L^2(\Omega_0)} (\sum_{\alpha,\beta} \|\varepsilon_{\alpha\beta}(u^\lambda)\|_{L^2(\Omega_0)}) \\[3mm] \displaystyle\left| \int_{\Omega_0} \lambda r_{\alpha\beta} u_{3,\alpha}^\lambda \bar{u}_{\beta,3}^\lambda d\Omega \right| \leq c\sqrt{\lambda} \|\varepsilon_{33}(u^\lambda)\|_{L^2(\Omega_0)} (\sum_{\alpha,\beta} \sqrt{\lambda} \|\varepsilon_{\alpha\beta}(\bar{u}^\lambda)\|_{L^2(\Omega_0)}) \end{array} \right\} $$

$$(6.4.66)$$

which constitute the crucial point of the proof. These estimates combined with (6.4.23a), (6.4.23b), (6.4.60) and with (6.4.62) lead directly to (6.4.58). Thus the proof is complete.   q.e.d.

To obtain an appropriate estimate for the integral

$$\int_{\Omega_0} f_i \bar{u}_i^\lambda d\Omega \qquad (6.4.67)$$

some additional requirements concerning $f = \{f_i\}$ are needed. To formulate them we shall use the notations

$$B_{\alpha\beta\gamma\mu} = C_{3333} C_{\alpha\beta\gamma\mu} - C_{33\alpha\beta} C_{33\gamma\mu}. \qquad (6.4.68)$$

Compatibility Condition (C.2): Suppose that the following hypotheses hold:

  (i) $f_3 = 0$ in Ω_0;   (6.4.69a)
  (ii) There exists $v^0 = (v_1^0, v_2^0, 0) \in [H_0^1(\Omega_0)]^3$ such that the equalities

$$\left(\frac{1}{C_{3333}} B_{\alpha\beta\gamma\mu} \varepsilon_{\gamma\mu}(v^0) \right)_{,\beta} + f_\alpha = 0 \quad \text{in } \Omega_0 \qquad (6.4.69b)$$

hold in the sense of distributions.

Both (C.1) and (C.2) will be referred to as the safe load conditions for the delamination problem under consideration.

Lemma 6.9 Suppose that (C.1) and (C.2) are satisfied. Moreover, let the following regularity conditions hold

$$\left(\frac{C_{\alpha\beta33}}{C_{3333}}\varepsilon_{\alpha\beta}(v^0)\right)_{,\mu} \in L^2(\Omega_0). \tag{6.4.70}$$

Then

$$|\int_{\Omega_0} f_\alpha \bar{u}_\alpha^\lambda d\Omega| \le c||u^\lambda||_{L^2(\Omega_0)}. \tag{6.4.71}$$

Proof. Let us set

$$H_{\alpha\beta}^\lambda \zeta_\alpha \xi_\beta = \frac{\gamma_T}{k(1-\lambda)}(k(1)\zeta_\alpha\xi_\alpha + 2\lambda\gamma_T C_{\alpha3\beta3}\zeta_\alpha\xi_\beta), \quad \forall\zeta_\alpha, \xi_\beta \in R. \tag{6.4.72}$$

Putting $v = v^0$ into (6.4.49) we get

$$\int_{\Omega_0} w_\alpha^\lambda v_\alpha^0 d\Omega = \alpha_\lambda(\bar{u}^\lambda, v^0) = \int_{\Omega_0^{\lambda-}} [C_{\alpha\beta33}\varepsilon_{33}(\bar{u}^\lambda) + C_{\alpha\beta\gamma\mu}\varepsilon_{\gamma\mu}(\bar{u}^\lambda)]\varepsilon_{\alpha\beta}(v^0)d\Omega$$

$$+ \int_{\Omega_0^{\lambda-}} \frac{\lambda\gamma_N}{\lambda\gamma_N + C_{3333}}[C_{\alpha\beta33}\varepsilon_{33}(\bar{u}^\lambda) + C_{\alpha\beta\gamma\mu}\varepsilon_{\gamma\mu}(\bar{u}^\lambda)]\varepsilon_{\alpha\beta}(v^0)d\Omega \tag{6.4.73}$$

$$+ \int_{\Omega_0^{\lambda+}} \frac{1}{\lambda\gamma_N + C_{3333}}B_{\alpha\beta\gamma\mu}\varepsilon_{\alpha\beta}(\bar{u}^\lambda)\varepsilon_{\gamma\mu}(v^0)d\Omega + \int_{\Omega_0} \lambda H_{\alpha\beta}^\lambda \bar{u}_{\alpha,3}^\lambda v_{\beta,3}^0 d\Omega.$$

The integral over $\Omega_0^{\lambda-}$ is transformed by applying the identity

$$[C_{\alpha\beta33}\varepsilon_{33}(\bar{u}^\lambda) + C_{\alpha\beta\gamma\mu}\varepsilon_{\gamma\mu}(\bar{u}^\lambda)]\varepsilon_{\alpha\beta}(v^0) \equiv \frac{1}{C_{3333}}B_{\alpha\beta\gamma\mu}\varepsilon_{\alpha\beta}(\bar{u}^\lambda)\varepsilon_{\gamma\mu}(v^0)$$

$$+ \frac{C_{\alpha\beta33}}{C_{3333}}[C_{3333}\varepsilon_{33}(\bar{u}^\lambda) + C_{33\gamma\mu}\varepsilon_{\gamma\mu}(\bar{u}^\lambda)]\varepsilon_{\alpha\beta}(v^0), \tag{6.4.74}$$

which allows us to put (6.4.73) in the form

$$\int_{\Omega_0^{\lambda+}} \frac{1}{\lambda\gamma_n + C_{3333}}B_{\alpha\beta\gamma\mu}\varepsilon_{\alpha\beta}(\bar{u}^\lambda)\varepsilon_{\gamma\mu}(v^0)d\Omega + \int_{\Omega_0^{\lambda-}} \frac{1}{C_{3333}}B_{\alpha\beta\gamma\mu}\varepsilon_{\alpha\beta}(\bar{u}^\lambda)\varepsilon_{\gamma\mu}(v^0)d\Omega$$

$$+ \int_{\Omega_0^{\lambda-}} \frac{C_{\alpha\beta33}}{C_{3333}}[C_{3333}\varepsilon_{33}(\bar{u}^\lambda) + C_{33\gamma\mu}\varepsilon_{\gamma\mu}(\bar{u}^\lambda)]\varepsilon_{\alpha\beta}(v^0)d\Omega$$

$$= \int_{\Omega_0} w_\alpha^\lambda v_\alpha^0 d\Omega - \int_{\Omega_0^{\lambda+}} \frac{\lambda\gamma_N}{\lambda\gamma_N + C_{3333}}[C_{\alpha\beta33}\varepsilon_{33}(\bar{u}^\lambda) + C_{\alpha\beta\gamma\mu}\varepsilon_{\gamma\mu}(\bar{u}^\lambda)]\varepsilon_{\alpha\beta}(v^0)d\Omega$$

$$- \int_{\Omega_0} \lambda H_{\alpha\beta}^\lambda \bar{u}_{\alpha,3}^\lambda v_{\beta,3}^0 d\Omega. \tag{6.4.75}$$

Due to (6.4.48) and (6.4.60) the right hand side of the foregoing equality is bounded. Indeed, we have that

$$|\int_{\Omega_0^{\lambda+}} \frac{\lambda\gamma_N}{\lambda\gamma_N + C_{3333}}[C_{\alpha\beta33}\varepsilon_{33}(\bar{u}^\lambda) + C_{\alpha\beta\gamma\mu}\varepsilon_{\gamma\mu}(\bar{u}^\lambda)]\varepsilon_{\alpha\beta}(v^0)d\Omega| \le c||u^\lambda||_{L^2(\Omega_0)},$$

$$|\int\limits_{\Omega_0} \lambda H^\lambda_{\alpha\beta} \bar{u}^\lambda_{\alpha,3} v^0_{\beta,3} d\Omega| \leq c||u^\lambda||_{L^2(\Omega_0)}, \tag{6.4.76}$$

$$|\int\limits_{\Omega_0} w^\lambda_\alpha v^0_\alpha d\Omega| \leq c||u^\lambda||_{L^2(\Omega_0)}.$$

Hence

$$|\int\limits_{\Omega^{\lambda-}_0} \frac{1}{C_{3333}} B_{\alpha\beta\gamma\mu} \varepsilon_{\alpha\beta}(\bar{u}^\lambda) \varepsilon_{\gamma\mu}(v^0) d\Omega \tag{6.4.77}$$

$$+ \int\limits_{\Omega^{\lambda+}_0} \frac{1}{\lambda\gamma_N + C_{3333}} B_{\alpha\beta\gamma\mu} \varepsilon_{\alpha\beta}(\bar{u}^\lambda) \varepsilon_{\gamma\mu}(v^0) d\Omega$$

$$+ \int\limits_{\Omega^{\lambda-}_0} \frac{C_{\alpha\beta33}}{C_{3333}} [C_{3333}\varepsilon_{33}(\bar{u}^\lambda) + C_{33\gamma\mu}\varepsilon_{\gamma\mu}(\bar{u}^\lambda)] \varepsilon_{\alpha\beta}(v^0) d\Omega| \leq c||u^\lambda||_{L^2(\Omega_0)}.$$

Using

$$|\int\limits_{\Omega^{\lambda+}_0} \frac{\lambda\gamma_N}{C_{3333}(\lambda\gamma_N + C_{3333})} B_{\alpha\beta\gamma\mu} \varepsilon_{\alpha\beta}(\bar{u}^\lambda) \varepsilon_{\gamma\mu}(v^0) d\Omega| \leq c||u^\lambda||_{L^2(\Omega_0)}, \tag{6.4.78}$$

and (6.4.77) we obtain the estimate

$$|\int\limits_{\Omega_0} \frac{1}{C_{3333}} B_{\alpha\beta\gamma\mu} \varepsilon_{\alpha\beta}(\bar{u}^\lambda) \varepsilon_{\gamma\mu}(v^0) d\Omega \tag{6.4.79}$$

$$+ \int\limits_{\Omega^{\lambda-}_0} \frac{C_{\alpha\beta33}}{C_{3333}} [C_{3333}\varepsilon_{33}(\bar{u}^\lambda) + C_{33\gamma\mu}\varepsilon_{\gamma\mu}(\bar{u}^\lambda)] \varepsilon_{\alpha\beta}(v^0) d\Omega| \leq c||u^\lambda||_{L^2(\Omega_0)}.$$

Let us recall that by virtue of (6.4.69b)

$$\int\limits_{\Omega_0} \frac{1}{C_{3333}} B_{\alpha\beta\gamma\mu} \varepsilon_{\alpha\beta}(\bar{u}^\lambda) \varepsilon_{\gamma\mu}(v^0) d\Omega = \int\limits_{\Omega_0} f_\alpha \bar{u}^\lambda_\alpha d\Omega; \tag{6.4.80}$$

thus it is clear that in order to prove the result we have to determine an appropriate bound for the integral over $\Omega^{\lambda-}_0$ in (6.4.79). To this end let us define

$$z(x_1, x_2, x_3) = \int\limits_{-h}^{x_3} \frac{C_{\alpha\beta33}}{C_{3333}} \varepsilon_{\alpha\beta}(v^0(x_1, x_2, s)) ds \tag{6.4.81}$$

with h large enough so that

$$\{x \in R^3 : x_3 = -h\} \cap \bar{\Omega}_0 = \emptyset. \tag{6.4.82}$$

Now, let $\phi \in C^\infty_0(\Omega_0)$ be such that $\phi(x) = 1$ for $x \in \bar{\Omega}'_0$. Thus $\bar{z} = \phi z$ is an element from $H^1_0(\Omega_0)$ having the following properties

$$\bar{z}_{,3} = \frac{C_{\alpha\beta33}}{C_{3333}}\varepsilon_{\alpha\beta}(v^0) \quad \text{in } \bar{\Omega}_0' \tag{6.4.83a}$$

$$||\bar{z}_{,\alpha}||_{L^2(\Omega_0')} \leq c(\sum_{\mu} ||(\frac{C_{\alpha\beta33}}{C_{3333}}\varepsilon_{\alpha\beta}(v^0))_{,\mu}||_{L^2(\Omega_0)}). \tag{6.4.83b}$$

Now let us substitute $v = (0,0,\bar{z})$ into (6.4.49). It yields

$$\int_{\Omega_0^{\lambda+}} \frac{\lambda\gamma_N}{\lambda\gamma_N + C_{3333}}[C_{3333}\varepsilon_{33}(\bar{u}^\lambda) + C_{33\gamma\mu}\varepsilon_{\gamma\mu}(\bar{u}^\lambda)]\frac{C_{\alpha\beta33}}{C_{3333}}\varepsilon_{\alpha\beta}(v^0)d\Omega$$

$$+ \int_{\Omega_0^{\lambda-}} [C_{3333}\varepsilon_{33}(\bar{u}^\lambda) + C_{33\gamma\mu}\varepsilon_{\gamma\mu}(\bar{u}^\lambda)]\frac{C_{\alpha\beta33}}{C_{3333}}\varepsilon_{\alpha\beta}(v^0)d\Omega \tag{6.4.84}$$

$$+ \int_{\Omega_0} \lambda H_{\alpha\beta}^\lambda \bar{u}_{3,\alpha}^\lambda \bar{z}_{,\beta}d\Omega = \int_{\Omega_0} w_3^\lambda \bar{z}d\Omega.$$

Hence, by means of (6.4.70) and (6.4.60), (6.4.48) and (6.4.83b) we obtain that

$$|\int_{\Omega_0^{\lambda-}} \frac{C_{\alpha\beta33}}{C_{3333}}[C_{3333}\varepsilon_{33}(\bar{u}^\lambda) + C_{33\gamma\mu}\varepsilon_{\gamma\mu}(\bar{u}^\lambda)]\varepsilon_{\alpha\beta}(v^0)d\Omega| \leq c||u^\lambda||_{L^2(\Omega_0)}, \tag{6.4.85}$$

which due to (6.4.79) and (6.4.80) leads to the conclusion that

$$|\int_{\Omega_0} f_\alpha \bar{u}_\alpha^\lambda d\Omega| \leq c||u^\lambda||_{L^2(\Omega_0)}. \tag{6.4.86}$$

Thus the proof is complete. q.e.d.

Now we are in the position to establish the until now lacking *a priori* estimates for u^λ.

Proposition 6.10 Suppose that the assumptions of Lemma 6.8 and of Lemma 6.9 hold. Then

$$||u^\lambda||_{L^2(\Omega)} \leq c. \tag{6.4.87}$$

Proof. Let us apply (6.4.58) and (6.4.71) to (6.4.54). This implies that the estimate

$$|\int_{\Omega_0} u^\lambda \cdot w^\lambda d\Omega| \leq c||u^\lambda||_{L^2(\Omega_0)} \tag{6.4.88}$$

holds for any $w^\lambda \in [C_0^\infty(\Omega_0)]^3$ with

$$||w^\lambda||_{L^2(\Omega_0)} \leq ||u^\lambda||_{L^2(\Omega_0)}. \tag{6.4.89}$$

Accordingly we can determine a sequence $\{w_m^\lambda\} \subset [C_0^\infty(\Omega_0)]^3$ converging strongly to u^λ in $[L^2(\Omega_0)]^3$ as $m \to \infty$, such that

$$||w_m^\lambda||_{L^2(\Omega_0)} \leq ||u^\lambda||_{L^2(\Omega_0)} \tag{6.4.90}$$

Hence

$$\left| \int_{\Omega_0} u^\lambda \cdot w_m^\lambda d\Omega \right| \le c \|u^\lambda\|_{L^2(\Omega_0)}. \tag{6.4.91}$$

and consequently, letting $m \to \infty$ yields

$$\int_{\Omega_0} |u^\lambda|^2 d\Omega \le c \|u^\lambda\|_{L^2(\Omega_0)}. \tag{6.4.92}$$

The foregoing estimate implies the uniform boundedness of $\{\|u^\lambda\|_{L^2(\Omega_0)}\}$. To get the boundedness of $\{\|u^\lambda\|_{L^2(\Omega\setminus\Omega_0)}\}$ it is enough to recall the estimates (6.4.23b) from which it follows that the sequence $\{\|u^\lambda\|_{H^1(\Omega\setminus\Omega_0)}\}$ is bounded. This fact leads easily to the result.   q.e.d.

Now we will complete our discussion concerning the existence of solutions of (P). The *a priori* estimates (6.4.13), (6.4.23) and (6.4.87) suggest the introduction of the following space for the displacements

$$P(\Omega; \Omega_0) = \{v \in [L^2(\Omega)]^3 : \varepsilon_{\alpha\beta}(v) \in L^2(\Omega), \ \varepsilon_{i3}(v) \in L^2(\Omega \setminus \Omega_0),$$

$$\varepsilon_{33}(v) \in M(\Omega_0), \ v_{|\Gamma_U} = 0, \ i = 1, 2, 3, \ \alpha, \beta = 1, 2\}, \tag{6.4.93}$$

where $M(\Omega_0)$ is the space of Radon measures with bounded total variation in Ω_0; moreover, we keep the space Y for the stresses.

Let us pay attention to some evident properties of $P(\Omega; \Omega_0)$.

1. $P(\Omega; \Omega_0)$ is a Banach space when endowed with the norm

$$\|v\|_{P(\Omega,\Omega_0)} = \|v\|_{L^2(\Omega)} + \sum_{\alpha,\beta} \|\varepsilon_{\alpha\beta}(v)\|_{L^2(\Omega)} + \sum_i \|\varepsilon_{i3}(v)\|_{L^2(\Omega\setminus\Omega_0)} + \|\varepsilon_{33}(v)\|_{M(\Omega_0)}. \tag{6.4.94}$$

2. Traces on Γ_U and tractions on Γ_S are well defined for any $v \in P(\Omega; \Omega_0)$ since

$$\Gamma_U \cup \Gamma_S \subset \partial(\Omega \setminus \bar{\Omega}_0) \quad \text{and} \quad v \in [H^1(\Omega \setminus \bar{\Omega}_0)]^3. \tag{6.4.95}$$

3. $P(\Omega; \Omega_0)$ admits displacement discontinuities along $\partial\Omega_0$.

To complete the proof we shall apply a compactness argument. Taking into account (6.4.13), (6.4.23), (6.4.40) and (6.4.87) we can extract a subsequence $(u^{\lambda_n}, \sigma^{\lambda_n}) \equiv (u^n, \sigma^n)$ of $(u^\lambda, \sigma^\lambda)$, where $\lambda_n \to 0$, such that

$$
\begin{array}{lll}
u^n \to u & \text{weakly in } [L^2(\Omega)]^3 & \\
\varepsilon_{\alpha\beta}(u^n) \to \varepsilon_{\alpha\beta}(u) & \text{weakly in } L^2(\Omega), & 1 \le \alpha, \beta \le 2 \\
\varepsilon_{i3}(u^n) \to \varepsilon_{i3}(u) & \text{weakly in } L^2(\Omega \setminus \Omega_0), & 1 \le i \le 3 \\
\varepsilon_{33}(u^n) \to \varepsilon_{33}(u) & \text{weakly}_* \text{ in } M(\Omega_0) & \\
\sigma_{ij}^n \to \sigma_{ij} & \text{weakly in } L^2(\Omega), & 1 \le i, j \le 3 \\
\sigma_{33}^{n+} \to \sigma_{33}^+ & \text{weakly in } L^2(\Omega \setminus \Omega_0) & \\
\frac{1}{\lambda_n}\sigma_{33}^{n+} \to m \ge 0 & \text{weakly}_* \text{ in } M(\Omega_0) &
\end{array} \tag{6.4.96}
$$

for some $u \in P(\Omega; \Omega_0)$, $\sigma \in Y$ and $m \in M(\Omega_0)$. We will now show that the triple (u, σ, m) is a solution of the original problem. From (6.4.9a) we have

$$L^* \sigma^n + J_n'(u^n) - g = 0, \tag{6.4.97}$$

where $J_n'(u^n)$ is given by the formula

$$\langle J_n'(u^n), v \rangle_V = \int_{\Gamma_S} j_n'(u^n) \cdot v d\Gamma, \quad \forall v \in V, \ (j_n' \equiv j_{\lambda_n}'). \tag{6.4.98}$$

Let us recall that $\Gamma_S \subset \partial(\Omega \setminus \bar{\Omega}_0)$ and $\{u^n\}$ is bounded in $[H^1(\Omega \setminus \bar{\Omega}_0)]^3$. Thus without loss of generality we may assume that $\{u^n\}$ converges weakly to u in $[H^1(\Omega \setminus \bar{\Omega}_0)]^3$. But then the Rellich theorem ensures the strong convergence of $\{u^n\}$ to u in $[L^2(\Gamma_S)]^3$. By taking a suitable subsequence (again denoted by the same symbol) it is possible to suppose that

$$u^n \to u \quad \text{a.e. on } \Gamma_S. \tag{6.4.99}$$

On the other hand, from the hypothesis that

$$|j(x, \xi) - j(x, \eta)| \leq k(x)|\xi - \eta| \quad \forall \xi, \eta \in R^3, \ \text{with } k \in L^2(\Gamma_S) \tag{6.4.100}$$

we deduce easily the boundedness of $\{j_n'(u^n)\}$ in $[L^2(\Gamma_S)]^3$ (see (6.4.8)). Accordingly, we may assume that $\{j'(u^n)\}$ converges weakly to some $\chi \in [L^2(\Gamma_S)]^3$. To assert that $\chi(x) \in \partial j(u(x))$ for a.e. $x \in \Gamma_S$ it is enough to proceed along the lines of the corresponding part of the proof of Theorem 5.4. Therefore the details will be omitted here. Accordingly, letting n converge to infinity in (6.4.97) and taking into account the weak convergence of $\{\sigma^n\}$ to σ in Y we obtain the equilibrium condition in the form

$$\left. \begin{array}{l} \displaystyle \int_\Omega \sigma \cdot \varepsilon(v) d\Omega - \int_\Omega f_i v_i d\Omega - \int_{\Gamma_F} F_i v_i d\Gamma + \int_{\Gamma_S} \chi \cdot v d\Gamma = 0 \quad \forall v \in V \\ \qquad \qquad \chi(x) \in \partial j(x, u(x)) \quad \text{for a.e. } x \in \Gamma_S \end{array} \right\} \tag{6.4.101}$$

which can be written equivalently as the following hemivariational inequality

$$\int_\Omega \sigma \cdot \varepsilon(v) d\Omega - \int_\Omega f_i v_i d\Omega - \int_{\Gamma_F} F_i v_i d\Gamma + \int_{\Gamma_S} j^0(u; v) d\Omega \geq 0 \quad \forall v \in V. \tag{6.4.102}$$

Moreover, from (6.4.13) we easily obtain that

$$\left. \begin{array}{l} \sigma_{\alpha 3} = 0 \\ \sigma_{33} \leq 0 \end{array} \right\} \quad \text{in } \Omega_0. \tag{6.4.103}$$

Now, it will be proved that

$$\sigma^n \to \sigma \text{ strongly in } Y. \tag{6.4.104}$$

Indeed, from (6.4.9b) and (6.4.101) the following inequality can be obtained

$$\begin{aligned} \Psi^n(\sigma) - \Psi^n(\sigma^n) &\geq (D\sigma^n, \sigma^n - \sigma)_{L^2(\Omega)} + (Lu_n, \sigma - \sigma^n)_{L^2(\Omega)} \tag{6.4.105} \\ &= (D\sigma^n, \sigma^n - \sigma)_{L^2(\Omega)} + \langle J_n'(u^n) - \chi, u^n \rangle_{H^1(\Omega)} \\ &= (D\sigma^n, \sigma^n - \sigma)_{L^2(\Omega)} + \int_{\Gamma_S} (j_n'(u^n) - \chi) \cdot u^n d\Gamma. \end{aligned}$$

Let us note that due to (6.4.103) we have $\Psi^n(\sigma) = \Psi^\delta(\sigma)$. Thus it is not difficult to observe that

$$\Psi^n(\sigma^n) \geq \int\limits_{\Omega\setminus\Omega_0} \frac{1}{1-\delta}\{\frac{1}{2\gamma_N}(\sigma^n_{33+})^2 + \frac{1}{2\gamma_T}[(\sigma^n_{13})^2 + (\sigma^n_{23})^2]\}d\Omega. \qquad (6.4.106)$$

The weak convergence of $\{\sigma_n\}$ to σ yields that

$$\liminf \Psi^n(\sigma^n) \geq \Psi^\delta(\sigma). \qquad (6.4.107)$$

This fact together with

$$\int\limits_{\Gamma_S}(j'_n(u^n) - \chi)\cdot u^n d\Gamma \to 0, \qquad (6.4.108)$$

$(j'_n(u^n) \to \chi$ weakly in $[L^2(\Gamma_S)]^3$ and $u^n \to u$ strongly in $[L^2(\Gamma_S)]^3$) allows us to derive from (6.4.105) the estimate

$$\limsup(D\sigma^n, \sigma^n - \sigma)_{L^2(\Omega)} \leq 0. \qquad (6.4.109)$$

which due to the strong ellipticity of D implies the strong convergence of $\{\sigma^n\}$ to σ.

By making use of the same argument as for Γ_S we obtain that

$$u^n \to u \text{ strongly in } [L^2(\Gamma_F \cap \text{supp}(F))]^3. \qquad (6.4.110)$$

The foregoing results lead to

$$(\sigma^n, Lu^n)_{L^2(\Omega)} = \langle g, u^n\rangle_{H^1(\Omega)} - \int\limits_{\Gamma_S} j'_n(u^n)\cdot u^n d\Gamma$$

$$\to \int\limits_{\Omega} f_i u_i d\Omega + \int\limits_{\Gamma_F} F_i u_i d\Gamma - \int\limits_{\Gamma_S} \chi\cdot u d\Gamma, \qquad (6.4.111)$$

and

$$(D\sigma^n, \sigma^n)_{L^2(\Omega)} \to (D\sigma, \sigma)_{L^2(\Omega)}. \qquad (6.4.112)$$

Further, we introduce the set

$$Y_0 = \{\tau \in Y; \ \tau_{33|\Omega_0} \in C_0^\infty(\Omega_0), \ \tau_{\alpha3|\Omega_0} = 0, \ \alpha = 1, 2\}, \qquad (6.4.113)$$

and we see that for any $\tau \in Y_0$ the limit

$$\lim(Lu^n, \tau)_{L^2(\Omega)} = \int\limits_{\Omega\setminus\Omega_0} \varepsilon_{ij}(u)\tau_{ij}d\Omega + \int\limits_{\Omega_0} \varepsilon_{\alpha\beta}(u)\tau_{\alpha\beta}d\Omega + \int\limits_{\Omega_0} \tau_{33}\varepsilon_{33}(u) \ \forall\tau \in Y_0,$$

$$(6.4.114)$$

is also well defined. Before passing to the limit procedure we introduce the notation

$$\langle Lu, \tau \rangle_{P(\Omega;\Omega_0)} \equiv \int_{\Omega \backslash \Omega_0} \varepsilon_{ij}(u)\tau_{ij}d\Omega + \int_{\Omega_0} \varepsilon_{\alpha\beta}(u)\tau_{\alpha\beta}d\Omega + \int_{\Omega_0} \tau_{33}\varepsilon_{33}(u) \quad \forall \tau \in Y_0.$$

$$(6.4.115)$$

Then by means of (6.4.111), (6.4.112), (6.4.14) and (6.4.107), the inequality

$$(D\sigma^n, \tau - \sigma^n)_{L^2(\Omega)} - (Lu^n, \tau - \sigma^n)_{L^2(\Omega)} + \Psi^n(\tau) - \Psi^n(\sigma^n) \geq 0 \,\forall \tau \in Y_0 \quad (6.4.116)$$

implies for $n \to \infty$ the relation

$$(D\sigma, \tau - \sigma)_{L^2(\Omega)} - \langle Lu, \tau \rangle_{P(\Omega;\Omega_0)} + \langle g, u \rangle_{H^1(\Omega)} \qquad (6.4.117)$$
$$- \int_{\Gamma_S} \chi \cdot u d\Gamma + \Psi^\delta(\tau) - \Psi^\delta(\sigma) \geq 0 \quad \forall \tau \in Y_0.$$

This gives rise to the inequality

$$0 \geq \int_{\Omega \backslash \Omega_0} \varepsilon_{ij}(u)\sigma_{ij}d\Omega + \int_{\Omega_0} \varepsilon_{\alpha\beta}(u)\sigma_{\alpha\beta}d\Omega$$
$$+ \int_{\Omega_0} \tau_{33}\varepsilon_{33}(u) - \int_{\Omega_0} D_{33kl}\sigma_{kl}(\tau_{33} - \sigma_{33})d\Omega \qquad (6.4.118)$$
$$- \int_\Omega f_i u_i d\Omega - \int_{\Gamma_F} F_i u_i d\Gamma + \int_{\Gamma_S} \chi \cdot u d\Gamma, \quad \forall \tau_{33} \in C_0^\infty(\Omega_0)^-,$$

where $C_0^\infty(\Omega_0)^-$ denotes the class of all non-positive $C_0^\infty(\Omega_0)$-functions. Further, when comparing (6.4.96) with

$$\left. \begin{array}{lll} \varepsilon_{\alpha\beta}(u^n) - D_{\alpha\beta kl}\sigma_{kl}^n & = 0 & \text{a.e. in } \Omega \\[2mm] \varepsilon_{\alpha 3}(u^n) - D_{\alpha 3\mu 3}\sigma_{\mu 3}^n & = \dfrac{1}{2\gamma_T(1-\delta)}\sigma_{\alpha 3}^n & \text{a.e. in } \Omega \backslash \bar{\Omega}_0 \\[2mm] \varepsilon_{33}(u^n) - D_{33kl}\sigma_{kl}^n & = \dfrac{1}{\gamma_N(1-\delta)}\sigma_{33+}^n & \text{a.e. in } \Omega \backslash \bar{\Omega}_0 \\[2mm] \varepsilon_{33}(u^n) - D_{33kl}\sigma_{kl}^n & = \dfrac{1}{\gamma_N \lambda_n}\sigma_{33+}^n & \text{a.e. in } \Omega_0 \end{array} \right\} \quad (6.4.119)$$

we arrive at the following relations

$$\left. \begin{array}{lll} \varepsilon_{\alpha\beta}(u) - D_{\alpha\beta kl}\sigma_{kl} & = 0 & \text{a.e. in } \Omega \\[2mm] \varepsilon_{\alpha 3}(u) - D_{\alpha 3\mu 3}\sigma_{\mu 3} & = \dfrac{1}{2\gamma_T(1-\delta)}\sigma_{\alpha 3} & \text{a.e. in } \Omega \backslash \bar{\Omega}_0 \\[2mm] \varepsilon_{33}(u) - D_{33kl}\sigma_{kl} & = \dfrac{1}{\gamma_N(1-\delta)}\sigma_{33+} & \text{a.e. in } \Omega \backslash \bar{\Omega}_0 \\[2mm] \sigma_{33} & \leq 0 & \text{a.e. in } \Omega_0 \\[2mm] \sigma_{\alpha 3} & = 0 & \text{a.e. in } \Omega_0 \\[2mm] \varepsilon_{33}(u) & - D_{33kl}\sigma_{kl}d\Omega \in M^+(\Omega_0). \end{array} \right\} \quad (6.4.120)$$

Here $M^+(\Omega_0)$ denotes the set of all non-negative Radon measures on Ω_0. The relations (6.4.118) and (6.4.120) together with the hemivariational inequality (6.4.101) constitute the pointwise-variational formulation of the full system of the governing relations for the delamination problem under consideration. All previous results permit one to give the following definition for the solution of the problem studied: a triple $(u, \sigma, \chi) \in P(\Omega; \Omega_0) \times Y \times [L^2(\Gamma_S)]^3$ satisfying the foregoing conditions will be called a solution of the original problem (P). To justify this definition let us notice that for $u \in P(\Omega; \Omega_0)$ and $\sigma \in Y$ there is not a meaningful duality between the strain field component $\varepsilon_{33}(u) \in M(\Omega_0)$ and the corresponding stress field component $\sigma_{33} \in L^2(\Omega_0)$. In terms of Mechanics the work produced by σ_{33} for ε_{33} cannot be defined. Moreover, the strain field components $\varepsilon_{\alpha3}(u)$, $\alpha = 1, 2$, fail to exist in Ω_0. All these inconveniences result from the noncoercivity of the constitutive operator G^δ and can be overcome by the acceptance of the aforementioned definition which in the case of smooth enough displacement fields reduces easily to the variational inequality

$$\int_\Omega (D\sigma - \varepsilon(u)) \cdot (\tau - \sigma)d\Omega + \Psi^\delta(\tau) - \Psi^\delta(\sigma) \geq 0 \quad \forall \tau \in Y_0, \qquad (6.4.121)$$

which is a variational expression for the constitutive law (6.1.12).

To sum up our results we formulate the following existence theorem.

Theorem 6.11 Suppose that (C.1), (C.2), (6.4.57) and (6.4.70) hold. Then the problem (P) has at least one solution $(u, \sigma, \chi) \in P(\Omega; \Omega_0) \times Y \times [L^2(\Gamma_S)]^3$ such that (6.4.101), (6.4.118) and (6.4.120) hold.

Remark 6.12 Due to the positive definiteness of D, σ is determined uniquely, whereas u is not unique.

Remark 6.13 The safe load condition (C.2) is not as restrictive. Indeed, if $\Omega_0 = \Pi_0 \times (-h, h)$, $\Pi_0 \subset R^2$, $h > 0$, and if C_{3333} and $B = \{B_{\alpha\beta\gamma\delta}\}$ are independent of x_3, then for each $f_\alpha = f_\alpha(x_1, x_2)$, $f_\alpha \in L^2(\Pi_0)$, $\alpha = 1, 2$, there exists $v^0 = v^0(x_1, x_2)$, $v^0 \in [H^1(\Pi_0)]^2$, such that (6.4.69) holds. This fact is a consequence of the positive definiteness of the tensor B given by (6.4.68) (see (6.2.7b)) and the Lax-Milgram theorem. In particular, if $f = 0$, then (C.2) holds immediately. This case is also important, since, as experimental evidence shows, in the delamination processes the body forces are negligible compared to the boundary tractions.

Remark 6.14 The fact that the displacement field u does not have $\varepsilon_{\alpha3}$-strains well defined in the totally delaminated region Ω_0 has a clear physical interpretation. Namely, in the totally delaminated region interlayer slips are out of our control and cannot be determined by the data of the problem under consideration. Moreover, the existence result obtained shows that if the external forces are safe enough (the safe load conditions (C.1) and (C.2) hold) then the total debonding in Ω_0 does not lead to the destruction of the composite as a whole. The foregoing interpretation is physically acceptable and is consistent with the observed behaviour of laminated composite structures subjected to quasistatic loadings.

7. Constrained Problems for Nonconvex Star-Shaped Admissible Sets

This Section is devoted to the study of constrained problems in reflexive Banach spaces, in which the set of all admissible elements is nonconvex but star-shaped. Due to the nonconvexity of the admissible sets the corresponding variational formulations are no longer variational inequalities but, instead, take the form of hemivariational inequalities. Making use of hemivariational inequalities we prove the existence of solutions to the following type of nonconvex constrained problems: find $u \in C$ such that

$$\langle Au - f, v \rangle \geq 0 \quad \forall v \in T_C(u),$$

where the admissible set $C \subset V$ is a star-shaped set with respect to a certain ball and $T_C(u)$ denotes Clarke's tangent cone of C at $u \in C$. It is worth to emphasize that the approach to nonconvex constrained problems presented here is not related to the methods developed in [Ba88, Mot86, Dac82,89]; it is rather based on the penalization method. The main idea of the proof is to make use of a discontinuity property of the generalized Clarke's differential $d_C^0(\cdot, \cdot)$; $d_C(\cdot)$ denotes the distance function from the admissible sets under consideration. This property makes possible to determine the solution of the original problem by applying the penalization method but, what is of importance, without making the small parameter tend to 0. This fact allows to relax the assumption of convexity in our approach. The method proposed here is an extension of the one developed by the first author in [Nan94] (see also [Nan89a]). Applications to Mechanics and Engineering close this Chapter.

7.1 Distance Function for Star-Shaped Sets. Basic Properties

Let V be a Banach space and let $B_V(u_0, \rho)$ denote the closed ball in V with center u_0 and radius ρ, i.e.

$$B_V(u_0, \rho) = \{v \in V : \|u_0 - v\|_V \leq \rho\}, \quad \rho > 0. \tag{7.1.1}$$

Let us suppose that C is a closed subset of V, which is assumed to be star-shaped with respect to $B_V(u_0, \rho)$, i.e.,

$$v \in C \iff \lambda v + (1 - \lambda)w \in C \text{ for any } \lambda \in [0,1] \text{ and } w \in B_V(u_0, \rho). \quad (7.1.2)$$

Consider the distance function of C, that is, the function $d_C : V \to R$ defined by

$$d_C(v) = \inf_{w \in C} ||v - w||_V, \quad v \in V. \quad (7.1.3)$$

Proposition 7.1 Let C be a closed star-shaped with respect to $B_V(u_0, \rho)$ subset of V. Then for any $u \notin C$ the following estimate holds

$$d_C(\lambda u + (1 - \lambda)u_0) \le \lambda d_C(u) - (1 - \lambda)\rho \quad \text{for} \quad \frac{\rho}{d_C(u) + \rho} \le \lambda \le 1. \quad (7.1.4)$$

Proof. Let an $\varepsilon > 0$ be given. By definition, there is a point $w \in C$ such that $d_C(u) + \varepsilon > ||u - w||_V$. Since

$$u_0 + \rho \frac{u - w}{||u - w||_V} \in B_V(u_0, \rho), \quad (7.1.5)$$

the hypothesis implies that

$$\lambda w + (1 - \lambda)(u_0 + \rho \frac{u - w}{||u - w||_V}) \in C \quad \text{for } \lambda \in [0,1]. \quad (7.1.6)$$

Hence

$$\begin{aligned}
d_C(\lambda u + (1 - \lambda)u_0) &\le ||\lambda u + (1 - \lambda)u_0 - \lambda w - (1 - \lambda)(u_0 + \rho \frac{u - w}{||u - w||_V}||_V \\
&= |\lambda||u - w||_V - (1 - \lambda)\rho|. \quad (7.1.7)
\end{aligned}$$

Now, it suffices to observe that for

$$\lambda \ge \frac{\rho}{d_C(u) + \rho} \quad (7.1.8)$$

we have

$$0 \le \lambda d_C(u) - (1 - \lambda)\rho \le \lambda ||u - w||_V - (1 - \lambda)\rho. \quad (7.1.9)$$

Finally,

$$d_C(\lambda u + (1 - \lambda)u_0) \le \lambda ||u - w||_V - (1 - \lambda)\rho \le \lambda d_C(u) - (1 - \lambda)\rho + \lambda \varepsilon. \quad (7.1.10)$$

But ε was chosen arbitrarily. Thus the proof is complete.   q.e.d.

Lemma 7.2 We make the same assumptions as in Proposition 7.1. Then the following relations hold

$$d_C^0(u, u_0 - u) \le -d_C(u) - \rho \quad \forall u \notin C, \quad (7.1.11)$$

$$d_C^0(u, u_0 - u) = 0. \quad \forall u \in C. \quad (7.1.12)$$

Proof. Let $u \notin C$. For the difference quotient the following estimate (for sufficiently small h) holds:

$$\frac{d_C(u + h + \lambda(u_0 - u)) - d_C(u + h)}{\lambda}$$

$$= \frac{d_C(u + h + \lambda(u_0 - u)) - d_C((1 - \lambda)(u + h) + \lambda u_0)}{\lambda}$$

$$+ \frac{d_C((1 - \lambda)(u + h) + \lambda u_0) - d_C(u + h)}{\lambda}$$

$$\leq \|h\|_V + \frac{(1 - \lambda)d_C(u + h) - \lambda \rho - d_C(u + h)}{\lambda}$$

$$\leq \|h\|_V - d_C(u + h) - \rho. \tag{7.1.13}$$

Here we have used (7.1.4) together with the Lipschitz property of d_C (with 1 as a Lipschitz constant). Hence

$$d_C^0(u, u_0 - u) = \limsup_{\substack{h \to 0 \\ \lambda \to 0+}} \frac{d_C(u + h + \lambda(u_0 - u)) - d_C(u + h)}{\lambda} \leq -d_C(u) - \rho,$$

$$\tag{7.1.14}$$

which is the assertion (7.1.11). To prove (7.1.12) it is enough to notice that the fact, that C is star-shaped with respect $B_V(u_0, \rho)$ ensures that $u_0 - u$ is hypertangent (cf. Sect. 1.2) to C at u.   q.e.d.

From Lemma 7.2 it follows that the function $V \ni v \to d_C^0(v, u_0 - v)$ admits a jump along the boundary ∂C of the set C, which is not less than ρ. This discontinuity will be the crucial point in establishing existence results for the nonconvex constrained problems under consideration.

7.2 Constrained Hemivariational Inequalities

7.2.1 General Method

Let V be a reflexive Banach space and let C be a closed subset of V which is assumed to be star-shaped with respect to a ball $B(u_0, \rho)$, i.e., (7.1.2) holds. Let us recall the definitions (cf. Sect. 1.2)

$$T_C(u) = \{k \in V : \forall u_n \to u, u_n \in C, \quad \forall \lambda_n \to 0_+, \quad \text{there exists } k_n \to k$$

$$\text{with } u_n + \lambda_n k_n \in C\}, \tag{7.2.1}$$

of Clarke's tangent cone of C at u, and

$$N_C(u) = \{u^\star \in V^\star : \langle u^\star, k \rangle_V \leq 0 \quad \forall k \in T_C(u)\}, \tag{7.2.2}$$

of Clarke's normal cone to C at u.

Let $A : V \to V^\star$ be a pseudo-monotone mapping from a reflexive Banach space V into $V^\star$ and let $g \in V^\star$ be given. Our main task in this Section is to study the following problem.

Problem (P). Find $u \in C$ such that

$$\langle Au - g, k \rangle \geq 0 \quad \forall k \in T_C(u). \tag{7.2.3}$$

Due to nonconvexity of C the problem (P) is called a constrained hemivariational inequality. It is easily seen that in the case of C convex (7.2.3) reduces to the variational inequality

$$\langle Au - g, v - u \rangle_V \geq 0 \quad \forall v \in C, \quad u \in C, \tag{7.2.4}$$

which is well known in the literature.

Theorem 7.3 Let $A : V \to V^*$ be a pseudo-monotone operator from a reflexive Banach space V into its dual V^*, and $C \subset V$ a closed subset of V. Suppose that the following assumptions hold:

(H1) C is star-shaped with respect to a ball $B_V(u_0, \rho), \rho > 0$, for some $u_0 \in V$;

(H2)   either $d_C \in PM(V)$, or A has the $(S)_+$ property and $d_C \in QPM(V)$;

(H3)   There exists a function $c : R^+ \to R$ with $c(r) \to \infty$ as $r \to \infty$, such that for any $v \in V, \langle Av, v - u_0 \rangle_V \geq c(\|v\|_V)\|v\|_V$.

Then for any $g \in V^*$ the problem (P) has at least one solution.

Proof. The penalty method will be used. We proceed in two steps.
Step 1. The idea is to formulate the regularized problem for any $\lambda > 0$. It reads:

Problem (P_λ). Find $u_\lambda \in V$ such that

$$\langle Au_\lambda - g, v - u_\lambda \rangle_V + \frac{1}{\lambda} d_C^0(u_\lambda, v - u_\lambda) \geq 0 \quad \forall v \in V. \tag{7.2.5}$$

Notice that because of the hypotheses the sum $A + \frac{1}{\lambda} \partial d_C$, with $\lambda > 0$, is pseudo-monotone (see Propositions 2.4 and 2.20). For the coercivity of $A + \frac{1}{\lambda} \partial d_C$ we use Lemma 7.2 and (H1) from which it follows that for $v^* \in \frac{1}{\lambda} \partial d_C(v)$, $\langle v^*, u_0 - v \rangle_V \leq 0$. Hence $A + \frac{1}{\lambda} d_C$ obeys the same coercivity condition as A. Thus Theorem 4.20 implies the existence of a solution u_λ to (P_λ).

Step 2. We shall show that for sufficiently small $\lambda, u_\lambda \in C$. Indeed, suppose that this is not true, i.e., $u_\lambda \notin C$ for each $\lambda > 0$. By the substitution of $v = u_0$ into (7.2.5) one gets

$$\langle Au_\lambda - g, u_0 - u_\lambda \rangle_V + \frac{1}{\lambda} d_C^0(u_\lambda, u_0 - u_\lambda) \geq 0 \tag{7.2.6}$$

From Lemma 7.2 it follows that

$$d_C^0(u_\lambda, u_0 - u_\lambda) \leq 0. \tag{7.2.7}$$

Hence, due to the coercivity of A,

$$||g||_{V^*}||u_\lambda - u_0||_V \geq c(||u_\lambda||_V)||u_\lambda||_V. \qquad (7.2.8)$$

But this leads directly to the conclusion that $\{u_\lambda\}$ is bounded, i.e. that

$$||u_\lambda||_V \leq M \qquad (7.2.9)$$

where M is a constant not depending on λ. By making use of Lemma 7.2 we can get some further results. Indeed, using the notation

$$\inf_{r \geq 0} rc(r) = \beta, \quad \beta \in R, \qquad (7.2.10)$$

we obtain from (7.2.6) and (7.2.9) that

$$||g||_{V^*}||u_\lambda - u_0||_V \geq \beta + \frac{1}{\lambda}(d_C(u_\lambda) + \rho) \geq \beta + \frac{\rho}{\lambda} \qquad (7.2.11)$$

and consequently

$$||g||_{V^*}(M + ||u_0||_V) - \beta \geq \frac{\rho}{\lambda}. \qquad (7.2.12)$$

But the last inequality cannot hold for

$$\lambda < \frac{\rho}{||g||_{V^*}(M + ||u_0||_V) - \beta}, \qquad (7.2.13)$$

which leads to a contradiction. Accordingly, if λ is small enough, u_λ must lie in C.

Step 3. In order to obtain that

$$\langle Au_\lambda - g, k \rangle_V \geq 0 \quad \forall k \in T_C(u_\lambda) \qquad (7.2.14)$$

(with λ satisfying (7.2.13)) it suffices to recall that

$$k \in T_C(u_\lambda) \iff d_C^0(u_\lambda, k) = 0 \qquad (7.2.15)$$

and to apply it to (7.2.5). Thus the proof is complete.   q.e.d.

7.2.2 Union of a Finite Collection of Convex Sets

Now we consider the case in which C can be represented as a union of a finite collection of closed convex sets $C_n, n = 1, \ldots, N$, i.e.,

$$C = C_1 \cup \ldots \cup C_N. \qquad (7.2.16)$$

The following theorem holds.

Theorem 7.4 Let $A : V \to V^*$ be a pseudo-monotone operator from a reflexive Banach space V into V^*, $C_i, i = 1, \ldots, N$, are closed convex subsets of V. Suppose that the following assumptions hold:

(H1) A satisfies the $(S)_+$ condition;

(H2) There exists $u_0 \in \bigcap_i \operatorname{int} C_i$;

(H3) $\langle Av, v - u_0 \rangle \geq c(\|v\|)\|v\|_V, v \in V$, with $c(r) \to \infty$ as $r \to \infty$.

Then for C given by (7.2.16) the problem (P) has at least one solution, i.e., (7.2.3) holds.

Proof. Denote by $d_n : V \to R$ the distance functions of C_n, $n = 1, \ldots, N$. Due to (H2) and the convexity of C_n, the set C is star-shaped with respect to a certain ball with its center at u_0. Further, notice that the distance function of C is expresses as a pointwise minimum of convex functions d_n, i.e., $d_C = \min(d_1, \ldots, d_N)$. So owing to Corollary 4.13, $d \in QPM(V)$. Applying Theorem 7.3 we get the result.   q.e.d.

Now we pass to a case where A does not have the $(S)_+$ property. Our task now is to formulate conditions guaranteeing that N_C is generalized pseudo-monotone.

Proposition 7.5 Let $C_i, i = 1, \ldots, N$, be a finite collection of closed convex subsets of V and let C be given by (7.2.16). Suppose that the following two conditions hold

$$T_C(u) = T_{\bigcap_{i \in I(u)} C_i}(u) \quad \forall u \in \partial C, \tag{7.2.17}$$

where ∂C denotes the boundary of C, $I(u) = \{i : u \in C_i\}$, and

If for a sequence $\{u_n\}$ in $V, u_n \to u$ weakly, $u_n \in \partial C$, there exists $\left.\begin{array}{l}\text{a sequence } \{u_n^*\} \subset V^* \text{ with } u_n^* \in N_C(u_n), \ u_n^* \to u^* \text{ weakly}, \ u^* \neq 0, \\ \text{and such that } \limsup\langle u_n^*, u_n - u\rangle_V \leq 0, \quad \text{then } u \in \partial C.\end{array}\right\} \tag{7.2.18}$

Then N_C is generalized pseudo-monotone.

Proof. Let $\{u_i\}$ be a sequence in C converging weakly to u, and let the corresponding sequence $\{u_i^*\} \subset V^*$ with $u_i^* \in N_C(u_i)$ converge also weakly to u^*. We have to show that the inequality $\limsup\langle u_i^*, u_i - u\rangle_V \leq 0$ implies that: i) $u \in C$, ii) $\lim\langle u_i^*, u_i - u\rangle_V = 0$ and iii) $u^* \in N_C(u)$.

The first condition follows directly from the fact that C is weakly closed. For the second condition we argue as follows. Let $\{u_i\}$ be a sequence in C converging weakly to $u, \{u_i^*\}$ a sequence in V^* with $u_i^* \in N_C(u_i)$, converging weakly to u^*. Suppose that $\limsup\langle u_i^*, u_i - u\rangle_V \leq 0$. Without loss of generality we may assume that $I(u_i) = \mathcal{H}$ for some fixed subset $\mathcal{H}$ of $\{1, \ldots, N\}$ (by passing to a subsequence of $\{u_i\}$, if necessary). This means that

$$u_i \in \bigcap_{n \in \mathcal{H}} C_n \setminus \left(\bigcup_{n \notin \mathcal{H}} C_n \right). \tag{7.2.19}$$

Due to the weak closedness of each C_n one gets that $u \in \bigcap_{n \in \mathcal{H}} C_n$. Hence

$$u - u_i \in T_{\bigcap_{n \in \mathcal{H}} C_n}(u_i). \tag{7.2.20}$$

Further, it can be easily verified that from $\{u_i\} \subset \bigcap_{n \in \mathcal{H}} C_n$ it results that

$$T_{\bigcap_{n \in \mathcal{H}} C_n}(u_i) \subset T_{\bigcup_{n \in \mathcal{H}} C_n}(u_i) = T_C(u_i). \tag{7.2.21}$$

Accordingly $u - u_i \in T_C(u_i)$ and owing to $u_i^* \in N_C(u_i)$ we obtain the relation

$$\langle u_i^*, u - u_i \rangle_V \le 0. \tag{7.2.22}$$

Combining (7.2.22) with $\limsup \langle u_i^*, u_i - u \rangle_V \le 0$ we are led directly to the desired equality in ii).

In order to obtain the last condition $u^* \in N_C(u)$ it suffices to consider the case $u^* \ne 0$ only, because $0 \in N_C(u)$ for each $u \in C$. In such a case we must also suppose that $u_i \in \partial C$ since for $u_i \in \text{int } C$ we would obtain $u_i^* = 0$ and consequently $u^* = 0$. The assumption (7.2.18) guarantees that the limit point u belongs to the boundary ∂C. Now, let $v \in \bigcap_{n \in I(u)} C_n$. By the weak closedness of C_n one gets $I(u) \supset \mathcal{H}$. Therefore

$$\bigcap_{n \in I(u)} C_n \subset \bigcap_{n \in \mathcal{H}} C_n \tag{7.2.23}$$

and we are led to

$$v - u_i \in T_{\bigcap_{n \in \mathcal{H}} C_n}(u_i) \subset T_{\bigcup_{n \in \mathcal{H}} C_n}(u_i) = T_C(u_i) \tag{7.2.24}$$

from which it follows that

$$\langle u_i^*, v - u_i \rangle_V \le 0 \quad \forall v \in \bigcap_{n \in I(u)} C_n. \tag{7.2.25}$$

Passing to the limit as $n \to \infty$ we obtain

$$\langle u^*, v - u \rangle_V \le 0 \quad \forall v \in \bigcap_{n \in I(u)} C_n, \tag{7.2.26}$$

which owing to (7.2.17) is equivalent to $u^* \in N_C(u)$. The proof is complete. q.e.d.

Remark 7.6 From the first part of the proof of Proposition 7.5 it follows that we can have N_C quasi generalized pseudo-monotone even if conditions (7.2.17) and (7.2.18) are omitted.

Theorem 7.7 Let $A : V \to V^*$ be a pseudo-monotone operator from a reflexive Banach space V into $V^*, C_i, i = 1, \ldots, N$, a finite collection of closed convex subsets of V and let C be given by (7.2.16). Suppose that (7.2.17), (7.2.18) hold and moreover, that the following assumptions hold:

(H1) There exists $u_0 \in \bigcap_i \text{int } C_i$;

(H2) $\langle Av, v - u_0 \rangle_V \ge c(\|v\|_V)\|v\|_V$, $v \in V$, with $c(r) \to \infty$ as $r \to \infty$.

Then the problem (P) has at least one solution, i.e., (7.2.3) holds.

Proof. We apply the regularization method. Let $J : V \to V^*$ be the duality mapping from V into V^*, i.e., $||J(u)||_{V^*} = ||u||_V$ and $\langle J(u), u \rangle_V = ||u||_V^2$ $\forall u \in V$. Since V is reflexive, without loss of generality we may assume that J is single-valued and satisfies the $(S)_+$ condition. This can be achieved by renorming V so that V and V^* become locally uniformly convex (for details see cf. [Bro83a]). For any $\varepsilon > 0$ let us now define an operator $A + \varepsilon J$. It is not difficult to observe that under the assumptions of the Theorem the sum $A + \varepsilon J$ is a pseudo-monotone coercive operator having the $(S)_+$ property. Thus we can apply Theorem 7.4 to prove the existence of $u_\varepsilon \in C$ with $g \in Au_\varepsilon + \varepsilon Ju_\varepsilon + N_C(u_\varepsilon)$ for any $\varepsilon > 0$, i.e.,

$$\langle Au_\varepsilon + \varepsilon Ju_\varepsilon - g, v \rangle_V \geq 0 \quad \forall v \in T_C(u_\varepsilon). \tag{7.2.27}$$

The set C is star-shaped with respect to a certain ball $B_V(u_0, \rho)$. This fact implies that $u_0 - u_\varepsilon \in T_C(u_\varepsilon)$. Hence

$$\langle Au_\varepsilon + \varepsilon Ju_\varepsilon - g, u_0 - u_\varepsilon \rangle_V \geq 0 \tag{7.2.28}$$

from which we are led to the estimate

$$||g||_{V^*}(||u_0||_V + ||u_\varepsilon||_V) + \varepsilon ||u_\varepsilon||_V ||u_0||_V \geq c(||u_\varepsilon||_V)||u_\varepsilon||_V + \varepsilon ||u_\varepsilon||_V^2. \tag{7.2.29}$$

Since $c(r) \to +\infty$ as $r \to \infty$, the boundedness of $\{u_\varepsilon\}$ follows, i.e., there exists a positive constant M which does not depend on ε such that

$$||u_\varepsilon||_V \leq M \quad \varepsilon > 0. \tag{7.2.30}$$

We denote by $\{\varepsilon_i\}$ a sequence converging to 0 for which the corresponding subsequence $\{u_{\varepsilon_i}\}$ of $\{u_\varepsilon\}$ converges weakly to some u. The subsequence $\{u_{\varepsilon_i}\}$ is denoted as $\{u_i\}$. Without loss of generality we may also assume that $I(u_i) = \mathcal{H}$ for some fixed subset $\mathcal{H}$ of $\{1, \ldots, N\}$. Thus we obtain

$$u_i \in \bigcap_{n \in \mathcal{H}} C_n \setminus (\bigcup_{n \notin \mathcal{H}} C_n). \tag{7.2.31}$$

Due to the weak closedness of each C_n one gets that $u \in \bigcap_{n \in \mathcal{H}} C_n$. Hence

$$u - u_i \in T_{\bigcap_{n \in \mathcal{H}} C_n}(u_i). \tag{7.2.32}$$

Further, it can be easily verified that $\{u_i\} \subset \bigcap_{n \in \mathcal{H}} C_n$ implies the relation

$$T_{\bigcap_{n \in \mathcal{H}} C_n}(u_i) \subset T_{\bigcup_{n \in \mathcal{H}} C_n}(u_i) = T_C(u_i). \tag{7.2.33}$$

Thus $u - u_i \in T_C(u_i)$ and because $g - Au_i - \varepsilon_i Ju_i \in N_C(u_i)$ one gets

$$\langle Au_i + \varepsilon_i Ju_i - g, u_i - u \rangle \leq 0. \tag{7.2.34}$$

Taking into account that $\varepsilon_i Ju_i \to 0$ in V^*, we are led to the condition

$$\lim \sup \langle Au_i, u_i - u \rangle_V \leq 0. \tag{7.2.35}$$

Since A has been assumed to be pseudo-monotone, the estimate

$$\lim \inf \langle Au_i, u_i - v \rangle_V \geq \langle Au, u - v \rangle_V \tag{7.2.36}$$

holds for any $v \in V$, from which we deduce the equality $\lim \langle Au_i, u_i - u \rangle_V = 0$. Moreover, we claim that $\{Au_i\}$ is bounded. Indeed, from (7.2.35) and (7.2.36) we obtain easily that for any $v \in V$ there exists a natural number $N_1 = N_1(v)$ such that

$$\langle Au_i, u - v \rangle_V \geq \langle Au, u - v \rangle_V - 2, \quad i > N_1(v). \tag{7.2.37}$$

The fact that v is arbitrary implies that for each $w \in V$ a natural number $N_2 = N_2(w)$ exists with the property that

$$|\langle Au_i, w \rangle_V| \leq ||Au||_{V^*} ||w||_V + 2, \quad i > N_2(w). \tag{7.2.38}$$

Thus, using the uniformly boundedness theorem we conclude that the sequence $\{||Au_i||_{V^*}\}$ is bounded in V^*. To show that

$$Au_i \to Au \quad \text{weakly in } V^*, \tag{7.2.39}$$

it is sufficient to prove that each weakly converging subsequence of $\{Au_i\}$ has Au as its limit. But this follows directly from the fact that A is also generalized pseudomonotone (see Proposition 2.1). Therefore the sequence $u_i^* = g - Au_i - \varepsilon_i J u_i$ has the following properties: $u_i^* \in N_C(u_i)$, $u_i^* \to g - Au$ weakly in V^* and $\lim \sup \langle u_i^*, u_i - u \rangle_V \leq 0$. Thus by means of the generalized pseudo-monotonicity of N_C it results that $g - Au \in N_C(u)$. Finally, (7.2.3) holds and the proof is complete. q.e.d.

7.2.3 The Case of a Compact Operator

In this Subsection we deal with a case in which a compactness argument is involved. We suppose that a linear compact operator $L : V \to Y$ from a reflexive Banach space V into a Banach space Y is given. Let $C_Y \subset Y$ be a closed set in Y which is assumed to be star-shaped with respect to a ball $B_Y(Lu_0, \bar{\rho}) = \{y \in Y : ||y - Lu_0||_Y \leq \bar{\rho}\}$, $\bar{\rho} > 0$, for some $u_0 \in V$. Define $C \subset V$ by

$$C = \{v \in V : Lv \in C_Y\}. \tag{7.2.40}$$

and associate to C a tangent cone $T_C^Y(\cdot)$ by means of the relation

$$k \in T_C^Y(u) \Longleftrightarrow Lk \in T_{C_Y}(Lu), \quad u \in C, \tag{7.2.41}$$

where $T_{C_Y}(\cdot)$ denotes the usual Clarke's tangent cone of C_Y. The following theorem can be formulated.

Theorem 7.8 Let $A : V \to V^*$ be a pseudo-monotone operator from a reflexive Banach space V into V^* and let $L : V \to Y$ be a linear and compact operator from V into a Banach space Y. Suppose that C is defined by (7.2.40) and that the following conditions hold:

(H1) C_Y is star-shaped with respect to a ball $B_Y(Lu_0, \bar{\rho}) = \{y \in Y : \|y - Lu_0\|_Y \leq \bar{\rho}\}$, $\bar{\rho} > 0$;

(H2) $\langle Av, v - u_0 \rangle_V \geq c(\|v\|_V)\|v\|_V, v \in V$, with $c(r) \to \infty$ as $r \to \infty$.

Then for any $g \in V^*$ there exists $u \in C$ such that

$$\langle Au - g, v \rangle_V \geq 0 \quad \forall v \in T_C^Y(u). \tag{7.2.42}$$

Proof. The proof follows the same steps as the proof of Theorem 7.3. In the first step we prove the existence of solutions of the regularized hemivariational inequality

$$\langle Au_\lambda - g, v - u_\lambda \rangle_V + \frac{1}{\lambda} d_Y^0(Lu_\lambda, Lv - Lu_\lambda) \geq 0 \quad \forall v \in V, \tag{7.2.43}$$

by applying Theorem 4.23. In (7.2.43) $d_Y : Y \to R$ denotes the distance function of C_Y. In the second step we show that for sufficiently small λ the corresponding solution u_λ belongs to C. Finally, by virtue of

$$k \in T_C^Y(u) \iff d_Y^0(Lu, Lk) = 0 \tag{7.2.44}$$

we show that u_λ is a solution to (7.2.42). Due to the similarity with the previous case, the detailed proof is omitted here.   q.e.d.

Remark 7.9 Let us consider the case in which C_Y is ∂-regular in the sense of Clarke, i.e. the contingent cone of C_Y, K_{C_Y}, coincides with T_{C_Y}. Recalling that $y \in K_{C_Y}(w)$, $w \in C_Y$, if there exist $y_n \to y$ in Y and $\lambda_n \to 0_+$ with $w + \lambda_n y_n \in C_Y$, we obtain the inclusion $T_C(v) \subset T_C^Y(v)$ which is valid for each $v \in C$. This fact allows us to conclude that if u is a solution of (7.2.42) then it is also a solution of the problem (P), i.e. (7.2.3) holds.

In many constrained problems the assumption that $C_Y \subset Y$ is star-shaped with respect to a ball $B_Y(Lu_0, \bar{\rho})$, $\bar{\rho} > 0$, $u_0 \in V$, is very restrictive. Such a situation takes place, for instance, when the interior of C_Y in Y is empty. Instead, it may happen, that the admissible set $C \subset V$ given by (7.2.40) is star-shaped with rerspect to a certain ball $B_V(u_0, \rho), \rho > 0$, in V. Then we proceed as following: let a function $\tilde{d}_Y : Y \to R$ be given by

$$\tilde{d}_Y(y) = \inf\{\|y - Lv\|_Y : Lv \in C_Y \text{ and } v \in V\}, \quad y \in Y, \tag{7.2.45}$$

and let us set $d : V \to R$ as

$$d(u) = \tilde{d}_Y(Lu), \quad u \in V. \tag{7.2.46}$$

Now we claim that the function d has the jump property

$$\left. \begin{array}{ll} d^0(u; u_0 - u) \leq -d(u) - \bar{\rho} & \text{if } u \notin C \\[2mm] d^0(u; u_0 - u) = 0 & \text{if } u \in C \end{array} \right\} \tag{7.2.47}$$

where $\bar{\rho}$ is a positive constant, analogous to that of the usual distance function of a star-shaped set with respect to a ball (see Lemma 7.2). According to the proof of Lemma 7.2, in order to get (7.2.47) it is enough to show that for $u \notin C$

$$d(\lambda u + (1 - \lambda)u_0) \leq \lambda d(u) - (1 - \lambda)\bar{\rho} \quad \text{for} \quad \lambda_0 \leq \lambda \leq 1, \ \lambda_0 < 1. \quad (7.2.48)$$

Let $\varepsilon \geq 0$ be arbitrary and $u \notin C$. Due to (7.2.45) and (7.2.46), there exists a point $w \in C$ such that

$$d(u) = \tilde{d}_Y(Lu) \geq ||Lu - Lw||_Y - \varepsilon. \quad (7.2.49)$$

Since

$$u_0 + \rho \frac{u - w}{||u - w||_V} \in B_V(u_0, \rho), \quad (7.2.50)$$

by the assumption that C is star-shaped with respect to $B_V(u_0, \rho)$, we get that

$$\lambda Lw + (1 - \lambda)(Lu_0 + \rho \frac{Lu - Lw}{||u - w||_V}) \in C_Y \quad \text{for} \ \lambda \in [0, 1]. \quad (7.2.51)$$

Hence

$$d(\lambda u + (1 - \lambda)u_0) = \tilde{d}_Y(\lambda Lu + (1 - \lambda)Lu_0)$$

$$\leq ||\lambda Lu + (1 - \lambda)Lu_0 - \lambda Lw - (1 - \lambda)(Lu_0 + \rho \frac{Lu - Lw}{||u - w||_V})||_Y$$

$$= \frac{||Lu - Lw||_Y}{||u - w||_V}|\lambda||u - w||_V - (1 - \lambda)\rho|. \quad (7.2.52)$$

Now, it suffices to observe that due to $||u - w||_V > 0$, there exists $\lambda_0 < 1$ such that

$$0 \leq \lambda||u - w||_V - (1 - \lambda)\rho \quad (7.2.53)$$

for $\lambda \in [\lambda_0, 1]$. Thus from (7.2.52) we obtain

$$d(\lambda u + (1 - \lambda)u_0) \leq \lambda||Lu - Lw||_Y - (1 - \lambda)\rho \frac{||Lu - Lw||_Y}{||u - w||_V}$$

$$\leq \lambda||Lu - Lw||_Y - (1 - \lambda)\rho c, \quad (7.2.54)$$

where c results from the linearity and continuity of L, i.e., $||Lu||_Y < c||u||_V, u \in V$. Combining (7.2.49) with (7.2.54) implies that

$$d(\lambda u + (1 - \lambda)u_0) \leq \lambda d(u) - (1 - \lambda)\rho c + \lambda \varepsilon. \quad (7.2.55)$$

Since ε was chosen arbitrarily, the assertion (7.2.48) follows with $\bar{\rho} = \rho c$.

In the next step we formulate the regularized hemivariational inequality

$$\langle Au_\lambda - g, v - u_\lambda \rangle_V + \frac{1}{\lambda}d^0(u_\lambda; v - u_\lambda) \geq 0 \quad \forall v \in V. \quad (7.2.56)$$

and on the basis of Theorem 7.3 and Proposition 4.18 (d is a composition of $\tilde{d}_Y$ with L) we deduce the existence, for λ sufficiently small, of a $u_\lambda \in C$ satisfying

(7.2.56). From the obvious estimate $d^0(u; v) \leq \tilde{d}^0_Y(Lu; Lv), u, v \in V$, if follows that the u_λ is also a solution of the hemivariational inequality

$$\langle Au_\lambda - g, v - u_\lambda \rangle_V + \frac{1}{\lambda}\tilde{d}^0_Y(Lu_\lambda; Lv - Lu_\lambda) \geq 0 \quad \forall v \in V. \tag{7.2.57}$$

Now, the tangent cone $\tilde{T}^Y_C(u)$, $u \in C$, is introduced by the relation

$$k \in \tilde{T}^Y_C(u) \iff \tilde{d}^0_Y(Lu, Lk) = 0. \tag{7.2.58}$$

Thus the existence of solutions to the problem: find $u \in C$ ($C \subset V$ defined by (7.2.40)) such that

$$\langle Au - g, k \rangle_V \geq 0 \quad \forall k \in \tilde{T}^Y_C(u), \tag{7.2.59}$$

is guaranteed. The foregoing results can be summarized in the following Theorem.

Theorem 7.10 Let $A : V \to V^*$ be a pseudo-monotone operator from a reflexive Banach space V into V^*, and let $L : V \to Y$ be a linear compact operator from V into a Banach space Y. Suppose that C_Y is a closed subset of Y and $C \subset V$ is defined by (7.2.40). The following assumptions are assumed to hold:

(H1) C is star-shaped with respect to a ball $B_V(u_0, \rho)$, $\rho > 0$, for some $u_0 \in V$;

(H2)   either A has the $(S)_+$ property, or $\tilde{d}_Y$ is ∂-regular;

(H3)   There exists a function $c : R^+ \to R$ with $c(r) \to \infty$ as $r \to \infty$, such that for any $v \in V$, $\langle Av, v - u_0 \rangle_V \geq c(||v||_V)||v||_V$.

Then problem (7.2.59) has at least one solution.

7.3 Constrained Problems with Strongly Monotone Operators

We begin with the following lemma.

Lemma 7.11 Let us suppose that A is strongly monotone, i.e.

$$\langle Au - Av, u - v \rangle_V \geq m||u - v||^2_V \quad \forall u, v \in V, \tag{7.3.1}$$

where $m = \text{const} > 0, f : V \to R$ a locally Lipschitz function which is assumed to fulfill the following condition of relaxed monotonicity

$$\langle u^* - v^*, u - v \rangle_V \geq -a||u - v||^2_V \quad \forall u, v \in V, \tag{7.3.2}$$

for any $u^* \in \partial f(u)$ and $v^* \in \partial f(v)$, with a positive constant "a" satisfying the inequality

$$a < m. \tag{7.3.3}$$

Then the problem: find $u \in V$ such that

$$\langle Au - g, v - u \rangle_V + f^0(u, v - u) \geq 0 \quad \forall v \in V \tag{7.3.4}$$

has a solution.

Proof. It is not difficult to check that the mapping $A + \partial f$ is strongly monotone. Indeed, we obtain easily that

$$\langle u^* - v^*, u - v \rangle_V \geq m^* ||u - v||_V^2 \quad \forall u, v \in V, \tag{7.3.5}$$

where $u^* \in Au + \partial f(u)$, $v^* \in Av + \partial f(v)$ and $m^* = m - a > 0$. We claim that $A + \partial f$ is pseudo-monotone (in fact, maximal monotone). To show this it suffices to check that, if $\{u_i\}$ is a sequence in V converging weakly to u, and $u_i^* \in Au_i + \partial f(u_i)$ is such that

$$\limsup \langle u_i^*, u_i - u \rangle_V \leq 0, \tag{7.3.6}$$

then for each element $v \in V$ there exists $u^*(v) \in Au + \partial f(u)$ with the property that

$$\liminf \langle u_i^*, u_i - v \rangle_V \geq \langle u^*(v), u - v \rangle_V. \tag{7.3.7}$$

Notice that from (7.3.5) and (7.3.6) the strong convergence of $\{u_i\}$ to u results. In such a case the remaining requirements follow immediately, due to the upper semicontinuity of ∂f from V to V^* with the weak topology. Since the multivalued mapping $A + \partial f$ as strongly monotone, it is also coercive. Accordingly, the maximal monotonicity implies that for a given $g \in V^*$ there exists u such that (7.3.4) holds. The proof is complete.   q.e.d.

Theorem 7.12 Let $A : V \to V^*$ be a maximal monotone operator from a reflexive Banach space V into V^* fulfilling (7.3.1) and let $C \subset V$ be star-shaped with respect to a ball $B_V(u_0, \rho)$. Suppose that $Au_0 \neq g$ and that the distance function d_C satisfies the condition of relaxed monotonicity

$$\langle u^* - v^*, u - v \rangle_V \geq -a ||u - v||_V^2 \quad \forall u, v \in V, \tag{7.3.8}$$

for any $u^* \in \partial d_C(u)$ and $v^* \in \partial d_C(v)$, with a constant "a" fulfilling the inequality

$$0 < a < \frac{4m^2 \rho}{||Au_0 - g||_{V^*}^2}. \tag{7.3.9}$$

Then problem (P) has at least one solution.

Proof. Choose $\lambda > 0$ such that

$$\frac{a}{m} < \lambda < \frac{4m\rho}{||Au_0 - g||_{V^*}^2}. \tag{7.3.10}$$

The existence of such a λ follows from (7.3.9). Consider the multivalued mapping $A + \frac{1}{\lambda} \partial d_C$. It is not difficult to check that this mapping is strongly monotone, i.e. that

$$\langle u^* - u^*, u - v \rangle_V \geq m^* \|u - v\|_V^2 \quad \forall u, v \in V \tag{7.3.11}$$

for any $u^* \in Au + \frac{1}{\lambda} \partial d_C(u)$, $v^* \in Av + \frac{1}{\lambda} \partial d_C(v)$, where $m^* = m - \frac{a}{\lambda} > 0$. The assumptions of Lemma 7.11 hold. Accordingly, for $g \in V^*$ there exists u_λ such that

$$\langle Au_\lambda - g, v - u_\lambda \rangle_V + \frac{1}{\lambda} d_C^0(u_\lambda, v - u_\lambda) \geq 0 \quad \forall v \in V. \tag{7.3.12}$$

Thus it remains to show that u_λ is, in fact, a solution of the original problem (P). Indeed, suppose that this is not the case, i.e., $u_\lambda \notin C$. Then by substituting $v = u_0$ into (7.3.12) we obtain

$$\langle Au_\lambda - g, u_0 - u_\lambda \rangle_V + \frac{1}{\lambda} d_C^0(u_\lambda, u_0 - u_\lambda) \geq 0, \tag{7.3.13}$$

from which, by means of to Lemma 7.2, it results that

$$\langle Au_\lambda - g, u_0 - u_\lambda \rangle_V \geq \frac{d_C(u_\lambda) + \rho}{\lambda} \geq \frac{\rho}{\lambda}. \tag{7.3.14}$$

Because of (7.3.1)

$$\|Au_0 - g\|_{V*} \|u_\lambda - u_0\|_V \geq m\|u_\lambda - u_0\|_V^2 + \frac{\rho}{\lambda}, \tag{7.3.15}$$

which can be written equivalently as

$$m\|u_\lambda - u_0\|_V^2 - \|Au_0 - g\|_{V*} \|u_\lambda - u_0\|_V + \frac{\rho}{\lambda} \leq 0. \tag{7.3.16}$$

But the last inequality holds only if the discriminant

$$\Delta = \|Au_0 - g\|_{V*}^2 - 4m\frac{\rho}{\lambda} \geq 0. \tag{7.3.17}$$

This means that

$$\lambda \geq \frac{4m\rho}{\|Au_0 - g\|_{V*}^2} \tag{7.3.18}$$

which is a contradiction to (7.3.10). Thus we have proved that $u_\lambda \in C$. Now to complete the proof it is enough to combine (7.2.15) with (7.3.13). q.e.d.

Corollary 7.13 Let V be a Hilbert space and let $b : V \times V \to R$ be a bilinear symmetric continuous form satisfying the condition

$$b(v, v) \geq m\|v\|_V^2 \quad v \in V, \, m = \text{const} > 0. \tag{7.3.19}$$

Denote by $A : V \to V^*$ a linear bounded operator associated with the form b, i.e. $\langle Au, v \rangle_V = b(u, v) \, \forall u, v \in V$. Suppose that $C \subset V$ is a closed subset of V which is assumed to be star-shaped with respect to a ball $B_V(u_0, \rho), \rho > 0$. Moreover, we suppose that $Au_0 \neq g$ and that

$$V \ni v \to d_C(v) + \frac{1}{2}a\|v\|_V^2 \tag{7.3.20}$$

is convex with a constant "a" fulfilling the condition

$$0 < a < \frac{4m^2\rho}{||Au_0 - g||_{V^*}^2}. \tag{7.3.21}$$

Let us introduce the function

$$\Pi(v) = \frac{1}{2}b(v, v) - \langle g, v \rangle_V + I_C(v), \quad v \in V, \tag{7.3.22}$$

where $I_C : V \to \bar{R}$ denotes the indicator of C. Note that Π is not convex due to nonconvexity of C. We claim that Π possesses at least one substationarity point. In other words, there exists $u \in C$ such that (cf. Sect. 1.2)

$$0 \in \partial\Pi(u), \tag{7.3.23}$$

where $\partial\Pi$ denotes the generalized gradient in the sense of Clarke and Rockafellar. Since the function $v \to \frac{1}{2}b(v, v) - \langle g, v \rangle_V$ is continuously differentiable,

$$\partial\Pi(\cdot) = \partial(\frac{1}{2}b(\cdot, \cdot) - \langle g, \cdot \rangle_V) + N_C(\cdot). \tag{7.3.24}$$

Thus in order to prove (7.3.23) it suffices to check that the assumptions of Theorem 7.12 hold. Indeed, we need only to notice that the strong monotonicity condition (7.3.1) follows from (7.3.19), whereas the relaxed monotonicity of ∂d_C, (7.3.8), is a consequence of the convexity of the function given in (7.3.20) and of (7.3.21).

Remark 7.14 It is worth noting that to obtain existence results for the posed nonconvex constrained problems we have applied the penalization method but, and this is important, without letting the small parameter tend to 0. This method based on the discontinuity of $d^0(\cdot, \cdot)$, permits to relax the condition of the convexity of C. It must be also emphasized that Theorem 7.12 is proved without imposing any compactness property on C.

7.4 Variational Inequalities with Nonconvex Domain

This Section is devoted to the study of variational problems involving convex lower semicontinuous functions and also the indicator functions of some closed nonconvex star-shaped sets. In other words, we deal with variational inequalities with a nonconvex domain.

Theorem 7.15 Let $A : V \to V^*$ be a pseudo-monotone operator from V into V^*, $\varphi : V \to \bar{R}$ a convex, lower semicontinuous function from V into $\bar{R}$ and let $C \subset V$ be a closed subset of V. Suppose that the following assumptions hold:

(H1) C is star-shaped with respect to a ball $B_V(u_0, \rho), \rho > 0$;

(H2) either the distance function d_C of C belongs to $PM(V)$, or $d_C \in QPM(V)$ and A has the $(S)_+$ property;

(H3) $u_0 \in D(\partial\varphi)$;

(H4) $\langle Av, v - u_0 \rangle_V \geq c(||v||_V)||v||_V, v \in V$, with $c(r) \to \infty$ as $r \to \infty$;

(H5) either A_{u_0} is quasi-bounded or $\partial\varphi_{u_0}$ is strongly quasi-bounded (for any operator T we denote by T_{u_0} the operator given by $T_{u_0}(v) = T(u_0 + v))$ $\forall v \in V$).

Then for any $g \in V^*$ the problem: find $u \in C$ such that

$$\langle Au - g, v - u \rangle_V + \varphi(v) - \varphi(u) \geq 0 \quad \forall v - u \in T_C(u) \tag{7.4.1}$$

has at least one solution.

Proof. In the first step we ensure the existence of a solution to the regularized problem: for a given $g \in V^*$ and $\lambda > 0$ find $u_\lambda \in V$ such that

$$\langle Au_\lambda - g, v - u_\lambda \rangle_V + \varphi(v) - \varphi(u_\lambda) + \frac{1}{\lambda}d_C^0(u_\lambda, v - u_\lambda) \geq 0 \quad \forall v \in V. \tag{7.4.2}$$

Along the lines of the proof of Theorem 7.3 we deduce the pseudo-monotonicity and coercivity of $A + \frac{1}{\lambda}\partial d_C$. By virtue of Theorem 2.12 the range of $A + \frac{1}{\lambda}\partial d_C + \partial\varphi$ coincides with the whole space V^*. Consequently, the existence of a solution for (7.4.2) results.

In the second step we shall show that for sufficiently small λ, $u_\lambda \in C$. Indeed, suppose that this is not true, i.e., that $u_\lambda \notin C$ for each $\lambda > 0$. By the substitution of $v = u_0$ into (7.4.2) one gets

$$\langle Au_\lambda - g, u_0 - u_\lambda \rangle_V + \varphi(u_0) - \varphi(u_\lambda) + \frac{1}{\lambda}d_C^0(u_\lambda, u_0 - u_\lambda) \geq 0. \tag{7.4.3}$$

Lemma 7.2 implies that

$$d_C^0(u_\lambda, u_0 - u_\lambda) \leq 0. \tag{7.4.4}$$

Since $u_0 \in D(\partial\varphi)$, there exists a constant $\alpha \in R^+$ such that

$$\varphi(u_0) - \varphi(v) \leq \alpha||v - u_0||_V \leq \alpha(||v||_V + ||u_0||_V) \quad \forall v \in D(\varphi). \tag{7.4.5}$$

Hence, from (7.4.4), (7.4.5) and (7.4.3) we get that

$$(||g||_{V^*} + \alpha)(||u_\lambda||_V + ||u_0||_V) \geq c(||u_\lambda||_V)||u_\lambda||_V. \tag{7.4.6}$$

Here the coercivity of A has also been used. It follows immediately that the sequence $\{u_\lambda\}$ is uniformly bounded, i.e. that

$$||u_\lambda||_V \leq M, \tag{7.4.7}$$

where the constant M does not depend on λ. Using (7.1.11) we can obtain some further results. Indeed using the notation

$$\inf_{r \geq 0} rc(r) = \beta, \quad \beta \in R, \tag{7.4.8}$$

and by means of (7.4.3) and (7.4.5) we are led to the relation

$$(\|g\|_{V^*} + \alpha)(\|u_\lambda\|_V + \|u_0\|_V) \geq \beta + \frac{1}{\lambda}(d_C(u_\lambda) + \rho) \geq \beta + \frac{\rho}{\lambda}, \tag{7.4.9}$$

which due to (7.4.7) can be written as

$$(\|g\|_{V^*} + \alpha)(M + \|u_0\|_V) - \beta \geq \frac{\rho}{\lambda}. \tag{7.4.10}$$

But the last inequality cannot be valid for

$$\lambda < \frac{\rho}{(\|g\|_{V^*} + \alpha)(M + \|u_0\|_V) - \beta}, \tag{7.4.11}$$

which leads to a contradiction. Accordingly, if λ is small enough, u_λ must belong to C. In order to obtain that

$$\langle Au_\lambda - g, v - u_\lambda \rangle_V + \varphi(v) + \varphi(u_\lambda) \geq 0 \quad \forall v - u_\lambda \in T_C(u_\lambda) \tag{7.4.12}$$

it suffices to use the relation

$$v \in T_C(u_\lambda) \iff d_C^0(u_\lambda, v) = 0 \tag{7.4.13}$$

and apply it to (7.4.3). The proof is complete.   q.e.d.

A direct consequence of Theorems 7.15, 7.4 and 7.7 is the following result concerning the family of nonconvex sets expressed by means of (7.2.16).

Theorem 7.16 Let $A : V \to V^*$ be a pseudo-monotone operator from V into V^*, $\varphi : V \to \bar{R}$ a convex, lower semicontinuous function from V into $\bar{R}$ and let $C_i, i = 1, \ldots, N$, be closed convex subsets of V. Suppose that the following four assumptions hold:

(H1) $u_0 \in \cap_i \text{int } C_i$;

(H2) $u_0 \in D(\partial\varphi)$;

(H3) $\langle Av, v - u_0 \rangle_V \geq c(\|v\|_V)\|v\|_V, v \in V$, with $c(r) \to \infty$ as $r \to \infty$;

(H4)   either A_{u_0} is quasi-bounded, or $\partial\varphi_{u_0}$ is strongly quasi-bounded.

Moreover, we suppose that either

(H5) $T_C(v) = T_{\cap_{i \in I(v)} C_i}(v) \quad \forall v \in \partial C$ (the boundary of C); and

(H6)   If for a sequence $\{u_n\}$ in V, $u_n \to u$ weakly, $u_n \in \partial C$, there exists a sequence $\{u_n^*\} \subset V^*$ with $u_n^* \in N_C(u_n)$, $u_n^* \to u^*$ weakly, $u^* \neq 0$, such that $\limsup \langle u_n^*, u_n - u \rangle_V \leq 0$, then $u \in \partial C$;

or, the condition

(H7) A has the $(S)_+$ property;

be satisfied. Then for C given by (7.2.16) the problem (7.4.1) has at least one solution.

The next result corresponds to the case involving a compact linear operator. We assume that $L : V \to Y$ is a linear compact operator from a reflexive Banach space V into a Banach space Y. Let $C_Y \subset Y$ be a closed set in Y which is assumed to be star-shaped with respect to a ball $B_Y(Lu_0, \bar{\rho}) = \{y \in Y : \|y\|_Y \leq \bar{\rho}\}, \bar{\rho} > 0$, for some $u_0 \in V$. Define $C \subset V$ by

$$C = \{v \in V : Lv \in C_Y\} \tag{7.4.14}$$

and associate with C a modified tangent cone $T_C^L(\cdot)$ by means of the formula

$$k \in T_C^L(u) \iff Lk \in T_{C_Y}(Lu), \quad u \in C, \tag{7.4.15}$$

where $T_{C_Y}(\cdot)$ denotes the usual Clarke's tangent cone to C_Y.

Theorem 7.17 Let $A : V \to V^*$ be a pseudo-monotone operator from a reflexive Banach space V into V^*, $L : V \to Y$ a linear compact operator from V into a Banach space Y and let $\varphi : V \to \bar{R}$ be a convex, lower semicontinuous function from V into $\bar{R}$. Let us define $C \subset V$ by means of (7.4.14). Suppose that the following assumptions hold

(H1) $C_Y \subset Y$ is a closed, star-shaped set with respect to a ball $B_Y(Lu_0, \bar{\rho})$, for some $u_0 \in V$ and $\bar{\rho} > 0$;

(H2) $u_0 \in D(\partial\varphi)$;

(H3) $\langle Av, v - u_0 \rangle \geq c(\|v\|_V)\|v\|_V, \ v \in V$, with $c(r) \to \infty$ as $r \to \infty$;

(H4) either A_{u_0} is quasi-bounded, or $\partial\varphi_{u_0}$ is strongly quasi-bounded.

Then for any $g \in V^*$ the problem: find $u \in C$ such that

$$\langle Au - g, v - u \rangle_V + \varphi(v) - \varphi(u) \geq 0 \quad \forall v - u \in T_C^L(u), \tag{7.4.16}$$

has at least one solution.

Proof. We proceed as in the proof of Theorem 7.15 with only a slight modification. The difference is that instead of the corresponding regularized problem (7.4.3) we consider a hemivariational inequality of the form

$$\langle Au_\lambda - g, v - u_\lambda \rangle_V + \varphi(v) - \varphi(u_\lambda) + \frac{1}{\lambda} d_Y^0(Lu_\lambda, Lv - Lu_\lambda) \geq 0 \quad \forall v \in V. \tag{7.4.17}$$

Here $d_Y : Y \to R$ is the distance function with respect to C_Y. Because of the similarity with the previous case the detailed study will be omitted here. q.e.d.

Taking into account Theorems 7.10 and 7.15 we can formulate the following result.

Theorem 7.18 Let $A : V \to V^*$ be a pseudo-monotone operator from a reflexive Banach space V into V^*, $L : V \to Y$ a linear compact operator from V into a Banach space Y and let $\varphi : V \to \bar{R}$ be a convex, lower semicontinuous function from V into $\bar{R}$. Let us define $C \subset V$ by means of (7.4.14). Suppose that the following hypotheses hold:

(H1) $C \subset V$ is a closed, star-shaped set with respect to a ball $B_V(u_0, \rho)$, for some $u_0 \in V$ and $\rho > 0$;

(H2) either A has the $(S)_+$ property, or $\tilde{d}_Y$ given by (7.2.45) is ∂-regular;

(H3) $u_0 \in D(\partial\varphi)$;

(H4) either A_{u_0} is quasi-bounded, or $\partial\varphi_{u_0}$ is strongly quasi-bounded;

(H5) $\langle Av, v - u_0 \rangle \geq c(\|v\|_V)\|v\|_V$, $\forall v \in V$, with $c(r) \to \infty$ as $r \to \infty$.

Then for any $g \in V^\star$ the problem: find $u \in C$ such that

$$\langle Au - g, v - u \rangle_V + \varphi(v) - \varphi(u) \geq 0 \quad \forall v - u \in \tilde{T}_C^Y(u), \tag{7.4.18}$$

where $\tilde{T}_C^Y(u)$ is given by (7.2.58), has at least one solution.

Now we consider the case with a strongly monotone operator A.

Theorem 7.19 Let $A : V \to V^\star$ be a strongly monotone operator from a reflexive Banach space V into $V^\star$, $\varphi : V \to \bar{R}$ a convex lower semicontinuous function with $u_0 \in D(\partial\varphi)$ and let $C \subset V$ be a closed subset of V, which is assumed to be star-shaped with respect to a ball $B(u_0, \rho), \rho > 0$. Suppose that $g \in V^\star$, $Au_0 \neq g$, and that (7.3.1) holds. Moreover, we suppose that the distance function of C, d_C, fulfills the condition of the relaxed monotonicity

$$\langle u^\star - v^\star, u - v \rangle_V \geq -a\|u - v\|_V^2 \tag{7.4.19}$$

for any $u, v \in V$ where $u^\star \in \partial d_C(u)$ and $v^\star \in \partial d_C(v)$, and that

$$0 < a < \frac{4m^2\rho}{(\|Au_0 - g\|_{V^\star} + \alpha)^2}, \tag{7.4.20}$$

where a and μ satisfy (7.4.5). Then (7.4.1) has at least one solution.

Proof. Choose $\lambda > 0$ such that

$$\frac{a}{m} < \lambda < \frac{4m\rho}{(\|Au_0 - g\|_{V^\star} + \alpha)^2}. \tag{7.4.21}$$

The existence of such λ follows from (7.4.20). Let us consider a multivalued mapping $A + \partial\varphi + \frac{1}{\lambda}\partial d_C$. It is not difficult to verify that this mapping is strongly monotone, i.e.

$$\langle u^\star - v^\star, u - v \rangle_V \geq m^\star\|u - v\|_V^2 \quad \forall u, v \in V, \tag{7.4.22}$$

where $u^\star \in Au + \partial\varphi(u) + \frac{1}{\lambda}d_C(u)$, $v^\star \in Av + \partial\varphi(v) + \frac{1}{\lambda}\partial d_C(v)$ and $m^\star = m - \frac{a}{\lambda} > 0$. Since $A + \frac{1}{\lambda}\partial d_C$ is a maximal monotone mapping with the effective domain $D(A + \frac{1}{\lambda}\partial d_C) = V$, the sum $A + \partial\varphi + \frac{1}{\lambda}d_C$ is maximal monotone, as well (see Lemma 7.11). Accordingly, for $g \in V^\star$ there exists u_λ such that

$$\langle Au_\lambda - g, v - u_\lambda \rangle_V + \varphi(v) - \varphi(u_\lambda) + \frac{1}{\lambda}d_C^0(u_\lambda, v - u_\lambda) \geq 0 \quad \forall v \in V, \tag{7.4.23}$$

holds. Now we show that for sufficiently small λ, u_λ is a solution of the original problem (7.4.1). Indeed, suppose the contrary, i.e., that $u_\lambda \notin C$ for all $\lambda > 0$. By substituting $v = u_0$ into (7.4.23), we obtain by means of (7.1.11) that

$$\langle Au_\lambda - g, u_0 - u_\lambda \rangle_V + \varphi(u_0) - \varphi(u_\lambda) \geq \frac{d(u_\lambda) + \rho}{\lambda} \geq \frac{\rho}{\lambda}. \tag{7.4.24}$$

Hence,

$$(||Au_0 - g||_{V^*} + \alpha)||u_\lambda - u_0||_V \geq m||u_\lambda - u_0||_V^2 + \frac{\rho}{\lambda}, \tag{7.4.25}$$

which can be written equivalently as

$$m||u_\lambda - u_0||_V^2 - (||Au_0 - g||_{V^*} + \alpha)||u_\lambda - u_0||_V + \frac{\rho}{\lambda} \leq 0. \tag{7.4.26}$$

The last inequality is valid only if the following inequality for the discriminant holds

$$\Delta = (||Au_0 - g||_{V^*} + \alpha)^2 - 4m\frac{\rho}{\lambda} \geq 0. \tag{7.4.27}$$

From (7.4.26) we obtain that

$$\lambda \geq \frac{4m\rho}{(||Au_0 - g||_{V^*} + a)^2} \tag{7.4.28}$$

which is a contradiction to (7.4.21). Accordingly, we have obtained that $u_\lambda \in C$. We use (7.2.15) and the proof is complete.   q.e.d.

7.5 Applications to the Plasticity Theory with Nonconvex Yield Condition

Let Ω be an open, bounded and connected subset of the three-dimensional Euclidean space R^3, with a Lipschitzian boundary Γ. Ω is occupied by a body in its undeformed state and is referred to a Cartesian coordinate system $0x_1x_2x_3$. We decompose Γ into two mutually disjoint parts Γ_U and Γ_F assuming that mes $\Gamma_U > 0$. By $n : \Gamma \rightarrow R^3$ we denote the outward unit normal vector field to Γ. For the space of 3×3 symmetric real-valued matrices we use as in Ch. 6 the symbol S^3. The inner product in R^3 and S^3 will be denoted by the same symbol "." if no ambiguity occurs. As usual, the well known notations $u : \Omega \rightarrow R^3$ and $\sigma : \Omega \rightarrow S^3$ will be introduced to denote the displacement vector field and the stress vector field, respectively.

Our considerations will be carried out within the small deformation theory and the holonomic theory of plasticity (i.e. without the consideration of time increments). Consequently, the interrelation between u and the corresponding strain tensor field $\varepsilon(u), \varepsilon(u) : \Omega \rightarrow S^3$, is given by

$$\varepsilon_{ij}(u) = \frac{1}{2}(u_{i,j} + u_{j,i}), \ i, j = 1, 2, 3. \tag{7.5.1}$$

The material properties of the body are determined by Hooke's tensor of elasticity $C = \{C_{ijkl}\}$ satisfying the symmetry property

$$C_{ijkl} = C_{jikl} = C_{klij}, \quad i, j, k, l = 1, 2, 3, \tag{7.5.2}$$

and the ellipticity condition

$$C_{ijkl}\varepsilon_{ij}\varepsilon_{kl} \geq \alpha\varepsilon_{ij}\varepsilon_{ij} \quad \text{in } \Omega, \quad \alpha = \text{const} > 0, \tag{7.5.3}$$

for any $\varepsilon \in S^3$. Denote by $D = C^{-1}$ the inverse of the elasticity tensor C. Suppose now that the body is submitted to external forces, i.e. to body forces $f : \Omega \to R^3$ and to surface tractions $F : \Gamma_F \to R^3$. On Γ_U the displacements $U : \Gamma \to R^3$ are supposed as given. Without loss of generality we may assume that $U = 0$. For the displacements we introduce the space $V = \{u = \{u_i\} : u_i \in H^1(\Omega), u_{i|\Gamma_U} = 0\}$, while for the stresses the space $Y = L^2(\Omega; S^3)$ will be used.

To formulate the stress constraints we consider a closed subset A of S^3. Let $\varepsilon > 0$ be a small parameter and let $\omega_\varepsilon : R^3 \to R$ be a mollifier corresponding to ε. We define the set

$$\Sigma_\varepsilon = \{\tau \in Y : \omega_\varepsilon \star \tau(x) \in A \quad \text{for all } x \in \Omega\}, \tag{7.5.4}$$

where

$$\omega_\varepsilon \star \tau(x) = \int_\Omega \omega_\varepsilon(x - y)\tau(y)d\Omega \quad \forall x \in \Omega. \tag{7.5.5}$$

Furthermore, we introduce a linear operator $L : V \to Y$ by the formula

$$Lu = \{\varepsilon_{ij}(u)\} = \{\tfrac{1}{2}(u_{i,j} + u_{j,i})\} \tag{7.5.6}$$

and the functional $g \in V^*$ as

$$\langle g, v \rangle_{H^1(\Omega)} = \int_\Omega f_i v_i d\Omega + \int_{\Gamma_F} F_i v_i d\Gamma \quad \forall v \in V. \tag{7.5.7}$$

We denote now by $d_\varepsilon : Y \to R$ the distance function with respect to Σ_ε. The tangent cone to Σ_ε at $\sigma \in \Sigma_\varepsilon$ is

$$T_{\Sigma_\varepsilon}(\sigma) = \{\tau \in Y : d_\varepsilon^0(\sigma, \tau) = 0\}, \tag{7.5.8}$$

while the normal cone to Σ_ε at $\sigma \in \Sigma_\varepsilon$ reads

$$N_{\Sigma_\varepsilon}(\sigma) = \{\eta \in Y : (\eta, \tau)_{L^2(\Omega)} \leq 0 \quad \forall \tau \in T_{\Sigma_\varepsilon}(\sigma)\}. \tag{7.5.9}$$

Denoting by $L^* : Y^* \to V^*$ the transpose of L we formulate the following problem.

Problem (P). Find $u \in V$ and $\sigma \in \Sigma_\varepsilon$ such that

$$\left. \begin{array}{l} L^*\sigma - g = 0 \\ Lu - D\sigma \in N_{\Sigma_\varepsilon}(\sigma). \end{array} \right\} \tag{7.5.10}$$

The foregoing system can be written equivalently as

$$\int_\Omega \sigma \cdot [\varepsilon(v) - \varepsilon(u)]d\Omega + \int_\Omega f_i(v_i - u_i)d\Omega - \int_{\Gamma_F} F_i(v_i - u_i)d\Gamma = 0 \quad \forall v \in V \quad (7.5.11)$$

$$\int_\Omega [D\sigma - \varepsilon(u)] \cdot (\tau - \sigma)d\Omega \geq 0 \quad \forall \tau - \sigma \in T_{\Sigma_\epsilon}(\sigma). \quad (7.5.12)$$

The following problem can be associated with (P):

Problem (P_σ): Find σ such that

$$0 \in D\sigma + \partial I_\Xi(\sigma) + N_{\Sigma_\epsilon}(\sigma), \quad (7.5.13)$$

where

$$\Xi = \{\tau \in Y : L^*\tau - g = 0\} \quad (7.5.14)$$

is the set of all statically admissible stresses.

Since Korn's inequality holds on V, we can write that $(\mathrm{Im}L = \text{image of } L)$

$$\partial I_\Xi(\sigma) = \begin{cases} \mathrm{Im}L = \{\eta \in Y : \eta = Lv \text{ for some } v \in V\}, & \text{if } \sigma \in \Xi \\ \emptyset & \text{otherwise.} \end{cases} \quad (7.5.15)$$

Therefore, if σ is a solution of (P_σ), then there exists u such that the pair (u, σ) is a solution of (P). Thus we have reduced (P) to the problem (P_σ) involving σ as its only unknown.

Our study will be carried out under the following assumptions. The generalized gradient of the distance function d_ϵ is supposed to satisfy the following condition of relaxed monotonicity

$$(\partial d_\epsilon(\eta) - \partial d_\epsilon(\tau), \eta - \tau)_{L^2(\Omega)} \geq -a\|\eta - \tau\|^2_{L^2(\Omega)} \quad \forall \eta, \tau \in Y, \quad (7.5.16)$$

where the positive constant $a > 0$ will be made more precise later. Denote by $m > 0$ a constant for which the ellipticity condition

$$D_{ijkl}\varepsilon_{ij}\varepsilon_{kl} \geq m\varepsilon_{ij}\varepsilon_{ij} \quad \forall \varepsilon \in S^3 \quad \text{in } \Omega. \quad (7.5.17)$$

holds. Concerning the functional $g \in V^*$ characterizing the external forces acting on the body it will be assumed the following

Safe load condition (G). There exists $\sigma^0 \in Y$ with the properties that

(a) $L^*\sigma^0 - g = 0$;

(b) Σ_ϵ is star-shaped with respect to a ball $B_Y(\sigma^0, \rho) = \{\tau \in Y : \|\tau - \sigma^0\|_Y \leq \rho\}$ with some positive ρ.

Now we can formulate the following existence result for (P).

Theorem 7.20 We make the assumptions (7.5.16), (7.5.17) and we assume the validity of the safe load condition (G). Moreover, let us suppose that

$$0 < a < \frac{4m^2\rho}{||D\sigma^0||^2_{L^2(\Omega)}}. \qquad (7.5.18)$$

Then (P) has at least one solution, i.e., (7.5.11) and (7.5.12) hold.

Proof. We choose $\lambda > 0$ such that

$$\frac{a}{m} < \lambda < \frac{4m\rho}{||D\sigma^0||^2_{L^2(\Omega)}}. \qquad (7.5.19)$$

The existence of such a λ follows from (7.5.18). Let us consider the multivalued mapping $D + \frac{1}{\lambda}\partial d_e$. It is not difficult to verify that this mapping is strongly monotone, i.e.

$$(\eta^* - \tau^*, \eta - \tau)_{L^2(\Omega)} \geq m^*||\eta - \tau||^2_{L^2(\Omega)} \quad \forall \eta, \tau \in Y. \qquad (7.5.20)$$

where $\eta^* \in D\eta + \frac{1}{\lambda}\partial d_e(\eta)$, $\tau^* \in D\tau + \frac{1}{\lambda}\partial d_e(\tau)$ and $m^* = m - \frac{a}{\lambda} > 0$, and maximal monotone (cf. the proof of Lemma 7.11). Accordingly, for $g \in V^*$ there exists $\sigma_\lambda \in \Xi$ such that

$$0 \in D\sigma_\lambda + \partial I_\Xi(\sigma_\lambda) + \frac{1}{\lambda}\partial d_e(\sigma_\lambda). \qquad (7.5.21)$$

Thus σ_λ satisfies the first relation in (7.5.10) and due to (7.5.15), there exists $u_\lambda \in V$ with the property that

$$(D\sigma_\lambda - Lu_\lambda, \tau - \sigma_\lambda)_{L^2(\Omega)} + \frac{1}{\lambda}d^0_e(\sigma_\lambda, \tau - \sigma_\lambda) \geq 0 \quad \forall \tau \in \Xi. \qquad (7.5.22)$$

Now it will be shown that $(u_\lambda, \sigma_\lambda)$ is, in fact, a solution of the original problem (P). First we claim that $\sigma_\lambda \in \Sigma_e$. Indeed, suppose that this is not true. Then by substituting $\tau = \sigma^0$ into (7.5.22) we get the inequality

$$(D\sigma_\lambda - Lu_\lambda, \sigma^0 - \sigma_\lambda) + \frac{1}{\lambda}d^0_e(\sigma_\lambda, \sigma^0 - \sigma_\lambda) \geq 0. \qquad (7.5.23)$$

Taking into account the safe load condition (G) and applying Lemma 7.2, we obtain the estimate

$$m||\sigma_\lambda||^2_{L^2(\Omega)} - ||D\sigma^0||_{L^2(\Omega)}||\sigma_\lambda||_{L^2(\Omega)} + \frac{\rho}{\lambda} \leq 0, \qquad (7.5.24)$$

which is valid only for

$$\lambda \geq \frac{4m\rho}{||D\sigma^0||_{L^2(\Omega)}}. \qquad (7.5.25)$$

This contradicts (7.5.19). Therefore $\sigma_\lambda \in \Sigma_e$ and consequently, from (7.5.23), the second relation in (7.5.10) results. Thus the pair $(u_\lambda, \sigma_\lambda)$ is a solution of (P). The proof is complete.   q.e.d.

Remark 7.21 The foregoing approach cannot be applied for the set of admissible stresses expressed by the formula

$$\Sigma = \{\tau \in Y : \tau(x) \in A \text{ for a. e. } x \in \Omega\}. \tag{7.5.26}$$

This is caused by the fact that Σ does not possess any internal points in the L^2-space. Thus we cannot verify the hypothesis (b) of the safe load condition (G), which is the crucial point of the proof. Indeed, the small parameter δ depends on ε in the sense that the condition $\varepsilon \to 0$ implies that $\delta \to 0$. Therefore the parameter "a" characterizing (via (7.5.16)) the nonconvexity of the admissible set Σ_ε cannot remain positive in the limit as $\varepsilon \to 0$. This method is appropriate when introducing the constraints on the average state of stresses but not on the stresses themselves.

7.6 Applications to the Theory of Elasticity

The results of this Chapter permit the formulation and solution of many new problems in the theory of elasticity. All the problems which we will study are connected with the treatment of linear elastic bodies whose displacement field or stress field is subjected to constraints expressed by means of nonconvex star-shaped and closed admissible sets. Such problems are studied here for the first time thoroughly and this only becomes possible by the theory developed in the previous Sections.

7.6.1 The Signorini Problem in Nonlinear Elasticity for Star-Shaped Closed Sets

Let $\Omega \subset R^3$ be an open bounded connected subset occupied by the nonlinear elastic body under consideration in its undeformed state. We put ourselves in the framework of (1.4.44) where now $u \in [W^{1,p}(\Omega)]^3$, $f_i \in L^p(\Omega)$ and we make the assumptions that (5.5.42) and (5.5.54) hold with $p > 3$. Moreover we assume that

$$u_i = 0, \quad i = 1, 2, 3 \quad \text{on} \quad \Gamma_U \tag{7.6.1}$$

and that on $\Gamma_S = \Gamma - \Gamma_U$, $-S_N$ is the force acting upon the body due to the frictionless contact of the points of Γ_S with the "internal side" of a closed star-shaped set K with respect to a ball in R^3. The tangential displacements u_T are prescribed to be zero on Γ_S. Then (Fig. 7.6.1)

$$-S_N \in N_K(u_N) = \partial I_K(u_N), \tag{7.6.2}$$

where N_K is the normal cone to the closed set K (see also Fig. 1.2.1)

Due to (1.2.28) relation (1.4.44) leads to the following problem for $f_i \in L^{p'}(\Omega)$, $C_{T_i} \in L^{q'}(\Gamma_S)$, $i = 1, 2, 3$, where p' and q' are defined as in subsection 5.5.4. Find $u \in V_0 = \{v : v \in [W^{1,p}(\Omega)]^3, u_{T_i} = 0, i = 1, 2, 3, \text{ on } \Gamma_S, v_i = 0 \text{ on } \Gamma_U\}$ with $u_N \in \tilde{K}$ such as to satisfy the hemivariational inequality

$$\int_\Omega [\frac{\partial w(\varepsilon(u))}{\partial \varepsilon}]_{ij}\varepsilon_{ij}(v)d\Omega + \int_{\Gamma_S} I_{T_K}(u_N)(v_N)d\Gamma \geq (l,v) \quad \forall v \in V_0, \qquad (7.6.3)$$

where

$$(l,v) = \int_\Omega f_i v_i d\Omega \qquad (7.6.4)$$

and

$$\tilde{K} = \{u \in V_0 : u_N(x) \in K \text{ for a.e. } x \in \Gamma_S\}. \qquad (7.6.5)$$

Thus (7.6.3) becomes: Find $u \in \tilde{K}$ such that

$$\int_\Omega [\frac{\partial w(\varepsilon(u))}{\partial \varepsilon}]_{ij}\varepsilon_{ij}(v)d\Omega \geq (l,v) \quad \forall v_N \in T_{\tilde{K}}(u_N). \qquad (7.6.6)$$

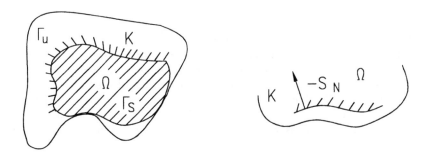

Fig. 7.6.1. The Signorini problem for star-shaped closed sets

Note first that the application $u \to u_N : [W^{1,p}(\Omega)]^3 \to L^\infty(\Gamma_S)$ is linear compact, because of the compact imbedding $W^{1,p}(\Omega) \subset C^0(\bar{\Omega})$ for $p > N = 3$. This fact implies that the definition of $\tilde{K}$ makes sense and that $\tilde{K}$ is a star-shaped closed set with respect to a ball in V_0. Moreover Korn's inequality (5.5.55) holds. Thus the application of Theorem 7.8 yields the existence of a solution of the nonclassical variational inequality (7.6.6).

Suppose further that the boundary Γ consists of Γ_U, where (7.6.1) holds, of Γ_S where (7.6.2) holds and of Γ_S' where monotone boundary conditions of the subdifferential type

$$-S \in \partial\varphi(u) \quad \text{on} \quad \Gamma_S' \qquad (7.6.7)$$

are given. Here φ is a convex, l.s.c and proper functional on R^3. Then we define the convex l.s.c and proper functional on $[W^{1,p}(\Omega)]^3$

$$\Phi(u) = \begin{cases} \int_{\Gamma_S'} \varphi(u)d\Gamma & \text{if } \varphi(u) \in L^1(\Omega) \\ \infty & \text{otherwise} \end{cases} \qquad (7.6.8)$$

and now the problem reads: Find $u \in \tilde{K}$ such that

$$\int_{\Omega} [\frac{\partial w(\varepsilon(u))}{\partial \varepsilon}]_{ij} \varepsilon_{ij}(v-u) d\Omega + \Phi(v) - \Phi(u) \geq (l, v - u) \quad \forall v_N - u_N \in T_{\tilde{K}}(u_N).$$

(7.6.9)

Here we apply Theorem 7.17 which implies the existence of a solution of the nonclassical variational inequality (7.6.9). Indeed $\mathrm{grad} w(\varepsilon(\cdot))$ is coercive and bounded and fulfills the S_+ property due to its strong monotonicity.

7.6.2 Constrained Skin Effects in Plane Nonlinear Elasticity

Let us consider a twodimensional nonlinear elastic body Ω obeying the law (5.5.42) and let us assume for the sake of simplicity that its boundary Γ is fixed, i.e.

$$u_i = 0 \quad i = 1, 2 \text{ on } \Gamma. \tag{7.6.10}$$

We assume further that the body forces $f = \{f_i\}$, $i = 1, 2$, are given in the form

$$f = \bar{f} + \bar{\bar{f}}, \quad -\bar{f} \in N_K(u) = \partial I_K(u). \tag{7.6.11}$$

Moreover let (5.5.54) hold with $p > 2$ and let $\bar{\bar{f}} \in [L^{p'}(\Omega)]^2$, $\frac{1}{p} + \frac{1}{p'} = 1$, be given. We assume that $\bar{f}$ is a force acting on the body and is caused (as a reaction force) by the fact that the skin displacements $u = \{u_i\}$, $i = 1, 2$, at each point $x \in \Omega$ are constrained to belong to a closed, star-shaped with respect to a ball in R^2 set K which may be different at each $x \in \Omega$. Here we make the additional assumption that the radii of all balls are bounbed from below by a positive number ρ_0. Thus we are led to the following variational formulation: find $u \in [\overset{0}{W}{}^{1,p}(\Omega)]^2$ with $u(x) \in \tilde{K}(x)$ such that

$$\int_{\Omega} [\frac{\partial w(\varepsilon(u))}{\partial \varepsilon}]_{ij} \varepsilon_{ij}(v) d\Omega + \int_{\Omega} I_{\tilde{K}}(u, v) d\Omega \geq (\bar{\bar{f}}, v) \quad \forall v \in [\overset{0}{W}{}^{1,p}(\Omega)]^2 \quad (7.6.12)$$

where

$$\tilde{K} = \{u : u \in [\overset{0}{W}{}^{1,p}(\Omega)]^2, u(x) \in K(x) \text{ a.e. on } \Omega\} \tag{7.6.13}$$

has a meaning, because $W^{1,p}(\Omega)$ is compactly imbedded into $C^0(\bar{\Omega})$, and is a closed star-shaped set with respect to a ball in $[\overset{0}{W}{}^{1,p}(\Omega)]^2$ and because of the previous assumption about the radii of the balls.

From (7.6.12) and (1.2.28) we obtain the following problem: find $u \in \tilde{K}$ such that

$$\int_{\Omega} [\frac{\partial w(\varepsilon(u))}{\partial \varepsilon}]_{ij} \varepsilon_{ij}(v) d\Omega \geq (\bar{\bar{f}}, v) \quad \forall v \in T_{\tilde{K}}(u). \tag{7.6.14}$$

This problem has a solution as one may conclude applying Theorem 7.8.

Let us assume in the previous application that Γ consists of Γ_U, where (7.6.10) holds and of $\Gamma_S = \Gamma - \Gamma_U$, where the boundary condition (7.6.7) holds.

Introducing (7.6.8) and on the assumption that (7.6.11) holds we are led to the following variational formulation: find $u \in \tilde{K}$, where $\tilde{K}$ is given by (7.6.13), such that

$$\int_{\Omega} [\frac{\partial w(\varepsilon(u))}{\partial \varepsilon}]_{ij}\varepsilon_{ij}(v-u)d\Omega + \Phi(v) - \Phi(u) \geq (\bar{\bar{f}}, v-u) \quad \forall v-u \in T_{\tilde{K}}(u) \quad (7.6.15)$$

We can apply Theorem 7.17 and we obtain the existence of at least one solution.

References

[Ant]   Antes, H., Panagiotopoulos, P.D.: The Boundary Integral Approach to Static and Dynamic Contact Problems. Equality and Inequality Methods. Birkhäuser Verlag, Basel, Boston 1992

[Anz]   Anzellotti, G.: Elasticity with Unilateral Constraints on the Stresses. Univ. Trento, Dip. Matematica, U.T.M. 158, 1984

[Argy65]   Argyris, J.H.: Three-Dimensional Anisotropic and Inhomogeneous Elastic Media. Matrix Analysis for Small and Large Displacements. Ing. Archiv 34 (1965) 33-55

[Argy66]   Argyris, J.H.: Continua and Discontinua, Proc. 1st Conf. Matrix Meth. Struct. Mech. Wright Patterson Air Force Base, Dayton, Ohio 1965, AFFDL TR 1966 66-80

[Argy69]   Argyris, J.H., Scharpf, D.W.: Some General Considerations on the Natural Mode Technique. Aeron. J. Royal Aeron. Soc. 73 (1969) 218-226 and 361-368

[Aub77]   Aubin, J.P.: Applied Abstract Analysis. J.Wiley and Sons, New York 1977

[Aub79]   Aubin, J.P., Clarke, F.H.: Shadow Prices and Duality for a Class of Optimal Control Problems. SIAM J. Control Optimization 17 (1979) 567-586

[Aub79a]   Aubin, J.P.: Applied Functional Analysis. J. Wiley and Sons, New York 1979

[Aub84]   Aubin, J.P., Ekeland, I.: Applied Nonlinear Analysis. Wiley-Inter-Science, N.York 1984

[Aub90]   Aubin, J.P., Frankovska, H.: Set-Valued Analysis. Birkhäuser Verlag, Basel, Boston 199(

[Aub91]   Aubin, J.P.: Viability Theory. Birkhäuser Verlag, Basel 1991.

[Avr]   Avriel, M.: R-convex functions Math. Programming 2 (1972) 309-323

[Ba84]   Baiocchi, C., Capelo, A.: Variational and Quasivariational Inequalities: Applications to Free Boundary Problems, J.Wiley and Sons, Chichester, 1984

[Ba86]   Baiocchi, C., Gastaldi, F., Tomarelli, F.: Some Existence Results on Noncoercive Variational Inequalities, Ann. Scuola Norm. Sup. Pisa cl. Sci., IV. 13 (1986) 617-659

[Ba88]   Baiocchi, C., Buttazzo, G., Gastaldi, F., Tomarelli, F.: General Existence Theorems for Unilateral Problems in Continuum Mechanics, Arch. Rational Mech. Anal. 100 (1988) 149-188

[Boc]   Boccardo, L., Giachetti, D., Murat, F.: A Generalization of a Theorem of H. Brézis & F.E. Browder and Applications to Some Unilateral Problems. Annales de l'Institut Henri Poincaré, Analyse Non-linéare, 7, (1990) 367-384

[Boi]   Boieri, P., Gastaldi, F., Kinderlehrer, D.: Existence, Uniqueness and Regularity Results for the Two Bodies Contact Problem, Appl. Math. Optim., 15 (1987) 251-277

[Bréz68]   Brézis, H.: Équations et inéquations non-linéaires dans les espaces véctorieles en dualité, Ann. Institut Fourier, Vol. 18 (1968) 115-176

[Bréz70]   Brézis, H., Crandall, M.G., Pazy, A.: Perturbations on Nonlinear Maximal Monotone Sets in Banach Spaces. Comm. Pure Appl. Math. 23 (1970) 123-144

[Bréz72]   Brézis, H.: Problèmes Unilatéraux. J. Math. Pures et Appl. 51 (1972) 1-168

[Bréz73]   Brézis, H.: Opérateurs Maximaux Monotones et Semigroupes de Contractions dans les Espaces de Hilbert. North-Holland Publ. Co., Amsterdam and American Elsevier Publ. Co., New York 1973

[Bréz82]   Brézis, H., Browder, F.E.: Some Properties of Higher Order Sobolev Spaces. J. Math. Pures et Appl., 61 (1982) 245-259

[Bro68]   Browder, F.E.: Nonlinear Eigenvalue Problems and Galerkin Approximations. Bull. Amer. Math. Soc., 74 (1968) 651-656

[Bro70a]   Browder, F.E.: Pseudo-Monotone Operators and the Direct Method of the Calculus of Variations. Arch. Rat. Mech. Anal., 38 (1970) 268-277

[Bro70b]   Browder, F.E.: Existence Theorems for Nonlinear Partial Differential Equations. Proceedings of Symposia in Pure Mathematics, Vol. 16, pp. 1-60, Amer. Math. Soc., Providence, 1970.

[Bro72]   Browder, F.E., Hess, P.: Nonlinear Mappings of Monotone Type in Banach Spaces. J. Funct. Anal., 11 (1972) 251-294

[Bro75]   Browder, F.E.: Nonlinear Equations of Evolution and Nonlinear Operators in Banach Spaces. Nonlinear Functional Analysis, Proceedings of Symposia in Pure Mathematics, Vol. 18, Part 2, Amer. Math. Soc. Providence, 1975

[Bro83]   Browder, F.E.: Fixed Point Theory and Nonlinear Problems. Bull. Amer. Math. Soc., 9, (1983) 1-39

[Brü90]   Brüning, E.: On the Variational Approach to Semilinear Elliptic Equations with Scale Covariance. J. Diff. Eq., Vol. 83 (1990) 109-144

[Brü92]   Brüning, E.: Weak K-Monotonicity in Minimization Problems, in "Dynamic of Complex and Irregular Systems", Proceedings of Symposium at the Center for Inter-Disciplinary Research, University of Bielefeld, December 1991; eds. S. Albeverio, Ph. Blanchard, World Scientific Publishing Co., 1992

[Ch]   Chang, K.C.: Variational Methods for Non-Differentiable Functionals and their Applications to Partial Differential Equations. J.Math.Anal. Appl. 80 (1981) 102-129

[Clar73]   Clarke, F.H.: Necessary Conditions for Nonsmooth Problems in Optimal Contorl and the Calculus of Variations. Ph.D.Thesis, University of Washington, Seattle 1973

[Clar75]   Clarke, F.H.: Generalized Gradients and Applications. Trans. A.M.S. 205 (1975) 247-262

[Clar81]   Clarke, F.H.: Generalized Gradients of Lipschitz Functionals. Advances in Math. 40 (1981) 52-67

[Clar83]   Clarke, F.H.: Optimization and Nonsmooth Analysis. Wiley, New York 1983

[Com]   Combini, A., Castagnoli, E., Martein, L., Mazzoleni, P., Schaible, S.(eds).: General Convexity and Fractional Programming with Economic Applications. Springer Verlag, New York 1990

[Crou]   Crouzeix, J.P, Ferland, J.A.: Criteria for Quasi-Convexity and Pseudomonotonicity: Relationships and Comparisons. Math. Programming 23 (1982) 193-205

[Dac82]   Dacorogna, B.: Weak Continuity and Weak Lower Semicontinuity of Nonlinear Functionals, Lectures Notes in Mathematics, Vol. 922, Springer-Verlag, Berlin, New York 1982

[Dac89]   Dacorogna, B.: Direct Methods in the Calculus of Variations, Springer-Verlag, Berlin, New York 1989.

[Dunf]   Dunford, N., Schwartz, J.T.: Linear Operators Part I: General Theory. Interscience Publishers, New York 1966

[Duv71]   Duvaut, G., Lions, J.L.: Un Problème d' élasticité avec frottement, J. de Mécanique 10 (1971) 409-420

[Duv72]   Duvaut, G., Lions, J.L.: Les Inéquations en Mécanique et en Physique. Dunod, Paris 1972

[Duv80]   Duvaut. G.: Equilibre d' un solide élastique avec contact unilateral et frottement de Coulomb. C.R. Acad. Sc. Paris 290 (1980) 263-265

[Eke]   Ekeland, I., Temam, R.: Convex Analysis and Variational Problems. North Holland, Amsterdam and American Elsevier, New York 1976

[Fich63]   Fichera, G.: The Signorini Elastostatics Problem with Ambiguous Boundary Conditions. Proc. Int. Conf. Application of the Theory of Functions in Continuum Mechanics, Vol. I, Tbilisi 1963

[Fich64]   Fichera, G.: Problemi Elastostatici con Vincoli Unilaterali: il Problema di Signorini con Ambigue Condizioni al Contorno. Mem. Accad. Naz. Lincei, VIII 7 (1964) 91-140

[Fich72]   Fichera, G.: Boundary Value Problems in Elasticity with Unilateral Constraints. In: Encyclopedia of Physics (ed. by S.Flügge) Vol. VI a/2. Springer-Verlag, Berlin 1972

[Fig91]   De Figueiredo, D.G., Miyagaki, O.H.: Semilinear Elliptic Equations with the Primitive of the Nonlinearity Away from the Spectrum, Nonlinear Analysis, Theory, Methods and Applications, 17 (1991) 1201-1219

[Flo]   Floegl, H., Mang, H.A.: Tension Stiffening Concept Based on Bond Slip. ASCE (ST 12) 108 (1982) 2681-2701

[Fré71]   Frémond, M.: Etude de structures viscoélastiques stratifiées soumises aux charges harmoniques et de solides élastiques reposant sur ces structures. Thèse de doctorat d'Etat, Univ. Paris VI, 1971.

[Frém82a]   Frémond, M.: Adhérence des solides, C. R. Acad. Sc. Paris, Série II, 295 (1982) 769-772

[Frém82b]   Frémond, M.: Equilibre de structures qui adhérent a leur support, C. R. Acad. Sc. Paris, Série II, 295 (1982) 913-915

[Frém82c]   Frémond, M., Ghidouche, H., Point, N.: Congélation d' un milieu poreux humide alimenté en eau, C. R. Acad. Sc. Paris Série II, 294 (1982) 711-714

[Frém83a]   Frémond, M.: Conditions unilatérales et non linéarité en calcul à la rupture. Matem. Aplicada e Computational, Brazil, 2 (1983) 237-256

[Frém83b]   Frémond, M.: Frost Action in Soils. In A. Fasano and M. Primicerio (eds), Free Boundary Problems, Theory and applications, Vol I, Pitman Res. Notes in Math. Vol 78, p.191-208 Pitman, London 1983

[Frém84]   Frémond, M.: Sur la fissuration, C. R. Acad. Sc. Paris Série II, 299 (1984) 487-490

[Frém85]   Frémond, M., Visintin, A.: Dissipation dans le changement de phase. Surfusion. Changement de phase irréversible, R. Acad. Sc. Paris Série II, 301 (1985) 1265-1268

[Frém87a]   Frémond, M.: Contact unilatéral avec adhérence: une théorie du premier gradient. Unilateral Problems in Structural Analysis - 2, (ed. by Del Pierro, G. and Maceri, F.), CISM Courses and Lectures 304, pp. 33-45, Springer-Verlag, N.York, Wien, 1987

[Frém87b   Frémond, M.: Matériaux à mémoire de forme, C. R. Acad. Sc. Paris, Série II, 304 (1987) 239-244

[Frém87c]   Frémond, M.: Adhérence des solides, J. Méc. Théor. Appl. 6 (1987) 383-407

[Frém88]   Frémond, M.: Contact with Adhesion. In Topics in Nonsmooth Mechanics (ed. by J.J.Moreau, P.D.Panagiotopoulos, G. Strang), pp. 157-185, Birkhäuser Verlag, Boston, Basel, Berlin 1988

[Frém90]   Frémond, M.: Shape Memory Alloys. A Thermomechanical Model, In K.H.Hoffmann and J. Sprekels (eds). Free Boundary Problems: Theory and Applications, Vol I, Pitman Res. Notes in Math. Vol 185 pp.295-306, Longman Scient. and Techn., Essex, 1990.

[Gas88a]   Gastaldi F.: Remarks on a Noncoercive Contact Problem with Friction in Elastostatics. Publ. No. 649, Istituto Anal. Num. C.N.R, Pavia 1988

[Gas88b]   Gastaldi, F. and Martins J.A.C.: A Noncoercive Steady-Sliding Problem with Friction. Publ. No. 650, Istituto Anal. Num. C.N.R. Pavia 1988

[Gia]   Gianquinta, M., Giusti, E.: Researches on the Equilibrium of Masonry Structures, Arch. Math. Mech. Anal. 88 (1985) 359-392

[Gir]   Girkmann, K.: Flächentragwerke. Springer-Verlag, Wien 1963

[Glow]   Glowinski, R., Lions J.L., Tremolières R.: Numerical Analysis of
            Variational Inequalities. Studies in Mathematics and its Applica-
            tions, Vol.8, North-Holland- Elsevier, Amsterdam-New York 1981

[Göp]   Göpfert, A.: Mathematische Optimierung in allgemeinen Vektorräu-
            men. B.G. Teubner, Leipzig 1973

[Ham]   Hamel, G.: Theoretische Mechanik. Springer-Verlag, Berlin 1967

[Has82]   Haslinger, J., Hlavaček, I.: Approximation of the Signorini Problem
            with Friction by a Mixed Finite Element Method. J.Math.Anal. 86
            (1982) 99-122

[Hasli89]   Haslinger, J., Panagiotopoulos, P.D.: Optimal Control of Hemivaria-
            tional Inequalities. In: Control of Boundaries and Stabilization, (ed.
            by J.Simon), Lect. Notes in Control and Information Sciences, Vol.
            1925 Springer-Verlag, N.York 1989, p.p. 128-139

[Hed]   Hedberg, L.I.: Two Approximation Problems in Function Spaces.
            Ark. Mat., 16 (1978) 51-81

[Hiri]   Hiriart-Urruty.: From Convex Optimization to Nonconvex Opti-
            mization. Necessary and Sufficient Conditions for Global Optimal-
            ity. In: Nonsmooth Optimization and Related Topics, (eds. F.H.
            Clarke, V.F. Dem'yanov and F. Giannessi), pp. 219-239 Plenum
            Press, New York-London 1989

[Hoff90a]   Hoffmann, K.H., Sprekels, J. (eds).: Free Boundary Problems: The-
            ory and Applications I,II, Pitman Res. Notes in Math. Vol. 185,
            Longman Scient. and Techn., Essex 1990

[Hoff90b]   Hoffmann, K.H., Sprekels, J. (eds).: Free Boundary Value Problems,
            ISNM Vol 95, Birkhäuser Verlag, Basel 1990

[Hult]   Hult J., Travniček L.: Carrying capacity of Fibre Bundles with Vary-
            ing Strength and Stiffness. J. Méc. Théor. et Appl. 2 (1983) 643-657

[Ja83]   Jarusek, J.: Contact Problems with Bounded Friction. Coercive
            Case. Czech. Math. J. 33 (1983) 254-278

[Ja84]   Jarusek, J.: Contact Problems with Bounded Friction. Semicoercive
            Case. Czech. Math. J. 34 (1984) 619-629

[Ka88]   Kalker, J.J.: Contact Mechanical Algorithms. Comm. in Applied
            Num. Methods 4 (1988) 25-32

[Ka90]   Kalker, J.J.: Three Dimensional Elastic Bodies in Rolling Contact.
            Kluwer Acad. Publ., Dordrecht 1990

[Kara]   Karamardian, S., Schaible, S.: Seven Kinds of Monotone Maps, J. Opt. Th. Appl. 66 (1990) 37-46

[Ken78]   Mc Kenna, P.J., Rauch, J.: Strongly Nonlinear Preturbations of Nonnegative Boundary Value Problems with Kernel. Journal of Differential Equations, 28 (1978) 253-265

[Kin]   Kinderlehrer, D.: Remarks About Signorini's Problem in Linear Elasticity. Ann. Scuola Norm. Sup. Pisa cl. Sci., IV, 8 (1981), 605-645.

[Kra]   Krasnoselskii, M.A.: Topological Methods in the Theory of Nonlinear Integral Equations Pergamon Press, Oxford, 1964

[Kuf]   Kufner, A., John, O. Fučik, S.: Function Spases, Noordhoff International Publ, Leyden, 1977

[Lan]   Lanczos, C.: The Variational Principles of Mechanics. University of Toronto Press, Toronto 1966

[Land]   Landesman, E.M., Lazer, A.C.: Nonlinear Perturbations of Linear Elliptic Boundary Value Problems at Resonance. J.Math.Mech. 19 (1970) 609-623

[Léné73]   Léné, F.: Sur les matériaux élastiques à énergié de déformation non quadratique. Thèse de $3^{\text{ème}}$ cycle, Université Paris VI, 1973

[Léné74]   Léné, F.: Sur les matériaux élastiques à énergié de déformation non quadratique. J. de Mécanique 13 (1974) 499-534

[Lio67]   Lions, J.L. and Stampacchia, G.S.: Variational Inequalities. Comm. Pure and Appl. Math. XX (1967) 493-519.

[Lio69]   Lions, J.L.: Quelques méthodes de résolution des problèmes aux limites non linéaires. Dunod/Gauthier-Villars, Paris 1969

[Lio71]   Lions, J.L.: Optimal control of systems governed by partial differential equations. Springer-Verlag, Berlin 1971

[Ma68]   Maier, G.: A Quadratic Programming Approach for Certain Classes of Nonlinear Structural Problems. Meccanica 3 (1968) 121-130

[Ma71]   Maier, G.: Incremental plastic analysis in the presence of large displacements and physical instabilizing. Int. J. Solids Structures. 7 (1971) 345-372

[Miet92]   Miettinen, M., Haslinger, J.: Approximation of Optimal Control Problems of Hemivariational Inequalities, Numer. Funct. Anal. and Optim. 13 (1992) 43-68

[Miet93]   Miettinen, M.: Approximation of Hemivariational Inequalities and Optimal Control Problems, Doct. Dissertation, Dept. of Mathematics, Univ. of Jyväskylä, Rep. 59, Jyväskyla, Finnland, 1993

[Mis92]   Mistakidis, E., Panagiotopoulos, P.D.: On the Numerical Treatment of Nonmonotone (zigzag) Friction and Adhesive Contact Problems with Debonding. Approximation by Monotone Subproblems, 1992. Int.J.Comp. and Struct. (to appear, 1994)

[Mis93]   Mistakidis, E., Panagiotopoulos, P.D.: On the Approximation of Nonmonotone Multivalued Problems by Monotone Subproblems, 1993. Comp. Meth. appl. Mech. Eng. (to appear, 1994)

[Mor67]   Moreau, J.J.: Fonctionnelles Convexes. Séminaire sur les équations aux dérivées partielles. Collège de France, Paris 1967

[Mor68]   Moreau, J.J.: La notion du surpotentiel et les liaisons unilatérales en élastostatique. C.R.Acad.Sci. Paris 167A (1968) 954-957

[Mor70]   Moreau, J.J.: Sur les lois de frottement, de plasticité et de viscosité. C.R. Acad. Sc. Paris 271A (1970) 608-611

[Mor86]   Moreau, J.J.: Une formulation du contact frottement sec; application au calcul numérique. C.R. Acad. Sci. Paris, Sér. II, 302 (1986) 799-801

[Mor88a]   Moreau, J.J., Panagiotopoulos, P.D.(eds): Nonsmooth Mechanics and Applications, CISM Vol. 302, Springer Verlag, Wien 1988

[Mor88b]   Moreau, J.J., Panagiotopoulos, P.D., Strang, G.(eds): Topics in Nonsmooth Mechanics, Birkhäuser Verlag, Basel, Boston 1988

[Mor88c]   Moreau, J.J.: Unilateral Contact and Dry Friction in Finite Freedom Dynamics. In: Nonsmooth Mechanics and Applications (ed. by J.J.Moreau and P.D.Panagiotopoulos),CISM Vol. 302, Springer Verlag, Wien, N.York 1988

[Mos]   Moser, K.: Faserkunststoffverbund, VDI Verlag, Düsseldorf 1992

[Mot86]   Motreanu, D.: Existence for minimization with Nonconvex Constraints, J.Math. Anal. Appl. 117 (1986) 128-137

[Mot93]   Motreanu, D., Panagiotopoulos, P.D.: Hysteresis: The Eigenvalue Problem for Hemivariational Inequalities. Models of Hysteresis (ed. by A. Visintin) Pitman Research Notes in Mathematics, Longman, Harlow (1993)

[Nan88]   Naniewicz, Z., Woźniak, C.: On the Quasi-Stationary Models of Debonding Processes in Layered Composites. Ing. Archiv 58 (1988) 403-412

[Nan89a]   Naniewicz, Z.: On Some Nonconvex Variational Problems Related to Hemivariational Inequalities. Nonlin. Anal. 13 (1989) 87-100

[Nan89b]   Naniewicz, Z.: On Some Nonmonotone Subdifferential Boundary Conditions in Elastostatics. Ing. Archiv 60 (1989) 31-40

[Nan92a]   Naniewicz, Z.: On the Pseudo-Monotonicity of Generalized Gradients of Nonconvex Functions. Applicable Analysis 47 (1992) 151-172

[Nan92b]   Naniewicz, Z.: On Some Noncoercive Problems related to Delamination in Layered Composites, in "Nonsmooth Optimization, Methods and Application", Ed. F. Giannessi, Gordon and Breach Science Publishers, Switzerland, USA, 1992.

[Nan93a]   Naniewicz, Z.: On the Existence of Solutions to the Continuum Model of Delamination. Nonl. Anal. Theory, Meth. Appl. 20 (1993) 481-507

[Nan93b]   Naniewicz, Z.: Hemivariational Inequalities with Functions Fulfilling Directional Growth Condition (to appear in Applicable Analysis)

[Nan94]   Naniewicz, Z.: Hemivariational Inequality Approach to Constrained Problems for Star-Shaped Admissible Sets. J. Opt. Theory Appl. 8, No. 1 (in press)

[Neit]   Neittaanmäki, P. (ed).: Numerical Methods for Free Boundary Problems, ISNM Vol 99, Birkhäuser Verlag, Basel 1991

[Neč]   Nečas, J., Jarusek, J., Haslinger, J.: On the Solution of the Variational Inequality to the Signorini Problem with Small Friction. Bulletino U.M.I. 17B (1980) 796-811

[Niez]   Niezgodka, M., Pawlow, I.: A Mathematical Model for Artificial Freezing of Geologic Formations, In: K.H.Hoffmann and J. Sprekels (eds) Free Boundary Problems: Theory and Applications, Vol I, Pitman Res. Notes in Math. Vol 185, pp.277-286, Longman Scient. and Techn., Essex, 1990.

[Oet]   Oettli, W., Prager, W.: Flow in Networks with Amplification and Coupling, Unternehmensforschung 10 (1966) 42-58

[Ond]   Ondracek G. (ed): Verbundwerkstoffe. Phasenverbindung und mechanische Eigenschaften, Deutsche Gesellschaft für Metallkunde, Band 1, 1985

[Orn]   Ornstein, D.: A Non-inequality for differential operators in the L^1-norm. Arch. Rat. Mech. Anal. 11 (1962) 40-49

[Pan75]   Panagiotopoulos, P.D.: A Nonlinear Programming Approach to the
            Unilateral Contact – and Friction – Boundary Value Problem in the
            Theory of Elasticity. Ing. Archiv 44 (1975) 421-432

[Pan76]   Panagiotopoulos, P.D.: Convex Analysis and Unilateral Static Prob-
            lems. Ing. Archiv 45 (1976) pp. 55-68

[Pan81]   Panagiotopoulos, P.D.: Non-Convex Superpotentials in the Sense of
            F.H. Clarke and Applications. Mech. Res. Comm. 8 (1981) 335-340

[Pan82]   Panagiotopoulos, P.D.: Non-Convex Energy Functionals. Applica-
            tion to Non-convex Elastoplasticity. Mech. Res. Comm. 9 (1982)
            23-29

[Pan82a]   Panagiotopoulos, P.D.: On a Method Proposed by W. Prager for the
            Nonlinear Network Flow Problem, Scient. Annual of the Faculty of
            Technology, Aristotle University Thessaloniki, Θ' (1982) 77-85

[Pan83]   Panagiotopoulos, P.D.: Nonconvex Energy Functions. Hemivaria-
            tional Inequalities and Substationarity Principles. Acta Mechanica
            42 (1983) 160-183

[Pan83a]   Panagiotopoulos, P.D.: A Boundary Integral Inclusion Approach to
            Unilateral B.V.Ps in Elastostatics. Mech. Res. Comm. 10 (1983)
            91-96

[Pan84]   Panagiotopoulos, P.D., Baniotopoulos, C.C.: A Hemivariational In-
            equality and Substationarity Approach to the Interface Problem.
            Theory and Prospects of Applications, Engineering Analysis 1 (1984)
            20-31

[Pan85]   Panagiotopoulos, P.D.: Inequality Problems in Mechanics and Ap-
            plications. Convex and Nonconvex Energy Functions. Birkhäuser
            Verlag, Basel, Boston 1985. (Russian Translation MIR Publ. Moscow
            1989)

[Pan87a]   Panagiotopoulos, P.D., Koltsakis, E.K.: Interlayer Slip and Delam-
            ination Effect: A Hemivariational Inequality Approach. Canadian
            Society for Mech. Engineering 11 (1987) 43-52.

[Pan87b]   Panagiotopoulos, P.D.: Ioffe's Fans and Unilateral Problems: A New
            Conjecture. In: Unilateral Problems in Structural Analysis 2, (ed. by
            G. del Piero, F.Maceri), CISM Courses and Lectures 304. Springer
            Verlag, Wien, N.York 1987

[Pan87c]   Panagiotopoulos, P.D., Koltsakis, E.K.: Hemivariational Inequali-
            ties for Linear and Nonlinear Elastic Materials. Meccanica 22 (1987)
            65- 75

[Pan88]   Panagiotopoulos, P.D.: Nonconvex Superpotentials and Hemivari-
            ational Inequalities. Quasidifferentiability in Mechanics. In: Nons-
            mooth Mechanics and Applications (ed. by J.J. Moreau, P.D. Pana-
            giotopoulos), CISM Courses and Lectures Nr. 302, Springer Verlag,
            Wien, N.York 1988

[Pan88a]   Panagiotopoulos, P.D., Stavroulakis, G.: A Variational-hemivariati-
            onal Inequality Approach to the Laminated Plate Theory under
            Subdifferential Boundary Conditions. Quart. of Appl. Math. XLVI
            (1988) 409-430

[Pan89a]   Panagiotopoulos, P.D.: Semicoercive Hemivariational Inequalities.
            On the Delamination of Composite Plates. Quart. of Appl. Math.,
            XLVII (1989) 611-629

[Pan89b]   Panagiotopoulos, P.D., Haslinger, J.: Optimal Control of Systems
            Governed by Hemivariational Inequalities. In: Mathematical Mod-
            els for Phase Change Problems, (ed. by J.F.Rodriques) Birkhäuser
            Verlag, Basel, Boston 1989

[Pan90]   Panagiotopoulos, P.D., Stavroulakis, G.: The Delamination Effect
            in Laminated von Karman Plates under Unilateral Boundary Con-
            ditions. A Variational-Hemivariational Inequality Approach. J. of
            Elasticity 23 (1990) 69-96

[Pan91]   Panagiotopoulos, P.D.: Coercive and Semicoercive Hemivariational
            Inequalities. Nonlin. Anal. 16 (1991) 209-231

[Pan92a]   Panagiotopoulos, P.D., Haslinger, J.: On the Dual Reciprocal Varia-
            tional Approach to the Signorini-Fichera Problem. Convex and Non-
            convex Generalizations. ZAMM 72 (1992) 497-506

[Pan92b]   Panagiotopoulos, P.D., Stavroulakis, G.: New Types of Variational
            Principles Based on the Notion of Quasidifferentiability. Acta Me-
            chanica 94 (1992) 171-194

[Pan92c]   Panagiotopoulos, P.D.: Adhesive Joints and Interfaces of Linear
            Elastic Bodies in Loading and Unloading. Semicoercive Hemivaria-
            tional Inequalities. J. of Elasticity 28 (1992) 29-54

[Pan93]   Panagiotopoulos, P.D.: Hemivariational Inequalities. Applications
            in Mechanics and Engineering, Springer Verlag, Berlin 1993

[Pana83]   Panagiotopoulos, P.D.: Nonconvex Energy Functions. Hemivaria-
            tional Inequalities and Substationarity Principles. Acta Mechanica
            42 (1983) 160-183

[Pana85a]   Panagiotopoulos, P.D.: Nonconvex Problems of Semipermeable Me-
            dia and Related Topics. ZAMM 65 (1985) 29-36

[Pana85b]   Panagiotopoulos, P.D.: Hemivariational Inequalites and Substation-
             rity in the Static Theory of von Kármán Plates. ZAMM 65 (1985)
             219-229

[Pana88a]   Panagiotopoulos, P.D.: Hemivariational Inequalities and their Ap-
             plications. In: Topics in Nonsmooth Mechanics (ed. by J.J.Moreau,
             P.D.Panagiotopoulos and G. Strang) Birkhäuser Verlag, Boston
             1980

[Pana88b]   Panagiotopoulos, P.D.: Nonconvex Superpotentials and Hemivari-
             ational Inequalities. Quasidifferentiablility in Mechanics. In: Nons-
             mooth Mechanics and Applications (ed. by J.J. Moreau and P.D.
             Panagiotopoulos) CISM Lect. Notes, Vol. 302, Springer Verlag,
             Wien, N.York 1988

[Pana88c]   Panagiotopoulos, P.D.: Variational Hemivariational Inequalities in
             Nolinear Elasticity. The Coercive Case. Aplikace Matematiky (now
             Applications of Mathematics) 33 (1988) 249-268

[Pana89]   Panagiotopoulos, P.D.: Boundary Integral Equations for Inequality
             Problems. The Nonconvex Case. Acta Mechanica 72 (1989) 152-168

[Pana91]   Panagiotopoulos, P.D.: The B.I.E.M. for Inequality Problems. Math.
             Comput. Modelling 15 (1991) 257-267

[Panag83]   Panagiotopoulos, P.D.: Optimal Control and Parameter Identifica-
             tion of Structures with Convex and Nonconvex Strain Energy Den-
             sity. Applications to Elastoplasticity and to Contact Problems. Solid
             Mech. Archives 8 (1983) 363-411

[Panag84]   Panagiotopoulos, P.D.: Optimal Control of Structures with Convex
             and Nonconvex Energy Densities and Variational and Hemivaria-
             tional Inequalities. Engng. Struct. 6 (1984) 12-18

[Panag89]   Panagiotopoulos, P.D., Haslinger, J.: Optimal Control of Systems
             governed by Hemivariational Inequalities. In: Math. Models for
             Phase Change Problems, (ed. by J.F.Rodrigues), Vol. 88, Birkhäuser
             Verlag, Basel Boston ISNM, 1989

[Panag90]   Panagiotopoulos, P.D.: Optimal Control of Systems Governed by
             Hemivariational Inequalities. Necessary Conditions. In: Int. Series
             of Num. Math. Vol. 95, (ed. by K.H.Hoffmann and J.Sprekels),
             Birkhäuser Verlag, Basel, Boston 1990

[Panag91]   Panagiotopoulos, P.D.: Optimal Control of Systems Governed by
             Variational-Hemivariational Inequalities. Proc. 4th U.P.S.A., Uni-
             lateral Problems in Structural Analysis, Capri, June 1989, (ed. by
             F.Maceri and G.Del Piero) Birkhäuser Verlag, Basel, Boston 1991

[Panag92]   Panagiotopoulos, P.D., Haslinger, J.: Optimal Control and Identification of Structures Involving Multivalued Nonmonotonicities. Existence and Approximation Results. European J.Mech. A/ Solids 11 (1992) 425-445

[Pot]   Potier-Ferry, M.: Problèmes semi-coercifs. Applications aux plaques de von Kármán. J. Math. pures et appl. 53 (1974) 331-346

[Prag57]   Prager, W.: On Ideal-locking Materials. Trans. Soc. Rheol. 1(1957) 169-175

[Prag58]   Prager, W.: Elastic Solids of Limited Compressibility. Proc. 9th Int. Congress Appl. Mech. Brussels, Vol. 5, 1958

[Prag]   Prager, W.: Problems of Network Flow. Z.A.M.P. 16 (1965) 185-193

[Rauch]   Rauch, J.: Discontinuous Semilinear Differential Equations and Multiple Valued Maps. Proc. Amer. Math. Soc. 64 (1977) 277-282

[Rock60]   Rockafellar, R.T.: Extension of Fenchel's Duality Theorem for Convex Functions. Duke Math. J. 33 (1960) 81-90

[Rock68]   Rockafellar, R.T.: Integrals which are Convex Functionals. Pacific J. Math. 24 (1968) 525-539

[Rock70]   Rockafellar, R.T.: Convex Analysis. Princeton Univ. Press, Princeton 1970

[Rock79]   Rockafellar, R.T.: La théorie des Sous-Gradients et ses Applications à l'optimization. Fonctions Convexes et Non-convexes, Les Presses de l' Université de Montréal, Montréal 1979

[Rock80]   Rockafellar, R.T.: Generalized Directional Derivatives and Subgradients of Non-convex Functions. Can.J.Math. XXXII (1980) 257-280

[Rod]   Rodrigues, J.F.(ed).: Mathematical Models for Phase Change Problems, ISNM Vol 88, Birkhäuser Verlag, Basel 1989

[Scha]   Schatzman, M.: Problèmes aux limites non-linéaires noncoercives. Ann. Sc. Norm. Sup. Pisa XXVII, III (1973) 641-686

[Schel]   Schellekens, J.C.J., De Borst, R.: Application of Linear and Nonlinear Fracture Mechanicm Options to Free Edge Delamination in Laminated Composites. Heron 36 (1991) 37-48

[Stav93a]   Stavroulakis, G.E.: Convex Decomposition for Nonconvex Energy Problems in Elastostatics and Applications. European J. of Mech. A /Solids 12 (1993) 1-20

[Stav93b]   Stavroulakis, G.E., Panagiotopoulos, P.D.: Convex Multivalued Decomposition Algorithms for Nonmonotone problems. Intern.J. for
Num. Meth. in Eng. (to appear)

[Struw]   Struwe, M.: Variational Methods, Applications to Nonlinear Partial
Differential Equations and Hamiltonian Systems. Springer-Verlag,
Berlin, Heidelberg 1990.

[Stu]   Stuart, C.A., Toland, J.F.: A Variational Method for Boundary
Value Probles with Discontinuous Nonlinearities. Journal of the
London Mathematical Society, 21 (1980) 319-328

[Suq]   Suqnet, P.M.: Plasticité et Homogénéisation. Thése de Doct. d' Etat
Univ. Paris VI, 1982

[Vis]   Visintin, A.: Stefan Problem with phase relaxation, J. Appl. Math.
I.M.A. 34 (1985) 225-246

[Web]   Webb, J.R.L.: Boundary Value Problems for Strongly Nonlinear Elliptic Equations, The Journal of the London Math. Soc., 21 (1980)
123-132

[Woź84]   Woźniak, C.: Materials with Generalized Constraints. Arch. Mech.,
36 (1984) 539-551

[Woź87]   Woźniak, C.: A Nonstandard Method of Modeling of Thermoelastic
Periodic Composites. Int. J. Eng. Sci., 25 (1987) 483-498

[Woź90]   Woźniak, C.: Discrete and Continuum Modeling of Delamination
Processes. Ingenieur-Archiv, Vol. 60 (1990) 335-344

[Woź91]   Woźniak, C., Kleiber, M.: Nonlinear Mechanics of Structures. PWN-
Polish Sci. Publ. Warszawa, Kluwer Academic Publishers, Dor-
drecht, London 1991

[Zeid]   Zeidler, E.: Nonlinear Functional Analysis and its Applications, Vol.
II 1A: Linear Monotone Operators, 1990; Vol. II 1B: Nonlinear
Monotone Operators, 1990, Vol. III: Variational Methods and Op-
timization, Springer-Verlag, New York, Tokyo, 1985

Index